...book is to be returned on or before
...date stamped below.

THE PAST THREE MILLION YEARS: EVOLUTION OF CLIMATIC VARIABILITY IN THE NORTH ATLANTIC REGION

THE PAST THREE MILLION YEARS: EVOLUTION OF CLIMATIC VARIABILITY IN THE NORTH ATLANTIC REGION

PROCEEDINGS OF A
ROYAL SOCIETY DISCUSSION MEETING
HELD ON 25 AND 26 FEBRUARY 1987

ORGANIZED AND EDITED BY
N. J. SHACKLETON, F.R.S., R. G. WEST, F.R.S.,
AND D. Q. BOWEN

LONDON
THE ROYAL SOCIETY
1988

Printed in Great Britain for the Royal Society
by the
University Press, Cambridge

ISBN 0 85403 348 3

First published in the *Philosophical Transactions of the Royal Society of London*,
series B, volume 318 (no. 1191), pages 409–688

Copyright

© The Royal Society and the authors of individual papers.

It is the policy of the Royal Society not to charge any royalty for the production of a single copy of any one article made for private study or research. Requests for the copying or reprinting of any article for any other purpose should be sent to the Royal Society.

British Library Cataloguing in Publication Data
The past three million years.
1. North Atlantic region
Climate. Changes
I. Shackleton, N.J. II. West, R.G.
III. Bowen, D.Q. (David Quentin), 1938–
IV. Royal Society
551.69′09182′1
ISBN 0-85403-348-3

LIBRARY
BULMERSHE
COLLEGE OF
HIGHER EDUCATION
READING
ACCESS No:
210570
CLASS No:
F 551.6909182

Published by the Royal Society
6 Carlton House Terrace, London SW1Y 5AG

PREFACE

The study of the Quaternary is to a large extent the study of the effects of climatic variability as recorded in the geological record. Although the Quaternary can be examined with any chosen spatial or temporal focus, the underlying climate can only be understood when it is examined on a global scale, because climate is controlled by the global atmospheric circulation. By compiling data gathered by a diverse range of Quaternary geologists over many years of investigation it has been possible, for example, to study the climate of the last glaciation on a global basis (CLIMAP 1976), and there have been many atmospheric general circulation model runs simulating the climate of that time (Williams *et al.* 1974; Gates 1976; Manabe & Hahn 1977). We can look forward to increasingly successful simulations of the climatic succession through the last glacial cycle with changes in the earth's orbital geometry as the external forcing function (see, for example, Prell & Kutzbach 1987).

When we look at the longer Quaternary record, we see an increasingly complex picture. Although we do apparently see a quasi-cyclicity with similar events occurring more than once, we also see a long-term trend towards conditions three million years ago which at least in Britain were more stable, and warmer, than the average for the last few cycles. The study of long records of oxygen isotopic composition in deep ocean sediment cores has begun to give us a picture of the long-term history of changing amplitude and frequency of Quaternary climatic variability, but this is far too simple a parameter to give us any understanding of the underlying causes of this change in pattern, which is not predicted by the variations in distribution of solar energy deriving from the Milankovitch mechanism. We are not yet in a position to describe this trend on a global basis, but since many aspects of climatic variability around the North Atlantic appears to be intimately linked because of the effect of changes in ocean surface circulation, it will be useful to examine the longer term record in that region. For European workers this concept was brought home by Ruddiman, Sancetta & McIntyre (1977), when at a Royal Society Discussion Meeting they introduced the picture of the Polar Front swinging to and fro between Iceland at its most interglacial position, and Portugal at its extreme glacial position. For this meeting in 1987 we have attempted to review the state of knowledge of the evolution of climatic variability around the whole of the North Atlantic.

At first sight it would appear that the major boundary conditions that constrain atmospheric circulation and hence climate have changed relatively little during the Quaternary. However, the changes have not by any means been immeasurable; major mountain ranges have probably changed dramatically and the final cessation of ocean surface water exchange across Panama may have been only about three million years ago. Changes in ocean productivity may have given rise to a significant change in the carbon dioxide concentration of the atmosphere. If we can calibrate the effect of any or all of these slow changes on global climate, we will provide a valuable service that will be appreciated well outside the confines of Quaternary geology. There has been a tendency for Quaternary specialists to distance themselves from phenomena that have characteristic times longer than a few tens of thousands of years, just as the rest of the geological community regard anybody investigating phenomena with characteristic times of less than a million years as a dangerous catastrophist. At this meeting we were forced to bridge that gap. The tectonics of central Europe, the relative uplift of the United States

Atlantic coastal plain, the extrusion of the Icelandic basaltic pile, all affect the character of the geological deposits that contain the history of changing climatic variability. At the same time they are all manifestations of the slow geological processes that may, together with more geographically distant processes such as the uplift of the Tibetan Plateau, be controlling that evolution of climatic variability.

References

CLIMAP project members 1976 The surface of the ice-age earth. *Science, Wash.* **191**, 1131–1137.

Gates, W. L. 1976 Modelling the ice age climate. *Science, Wash.* **191**, 1138–1144.

Manabe, S. & Hahn, D. G. 1977 Simulation of the tropical climate of an ice age. *J. geophys. Res.* **82**, 3889–3911.

Prell, W. L. & Kutzbach, J. E. 1987 A Monsoon variability over the past 150,000 years. *J. geophys. Res.* **92**, 8411–8425.

Ruddiman, W. F., Sancetta, C. D. & McIntyre, A. 1977 Glacial/Interglacial response rate of subpolar North Atlantic waters to climatic change: the record in oceanic sediments. *Phil. Trans. R. Soc. Lond.* B **280**, 119–142.

Williams, J. R. G., Barry, R. G. & Washington W. M. 1974 Simulation of the atmospheric circulation using the NCAR global circulation model with Ice Age boundary conditions. *J. appl. Met.* **13**, 305–317.

January 1988 N. J. SHACKLETON

CONTENTS

[Two plates]

Phil. Trans. R. Soc. Lond. B **318**, 411–430 (1988)
Printed in Great Britain

Northern Hemisphere climate régimes during the past 3 Ma: possible tectonic connections

BY W. F. RUDDIMAN AND M. E. RAYMO

Lamont-Doherty Geological Observatory of Columbia University, Palisades, New York 10964, *U.S.A.*; *Department of Geological Sciences, Columbia University, New York* 10027, *U.S.A.*

The 100 ka rhythm of orbital eccentricity has dominated large-amplitude climatic variations in the high-latitude North Atlantic during the Brunhes magnetic chron (0–0.735 Ma BP). Earlier, during the Matuyama chron (0.735–2.47 Ma BP), climatic variations in this region were lower in amplitude and concentrated mainly at the 41 ka rhythm of orbital obliquity. These rhythmic climatic responses to orbital forcing are evident both in stable isotopic ($\delta^{18}O$) indicators of ice volume or temperature and in biotic and lithologic indicators of local North Atlantic surface-ocean variability. The synchronous responses of these indicators are consistent with results from atmospheric general circulation models showing that the North American ice sheet directly controls North Atlantic surface-ocean responses via strong cold winds that are generated on the northern ice-sheet flanks and blow out across the ocean, chilling its surface. Before 2.47 Ma BP, smaller-scale quasiperiodic oscillations of the planktonic fauna and flora occurred, but the cause of these variations in the absence of significant ice sheets is unclear.

INTRODUCTION

Analyses of $\delta^{18}O$ (a rough approximation of global ice volume) and of indicators of local climate near Northern Hemisphere ice sheets (sea-surface temperature and ice-rafted debris in the North Atlantic ocean) have demonstrated that Plio-Pleistocene ice sheets were primarily driven by orbitally controlled variations in insolation (Hays *et al.* 1976; Imbrie *et al.* 1984). To first order, the Brunhes chron (0–0.735 Ma BP) was a time of large-amplitude variation at the 100 ka period, the Matuyama chron (0.735–2.47 Ma BP) was a time of moderate-amplitude variation at the 41 ka cycle, and the Gauss chron (2.47–3.40 Ma BP) was a time of very small or non-existent Northern Hemisphere ice sheets. We review here some of the evidence for these distinct climatic régimes, with greatest focus on non-isotopic indicators.

Not yet explained are the changes between these régimes: the initiation of Northern Hemisphere glaciation at or near the Gauss–Matuyama transition; and the intensification of glaciation in the upper Matuyama and into the lower Brunhes. We review possible explanations for these changes in prevailing climatic régime, and further develop a recently proposed tectonic theory.

EVIDENCE FROM THE NORTH ATLANTIC

Oceanic records contain three independent monitors of the amount of ice on land. Oxygen-isotope ratios ($\delta^{18}O$) from late Pleistocene cores in open-ocean areas are largely (*ca.* 70%) controlled by changes in global ice volume (Shackleton & Opdyke 1973), with North

American ice representing one half or more of the ice-volume component (Flint 1957). The amount of continental detritus ice-rafted into the high and middle latitudes of the North Atlantic is also a first-order measure of the size of ice sheets on surrounding continents, with North American ice dominating this signal because of geographic proximity and ocean circulation patterns (Ruddiman 1977). Estimates of sea-surface temperature (SST) in the high-latitude North Atlantic are a third indicator of ice volume; both geological evidence and climate modelling experiments isolate North American ice as the key control on this response as well (Ruddiman & McIntyre 1984; Manabe & Broccoli 1985). A full discussion of the rationale behind the use of percentages of $CaCO_3$, and estimated SST, as indicators of Northern Hemisphere (and particularly Laurentide) ice-sheet size is given by Ruddiman *et al.* (1986*b*).

The $\delta^{18}O$ record by no means entirely reflects changes in North American ice volume; it also reflects volumetric changes in other ice sheets, as well as variations in local temperature signals. Still, the great resemblance of the local % $CaCO_3$ and SST records in high-latitude North Atlantic cores to the $\delta^{18}O$ signal is strong confirmation that all three signals are good first-order measures of changes in North American ice volume during the Pleistocene and late Pliocene.

Initiation of Northern Hemisphere glaciation

The initiation of moderate-sized ice sheets in the Northern Hemisphere occurred at 2.40 Ma BP, although it was preceded by small increases in ice volume at *ca.* 2.55 Ma BP (Backman 1979; Shackleton *et al.* 1984; Zimmerman *et al.* 1985). This finding is based primarily on % $CaCO_3$ records from Deep-Sea Drilling Project (DSDP) Site 552 in the North Atlantic. The uniformly high carbonate values of the early Pliocene suddenly yielded to periodically much lower values as a result of influxes of ice-rafted continental debris (figure 1). In addition, the abrupt increase in $\delta^{18}O$ values at 2.40 Ma BP (figure 1) partly reflects the growth of ice sheets on land. Subsequent coring by the Deep-Sea Drilling Project and Ocean Drilling Project across the subpolar North Atlantic and in the Labrador and Norwegian Seas has confirmed 2.40 Ma BP as the age of onset of major ice rafting (Ruddiman, Kidd, Thomas *et al.* 1986; Eldholm, Thiede, Taylor *et al.* 1987; Srivastava, Arthur, Clement *et al.* 1987).

Other evidence suggests that the initiation of ice rafting to the North Atlantic Ocean at 2.40 Ma BP represents the culmination of a longer-term high-latitude cooling that began about 750 ka earlier. Progressive (but oscillatory) enrichment of $\delta^{18}O$ values in benthic foraminifera after 3.15 Ma BP (figure 1) suggests significant cooling of deep waters (Prell 1984; Weissert *et al.* 1984; Hodell *et al.* 1985; Keigwin 1986); depth-dependent decreases in $CaCO_3$ percentages in the North Atlantic during this interval (figure 2) have been attributed to increasingly corrosive deep waters (Ruddiman *et al.* 1986*c*), possibly due to the suppression of North Atlantic Deep water formation.

Progressive high-latitude cooling is also suggested by increasing percentages of cooler planktonic foraminifera (Loubere & Moss 1986; Raymo *et al.* 1986) and by decreasing discoaster abundances (Backman & Pestiaux 1986). Sea-surface temperature trends during this time interval are not easily reconstructed, however, because the assemblages are not analogous to modern faunas. A significant evolutionary turnover in planktonic foraminifera occurs in the late Pliocene near 2.40 Ma BP, probably in response to the effect of the earliest glaciations in cooling the surface North Atlantic and otherwise altering its structure (Raymo *et al.* 1986).

The history of ice-volume changes from 3.15 to 2.40 Ma BP is still controversial. Small-scale

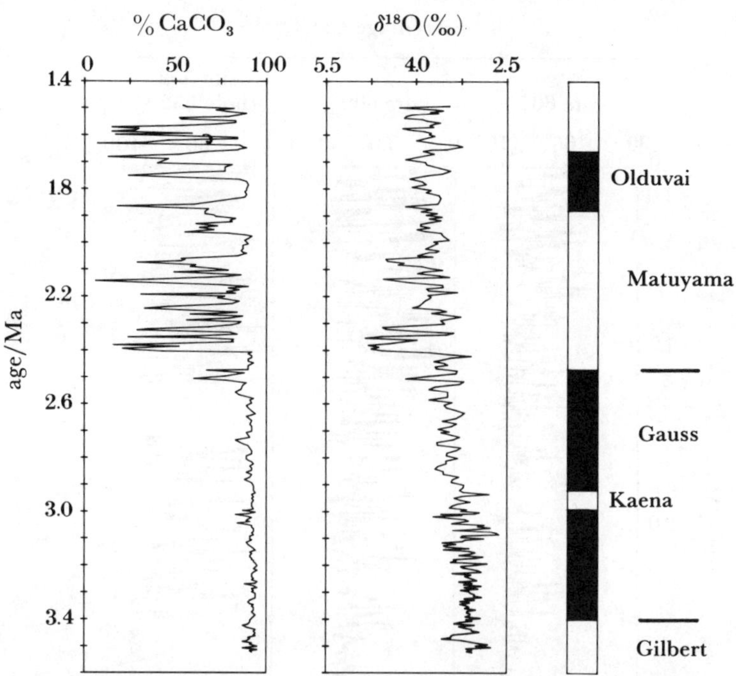

FIGURE 1. Late Pliocene and early Pleistocene % $CaCO_3$ and stable-isotope ($\delta^{18}O$) records at North Atlantic DSDP site 552 (after Shackleton *et al.* 1984). The abrupt decrease of % $CaCO_3$ values at 2.40 Ma BP marks initiation of deposition of ice-rafted sand. A correlative late-Pliocene increase in $\delta^{18}O$ values is superimposed on progressive long-term increase after 3.15 Ma BP.

glaciation began early in this interval on Iceland (McDougall & Wensink 1966) and in the Sierras (Curry 1966). Covariance of benthic and planktonic $\delta^{18}O$ values beginning at 2.90 Ma BP permits (but does not require) somewhat larger storage of ^{16}O-rich ice on land (Prell 1984; Keigwin 1986). It is plausible that Northern Hemisphere ice sheets periodically existed before 2.40 Ma BP but were too small to send ice-rafted detritus to the Atlantic. In any case, all the evidence is consistent with a progressive, but oscillatory, deterioration of Northern Hemisphere climate from 3.15 Ma BP (or earlier) until glaciations of substantial scale abruptly began at 2.40 Ma BP.

Recent evidence from North Atlantic DSDP sites 607 and 609 shows that Northern Hemisphere ice sheets fluctuated mainly at a period of 41 ka during the Matuyama (Ruddiman *et al.* 1986*b*). Variations in % $CaCO_3$, estimated sea-surface temperature, and benthic foraminiferal $\delta^{18}O$ were all dominated by the 41 ka rhythm of orbital obliquity (tilt) during the Matuyama magnetic chron (figures 3 and 4). Strong variance near 41 ka is evident from timescales based solely on linear interpolation between paleomagnetic datums (figure 4).

For the post-Olduvai (after 1.66 Ma BP) portion of the Matuyama chron at sites 607 and 609, we also developed an obliquity timescale by tuning the climatic indicators directly to the orbital obliquity signal. The obliquity timescale imposed age changes (relative to the magnetic time scale) no larger than 20 ka at any point in the records, and it predicted ages for the magnetic reversals (upper and lower Jaramillo and upper Olduvai) that agreed with the radiometric ages to within less than 1% of the absolute values. For the obliquity timescale, more than one half of the variance in these late-Matuyama records was narrowly centred at the 41 ka period (figure 4).

The obliquity rhythm is also prominent in the lower Matuyama (pre-Olduvai) portion of the

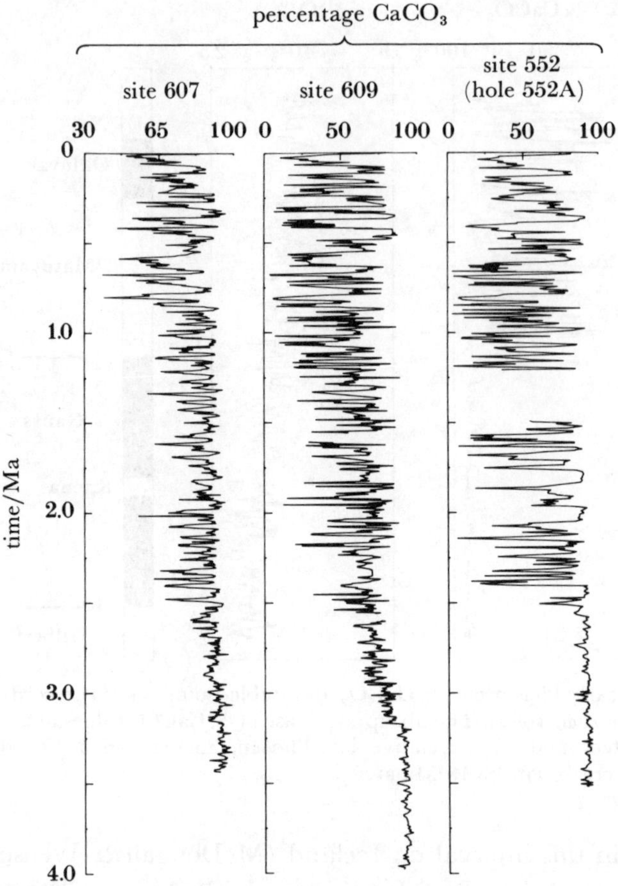

FIGURE 2. Plio-Pleistocene % $CaCO_3$ trends at North Atlantic DSDP sites 552, 607 and 609 (from Ruddiman *et al.* 1986*c*). All records show initiation of major glaciation by carbonate decreases at 2.40 Ma BP. Deeper cores (site 607, 3500 m; site 609, 3800 m) show progressive decrease of % $CaCO_3$ values between 3.15 and 2.40 Ma BP, indicating increasing dissolution with depth.

records from sites 607 and 609 (Ruddiman *et al.* 1986*b*). We did not, however, attempt to tune these records because of complications in assembling continuous records owing to coring problems. Our recent (unpublished) work confirms these early indications that the 41 ka rhythm was dominant in the early Matuyama as well.

Before 0.9 Ma BP, the dominance of 41 ka power in all signals is unequivocal and overwhelming (figure 3). After 0.9 Ma BP, the signals become more complex, with higher-amplitude fluctuations, and probably lower-frequency components as well. These changes mark the beginning of the shift into a different climatic régime during the Brunhes (Shackleton & Opdyke 1973; Pisias & Moore 1981; Prell 1982).

Because the climatic signals after 0.9 Ma BP differ from those earlier in the Matuyama, they complicate the spectral analysis shown in figure 4 by introducing effects of non-stationarity (changes in signal mean and variance). To assess this effect, we re-ran the spectral analyses of all four climatic indicators for the interval from the upper Olduvai boundary to 0.9 Ma BP. The only significant changes in relative concentration of spectral power were at very long periods (greater than 250 ka); reductions were minor for site 609 % $CaCO_3$ and site 607 $\delta^{18}O$, but larger (30–40 %) for site 607 % $CaCO_3$ and SST. Most of the low-frequency variance that

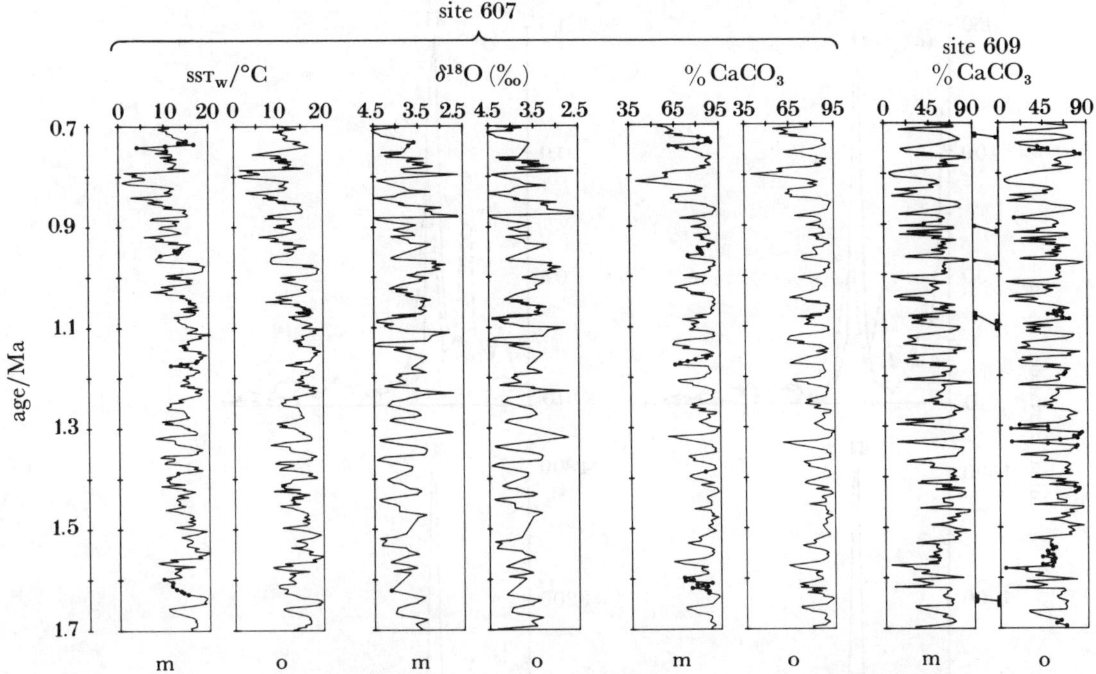

FIGURE 3. Early Pleistocene climatic trends at North Atlantic DSDP sites 607 (estimated winter sea-surface temperature (SST_w), $\delta^{18}O$, and % $CaCO_3$) and 609 (% $CaCO_3$); after Ruddiman *et al.* (1986*b*). Magnetic timescale based on linear interpolation between magnetic datums; obliquity timescale derived by tuning records to obliquity. Large dots are values spliced into main sequence from offset holes.

remained was concentrated at periods of 150 ka to 300 ka, apparently intermediate between the orbital eccentricity bands at 413 ka and 95–124 ka and thus of uncertain origin. For periods of 100 ka years or less, the late-Matuyama spectra were essentially unchanged, although the prominent 41 ka obliquity signal contains an even larger fraction of the total variance because of reduced power at the very low frequencies.

Intensification of Northern Hemisphere glaciation

After 0.9 Ma BP, changes in $\delta^{18}O$, % $CaCO_3$, and estimated sea-surface temperature increased in amplitude by *ca.* 50%; this increase suggests that ice sheets in the Northern Hemisphere during the late Matuyama and early Brunhes chrons grew to maximum volumes considerably larger than those attained during the early Matuyama. The first prominent $\delta^{18}O$ maximum occurred in isotopic stage 22 at 0.85–0.8 Ma BP (Shackleton & Opdyke 1973), an age correlating with the 'Nebraskan' glaciation in the magnetically reversed interval between 0.91 and 0.735 Ma BP (Boellstorff 1978; Rogers *et al.* 1985).

There is some regional variation in the SST and % $CaCO_3$ expression of the shift from the Matuyama to Brunhes climatic régimes. At mid-latitude site 607 (41° N), the largest SST and % $CaCO_3$ changes occurred at *ca.* 0.9–0.8 Ma BP (figure 3). At high-latitude site 552 (56° N), glacial SST minima also deepen at and after 0.9–0.8 Ma BP, but there is a further doubling in amplitude of SST fluctuations at a period of 95 ka in the middle of the Brunhes chron at 0.45 Ma BP (figure 5). $\delta^{18}O$ spectra show a similar doubling of variance at a period near 100 ka after 0.4 Ma BP (Imbrie 1985). Despite regional differences in SST response, all three climatic indicators (SST, % $CaCO_3$, and $\delta^{18}O$) show increasing variance after 0.9 Ma BP at

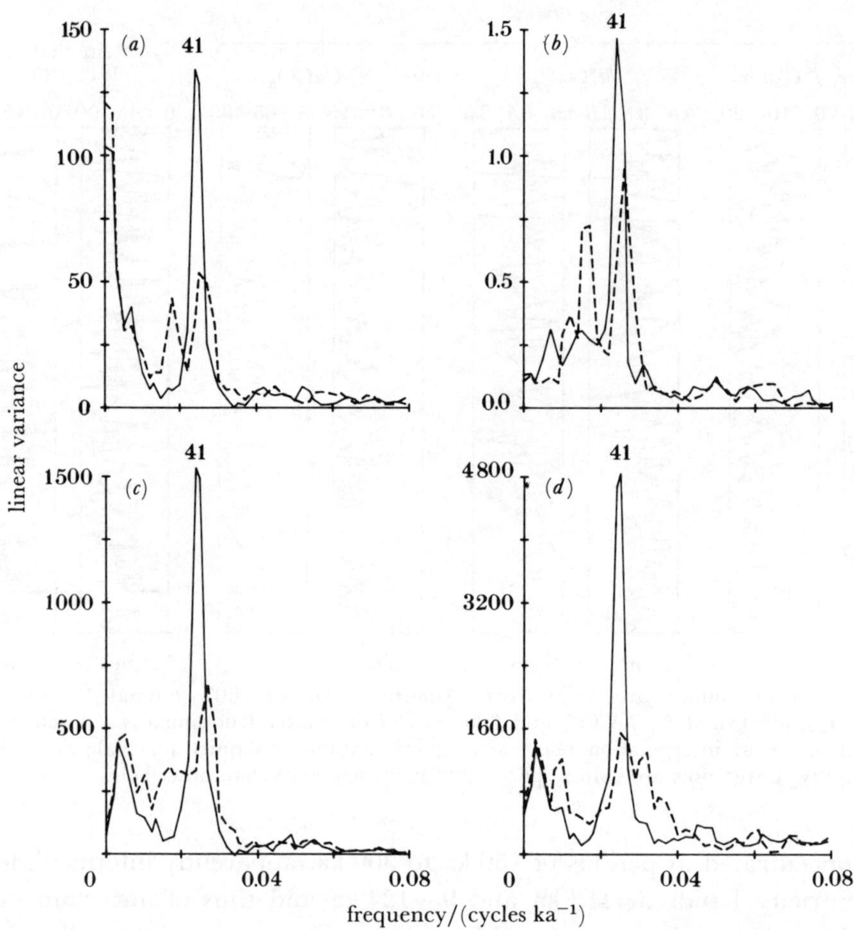

FIGURE 4. Spectral analysis of early Pleistocene records shown in figure 3 (from Ruddiman *et al.* 1986*b*). Spectra from both magnetic (broken lines) and obliquity timescales (solid lines) show prevalent 41 ka periodicity during the late Matuyama. (*a*) Site 607, SST_w; (*b*) site 607, $\delta^{18}O$; (*c*) site 607, % $CaCO_3$; (*d*) site 609, % $CaCO_3$; age range for all sites, 0.735–1.64 Ma BP.

periods of orbital eccentricity (near 100 ka) and precession (23 ka and 19 ka), and all indicate that the 100 ka rhythm attained strongest dominance in the late Pleistocene after 0.45 Ma BP.

Some uncertainty remains as to the form of the climatic response near the Brunhes–Matuyama boundary. SST records from site 552 suggest a brief interval (0.775–0.625 Ma BP) of prominent fluctuations at a period of 54 ka and a smaller coincident increase in 54 ka power also occurs in the SPECMAP $\delta^{18}O$ signal (Ruddiman *et al.* 1986*a*). It is not yet clear whether this 54 ka signal is real. It could have resulted from gaps in the single-HPC (hydraulic piston core) record cored at site 552, from incorrect correlation of the site 552 $\delta^{18}O$ record to the SPECMAP timescale, or from an error in the pre-Brunhes part of the SPECMAP timescale. More work is needed to resolve this important transition between the 41 ka (Matuyama) and 100 ka (Brunhes) régimes.

In summary, ice sheets large enough to send large amounts of ice-rafted debris to the open North Atlantic ocean did not exist before 2.40 Ma BP. There is evidence of Northern Hemisphere cooling after about 3.15 Ma BP, but major ice sheets only appeared by 2.40 Ma BP. The first 1.6–1.7 Ma of the Northern Hemisphere ice age (roughly the Matuyama

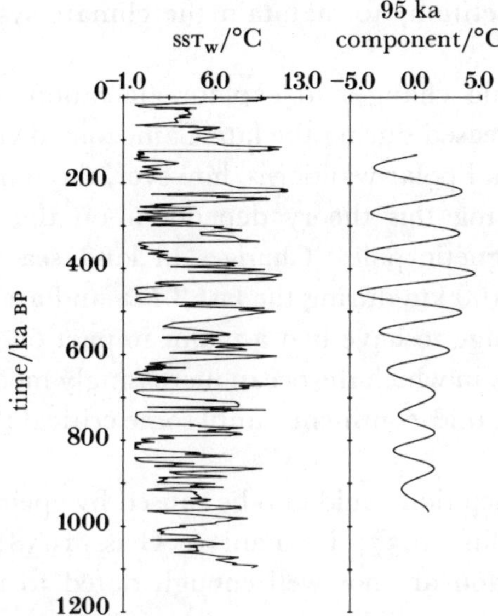

FIGURE 5. Late-Pleistocene record of estimated winter sea-surface temperature (SST$_w$) from spliced record of piston core K708-7 and DSDP site 552 (after Ruddiman *et al.* 1986*a*). Filtered record of dominant 95 ka SST component shows major increase in variance at 0.45 Ma BP.

chron) was largely a sequence of some 40 climatic cycles at the 41 ka rhythm of orbital obliquity. During the mid-Pleistocene (0.9–0.4 Ma BP), the obliquity rhythm was progressively superseded by an increased response at or near the 100 ka period of orbital eccentricity, and this rhythm dominated climatic responses throughout the Brunhes chron, particularly after 0.45 Ma BP. In addition, numerous glacial advances at periods of 41 ka and 23 ka were superimposed on the basic 100 ka cycle during the Brunhes chron.

This long-term perspective raises four fundamental questions: (1) Why did the Northern Hemisphere ice age begin? (2) Why did the Matuyama ice-sheet variations occur at a rhythm of 41 ka? (3) Why did the mid-Pleistocene shift in rhythmic response occur? (4) What is the origin of the 100 ka rhythm of the late Pleistocene? In this paper, we shall address a possible reason for the major changes in prevailing climatic régime during the late Pliocene and mid-Pleistocene (questions 1 and 3).

THEORIES OF ICE-AGE INITIATION AND INTENSIFICATION

Several explanations have been advanced to explain why glaciation in the northern hemisphere was restricted to the late Neogene. One group of theories looks to changes in atmospheric composition or in solar radiation. Some involve variations in parameters that are untestable, such as changes in solar strength (Opik 1958). Other plausible theories, such as changes in atmospheric CO_2, ozone or trace gases (Plass 1956), although also not yet tested, should be at least partly verifiable from geological data. Variations in orbitally driven insolation (Milankovitch 1941) are obviously critical on time scales of 20 to 100 (or 400) ka, but the mean long-term form and strength of this forcing (Berger 1984) did not change sufficiently to explain the glacial initiation or intensification discussed in the last section. Increased volcanism during the latest Cainozoic (Kennett & Thunell 1975) has also been suggested as a possible cause of glaciation, but vulcanism does not appear to provide persistent enough forcing (i.e. decadal repetition) to maintain the climate system in one 'régime' for intervals lasting a million years.

Other theories call on tectonic changes to explain glaciation. Ewing & Donn (1956) suggested that global albedo increased during the late Cainozoic, owing to relative movement of the poles and continents. Virtual polar wander is, however, thought to have been negligible (Schneider & Kent 1986), leaving this theory dependent on the unlikely possibility of a decoupling of the spin and magnetic poles. Changes in land–sea distribution by sea-floor spreading are of the order of 10–100 km during the last 3 Ma and are mostly latitude-parallel. This is probably too small a change to have had a major impact on climate, although North *et al.* (1983) suggest a possible link in which the ocean increasingly moderates summer radiative heating (and thus ablation) in the mid-continents, until some critical threshold or discontinuity is passed and glaciation begins.

Changes in susceptibility to glaciation could also be caused by epeirogenic uplift of highland regions in northern Canada (Flint 1957; Emiliani & Geiss 1958; Birchfield *et al.* 1982), although such changes in elevation are not well enough dated to relate unambiguously to glacial initiation and intensification. The consensus viewpoint today is probably that continental rearrangements and uplift have contributed to long-term cooling since the Cretaceous (Barron 1981) but are not very convincing as explanations for the more sudden initiation and intensification of glaciation in the Plio-Pleistocene, unless there are extremely sensitive thresholds of response in the climate system.

Another proposed tectonic explanation involves changes in oceanic heat and moisture fluxes (Emiliani *et al.* 1972) induced by shoaling or deepening of critical oceanic gateways such as the Panama Isthmus (Keigwin 1978, 1982) or the Bering Straits (Einarsson *et al.* 1967). Both of these tectonic changes occurred at 3.5–3.0 Ma BP, somewhat before the late-Pliocene initiation of significant glaciation at 2.40 Ma BP (figure 6), but very close to the beginning of the long-term late Pliocene climatic deterioration at 3.15 Ma BP. Neither of these tectonic changes could be causally involved in the mid-Pleistocene glacial intensification from 0.9–0.45 Ma BP.

We also believe that the changes in response of the climate system during the Pliocene and Pleistocene could not have been generated spontaneously within the internal physical variability of the ice–ocean–atmosphere system, but must be sought in tectonic changes that altered the solid boundary conditions within which that system operates. We discuss below two theories that ascribe Plio-Pleistocene climatic change to large-scale uplift in two key regions: Tibet and western North America. In both of these regions, there is abundant evidence of major late Cainozoic uplift that has continued through the Pleistocene (figure 6) and even into the last century. This evidence is summarized below.

Evidence for Plio-Pleistocene uplift in Tibet and North America

Although one geophysical interpretation holds that the Himalayas and Tibet have been high-standing since the Oligocene–Miocene orogeny because of continuous ('steady-state') underthrusting of the Indian continent (see, for example, Seeber & Armbruster 1983), other evidence favours the view that *ca.* 3 km of net uplift of the land surface has occurred during the Plio-Pleistocene (Liu & Menglin 1984). Early Pliocene pollen assemblages in both Tibet and the Himalayas contain mostly subtropical and warm deciduous flora indicative of elevations of 1.5 km or less (Hsu 1978); these deposits occur today at elevations of 4 km or more. Mammalian distributions (the *Hipparion* fauna) indicate that the Himalayas and Tibet

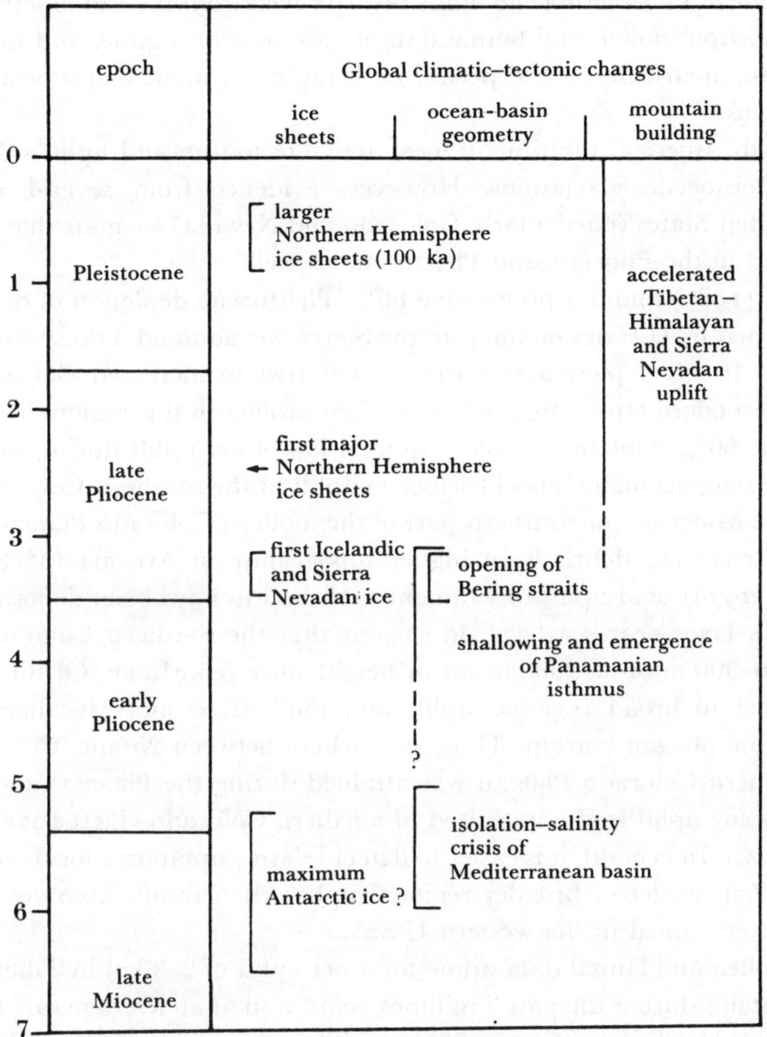

FIGURE 6. Timing of large-scale tectonic changes potentially important to Northern Hemisphere glaciation.

were not an effective barrier to north–south faunal exchange until the late Pliocene (West 1984). The Siwalik Formation at the southern foot of the Himalayas coarsens progressively upwards from early Pliocene silts to middle and late Pleistocene conglomerates, consistent with increasing uplift through the past 3 Ma.

The pollen data suggest mean rates of net uplift of just under 1 mm per year over the last 3 Ma. This agrees with the average rates of uplift in this century, determined from geodetic measurements in the Tibetan–Himalayan region (Mehta 1980). Radiometric data (Rb–Sr, K–Ar, and fission-track ages) also suggest mean rates of Himalayan uplift of 0.7–0.8 mm per year during the Miocene, Pliocene and Pleistocene (Mehta 1980), with some data suggesting progressively larger uplift rates during the Plio-Pleistocene (Saini *et al.* 1979) and, in some regions of the High Himalayas, rates as high as 3.5–10 mm per year during the late Pleistocene (Burbank & Johnson 1983).

Mineralogical data cannot be used to specify absolute elevations of the earth's surface in the past, because the competing effect of erosion in lowering the rising surface is not accurately known. Still, the mean Plio-Pleistocene uplift rates derived from the radiometric data support the rates estimated from pollen and faunal data if erosion is negligible, and the highest uplift rates estimated radiometrically would permit both rapid net uplift and substantial erosion of the High Himalayas.

In western North America, the time of most active tectonism and uplift substantially predates the Plio-Pleistocene glaciations. However, evidence from several regions in the Southwestern United States (particularly Colorado and Nevada) suggests that substantial net uplift has occurred in the Pliocene and Pleistocene as well.

Winograd *et al.* (1985) found a progressive 60‰ Pleistocene depletion of deuterium values in calcitic veins dated by U-series methods in the Sierra Nevada and Transverse Ranges of the Great Basin area. If this depletion is entirely attributed to increased elevation during the Pleistocene, and if modern lapse rates and deuterium profiles in this region are used as a basis for calibration, the 60‰ depletion would require 1 km of net uplift during the past 2 Ma.

Other evidence suggests major Plio-Pleistocene uplift of the southern Colorado Plateau. As recently as the late Miocene, the southern part of the modern Colorado Plateau was a relative topographic low, receiving debris from higher topography in Arizona (McKee & McKee 1972). Lucchitta (1979) used uplift of sediments, inferred to have been deposited in estuaries at sea level and in lakes near sea level, to suggest that the southern Colorado Plateau has attained some 800–900 m of its 2000 m mean height since 5.5 Ma BP. Of this total amount, 500 m is attributed to broad regional uplift and the rest to more localized fault-related movement along the plateau margin. Thus, somewhere between 25 and 45% of the present height of the southern Colorado Plateau was attained during the Pliocene and Pleistocene.

Significant Pliocene uplift is also indicated in northern Colorado (Izett 1975), but absolute values are not known. In general, it is easier to detect relative uplift on a local scale (by vertical offsets of strata) than to detect broader regional uplift. As a result, absolute late-Cainozoic uplift is not well constrained in the western U.S.A.

In summary, pollen and faunal data argue for a net uplift of 2–3 km in Tibet and the High Himalayan Mountains during the past 3 millions years, and at an average rate consistent with those measured geodetically over the past 70 years. Thus, as much as 75% of the net elevation of Tibet (an area approximately one third the size of the contiguous United States) may have been attained during the time of Northern Hemisphere glacial inception and intensification. Although significant net Plio-Pleistocene uplift (several hundred metres to 1 km) has also occurred in several regions of the southwest United States, these represent lesser alterations of a region with a long earlier history of high-standing mountains.

Taken together, these Plio-Pleistocene tectonic alterations may have had two effects on Northern Hemisphere climate that could be relevant to glacial inception and intensification. One involves major increases in zonal mean albedo (Birchfield & Weertman 1983); the other, large-scale rearrangement of the planetary wave structure (Ruddiman *et al.* 1986*b*).

Effects of uplift on zonal mean albedo

Using numerical experiments with a zonally averaged energy-balance model, Birchfield & Weertman (1983) proposed that uplift in Asia and North America would enhance albedo–temperature feedback on a globally significant scale. The cold air over the large, newly created region of high-standing topography at middle latitudes provides a temperature discontinuity that accelerates the normal southward movement of the snowline in autumn–winter. This enlarged high-albedo surface in turn leads to global cooling during the transitional and winter seasons.

Birchfield & Weertman did not attempt to resolve fully the sensitivity of the climate system to this albedo–temperature feedback; their experiments used numerous simplifying assumptions and could not simulate important geography-dependent features of the atmospheric circulation, such as the Asian monsoon. Their results thus pertain only to zonal mean climate, and it is not clear what specific effects these changes would have in the regions of North American and Eurasian glaciation. Nevertheless, the proposed albedo–temperature feedback is a plausible link between mountain uplift in Asia and North America and glaciation in the Northern Hemisphere.

Effects of uplift on planetary waves

Ruddiman *et al.* (1986*b*) suggested a second explanation for the initiation and intensification of Northern Hemisphere glaciation. This explanation specifically links changes in mid-latitude orography to climate in the glaciated regions of North America and Eurasia. They proposed that increased elevation in the Tibetan–Himalayan and Sierran–Coloradan regions have altered the planetary wave structure in such a way as to cool the North American and European land masses and increase their sensitivity to orbitally driven insolation changes. The concept is further developed and evaluated here.

At the simplest possible level, both the late-Pliocene initiation of glaciation and the mid-Pleistocene intensification require the same thing: some means of permitting larger ice sheets to grow, during favourable orbital configurations, than had previously been possible. We avoid here detailed discussion of the complex mechanisms involved in explaining the dominance of specific orbital rhythms during specific intervals. Instead, we focus on possible tectonic alterations of the planetary wave pattern that would permit faster (and thus greater) ice growth during glacial cycles.

Orographic control of the planetary waves

Maps of geopotential height or pressure distribution in the upper troposphere show that the middle-latitude westerlies are concentrated in narrow bands (including the jet stream) that are perturbed into large, wave-like meanders (the 'planetary waves'). The mean positions of the modern planetary waves in winter are shown in figure 7 (top left). Prominent southward meanders occur over east-central North America, western Europe, and the western North Pacific in winter (Bolin 1950); these waves continue in similar positions into the spring season. Because the southward meanders over North America and Europe match regions of Northern Hemisphere mid-latitude glaciation, they obviously have the potential to be relevant to the glaciation process. A weaker southward meander of the planetary waves occurs over eastern North America in summer and continues into autumn.

There has long been interest in what factors control the planetary long-wave structure. One view is that the waves are largely created by thermal contrasts at the earth's surface (Sutcliffe 1951; Smagorinsky 1953). Differential heating of land and sea surfaces sets up, via different vertical energy exchanges, longitudinal contrasts in the distribution of upper-tropospheric mass and pressure. Relative warmth over the heated land surface in summer and over the heat-

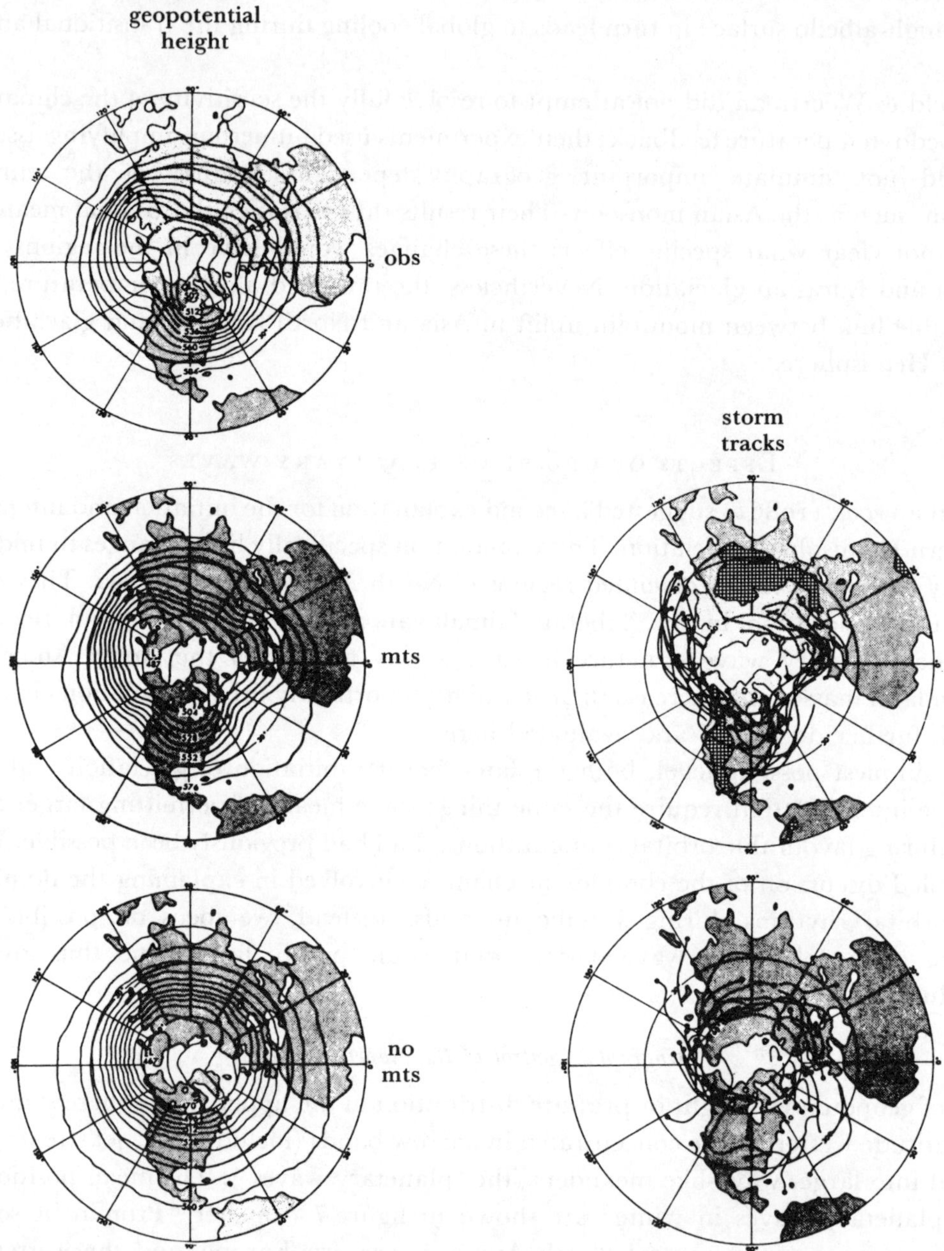

FIGURE 7. (*a*) Geopotential height (in decametres) at 500 mb for modern circulation and for 'mountain' and 'no-mountain' experiments; (*b*) storm tracks for 'mountain' and 'no-mountain' experiments (from Manabe & Terpstra 1974).

retaining ocean in winter promotes strong vertical fluxes from the lower to the upper troposphere compared with the weaker fluxes from the cooler summer ocean and colder winter land surface. These longitudinal contrasts then perturb the otherwise zonal upper westerlies into long-wave meanders.

The other view holds that orography controls the planetary wave meanders (Charney & Eliassen 1949; Bolin 1950). In some areas, the wave meanders tend to maintain similar longitudinal positions in both winter and summer, even though the land–sea temperature difference reverses sign between the seasons. This suggests that orography, which is constant between the seasons, is a stronger control on the waves than thermal contrasts. From a dynamical analysis of two-dimensional models, Bolin proposed that only two regions in the Northern Hemisphere are capable of controlling the planetary waves: mountains in the interior of Asia (Tibetan Plateau and Himalayas) and mountains in western North America (primarily the Rockies).

Control of the planetary waves is probably a combination of orographic and thermal factors (Trenberth 1983). Whereas thermal contrasts at the earth's surface are important determinants of many surface features, such as sea-level pressure, orography increasingly determines the distribution of pressure and mass (and thus the planetary waves) with altitude. There is also basic agreement that winter circulation is relatively dominated by dynamics and thus orography, whereas in summer the thermal control becomes stronger.

Bolin's analysis indicated that orography is particularly effective in positioning the planetary wave meanders over North America and the North Atlantic. In both the winter and summer seasons, the axis of the middle-tropospheric westerlies intercepts the long, linear Rocky Mountain chain and forms a downstream meander, despite shifting 10° in latitude. Over Asia, the orographic effect is strongest in winter when the westerly axis directly intercepts the Tibetan Plateau at 30°–40° N. In summer, the axis retreats far northward to latitudes where the mountains are lower and have less influence on the wave structure. Thus, thermal effects presumably are larger in Asia in summer.

The strongest confirmation of the orographic influence on the planetary waves in winter comes from general circulation model (GCM) modelling experiments by Manabe & Terpstra (1974). In one model run, present-day mountains were included in the smoothed form required by model grid-point constraints (figure 8); in the other run, mountains were omitted. The difference between the two runs showed that removing the mountains almost completely suppressed winter planetary waves in the Northern Hemisphere (figure 7). In contrast, GCM experiments by Kasahara & Washington (1971) indicated a smaller role for orography and a larger role for thermal effects.

In summary, orography appears to be the major factor determining the mean location and amplitude of the winter planetary waves. Land–sea thermal contrasts at the earth's surface are a secondary factor in winter but may be primary in summer.

Proposed link between planetary waves and glaciation

The prevailing climatological view since Hays *et al.* (1976) decisively confirmed the orbital-climate theory of Milankovitch (1941) is that the most critical factor favouring accumulation of ice sheets is decreased ablation over land, particularly in summer. This allows larger fractions of winter snowfall to survive as permanent ice. The present configuration of the planetary waves (figure 7) favours diminished ablation over North America in both seasons (Bolin 1950) because the southward meanders of the upper-level waves mark regions of increased outbreaks of lower-tropospheric polar air masses that chill the continental interiors. The wave configuration thus promotes glaciation via reduced ablation.

Other, partly conflicting, views suggest that the ocean is also critical to glaciation. One view holds that increased (or at least undiminished) snowfall in the cold season is essential to glaciation, particularly during intervals of rapid ice growth indicated by $\delta^{18}O$ data (Ruddiman & McIntyre 1981). This view calls on a warm, or at least ice-free, winter ocean as a moisture source during ice growth.

More recently, it has been suggested that the ocean may be critical to ablation, by

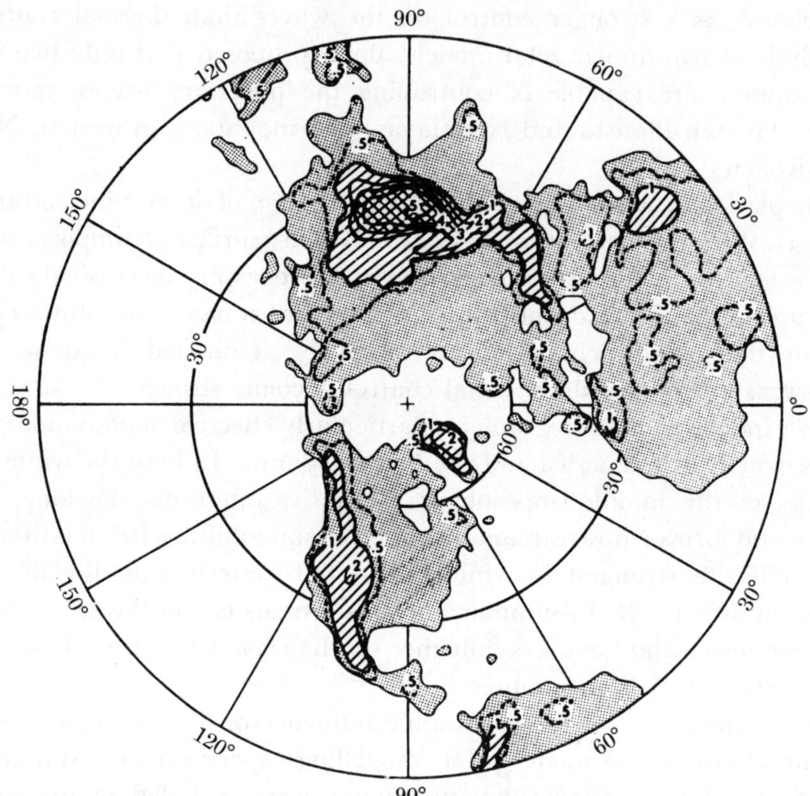

FIGURE 8. Topography entered into 'mountain' experiment of Manabe & Terpstra (1974); mountains of Tibet and western U.S.A. are main features in spectrally smoothed representation. Elevations are contoured in kilometres. 'No-mountain' experiment dropped all topography to sea level.

moderating the radiative summer heating of the continental interiors (North *et al.* 1983). In this view, a cooler summer ocean promotes glaciation.

At present, it is not possible to say which of these two effects dominates long-term climatic changes. The northward meander of the modern planetary waves over the North Atlantic (figure 7) is linked to prevailing southwesterly winds that drive warm, ice-free waters far to the north; this configuration promotes glaciation via the cold-season moisture flux but opposes it via the warm-season heat flux.

In any case, diminished ablation over the continental interiors, owing to colder temperatures, is widely considered the key factor in glaciation. Thus the modern planetary-wave pattern, with southward meanders over both eastern North America and western Europe in winter and over eastern North America in summer, favours glaciation by promoting influxes of colder air from the north.

Given that the modern waves in the upper westerlies affect the lower-tropospheric circulation in ways favourable to glaciation over North America and Europe, could the form of the planetary waves have been different from that of today during the unglaciated early Pliocene and smaller glaciations of the late Pliocene and early Pleistocene? Although this question cannot be answered without GCM modelling experiments, we suggest that the planetary-wave positions shown in figure 7 may indicate at least the direction of change from the Gauss magnetic chron (before Northern Hemisphere glaciation) to the large glaciations of the Brunhes chron.

Before late-Pliocene mountain uplift, early-Pliocene cold air masses were probably more confined to the northernmost parts of North America and Europe, as in the no-mountains experiment (figure 7). Because mountains in western North America were already high-standing, presumably some kind of southwards meander of the planetary wave was present. We suggest that the degree of southward penetration of the waves over North America and Europe gradually deepened during the late Pliocene and Pleistocene, possibly changing more rapidly during the major transitions in Northern Hemisphere climatic régime at 3.15–2.40 Ma BP and at 0.9–0.45 Ma BP. The full development of the modern planetary-wave pattern may only have occurred during the 0.9–0.45 Ma BP interval of the late Pleistocene.

These changes in the winter position of the planetary waves may also have occurred for much of the late-autumn and early-spring transitional seasons, which today are characterized by the winter wave patterns. If so, the Gauss-to-Matuyama-to-Brunhes trend in circulation should be one of decreasing air temperatures and decreasing ablation of snow and ice through the winter and large parts of the transitional seasons.

The modern southward wave over eastern North America is dynamically linked to the region of increased cyclogenesis in the Atlantic off the coast of North America, as indicated by the model results (figure 7) of Manabe & Terpstra (1974). Gradual deepening of this wave through the late Pliocene and Pleistocene would have increased the presence of cyclonic storms moving to the north and northeast along the east coast of North America. One likely consequence would have been increased precipitation in maritime regions of eastern Canada, although the extent of inland penetration of these storms was probably limited.

Finally, we add the caveat that the planetary-wave theory is mainly applicable when ice sheets are relatively small in size, particularly in the early parts of ice-growth phases. When ice sheets become large, their own orography has a large effect on the planetary-wave structure (Manabe & Broccoli 1985) that persists until the ice again shrinks under orbital forcing. The proposed contribution of mountain orography is to permit faster growth (or slower decay) during intervals when ice sheets are relatively small.

Discussion

The tectonic–climate connections proposed both by Birchfield & Weertman (1983) and Ruddiman *et al.* (1986 *b*) have the appeal of calling on tectonic changes that are (1) capable of affecting large-scale climate and (2) known to be occurring during the times of largest climatic change. Both hypotheses, however, have key features that need testing by climate-modelling experiments, as outlined below.

Both the albedo and orographic theories have in common one potentially major weakness. There is wide acceptance of the view of Milankovitch (1941) that glacial variations at the orbital periods are most sensitive to variations in summer insolation. Both of the above theories, however, have their largest effect on climate during the winter and transitional seasons, when orographic control of the waves is strongest. Hartman & Short (1979) have argued that the climatic effects of planetary-wave asymmetries should be greatest during summer and the transitional seasons.

Although summer insolation is the critical control for climatic changes at orbital timescales, it is not necessarily the only important ablation season for climatic change. Today, over North America, ablation does not only occur in summer; spring is a particularly important time for snowmelt, although this happens at lower rates than in midsummer. Reducing spring (and autumn) ablation must make some annual contribution to ice-sheet mass balance, and shortening the length of the high-ablation season by expanding the interval of winter-like planetary waves would further diminish annual ablation.

In addition, there is a small southward meander of the planetary waves over east-central North America even in summer, suggesting that orography continues to play some role in guiding planetary waves in that season. There may thus be a link between orography, planetary waves and ablation even in summer, although probably a weaker one than in other seasons. Ultimately, it will require GCM experiments to test whether integrated annual ablation over the glaciated regions is sufficiently affected by tectonic changes to enhance preservation of snow and ice.

A second issue is specific to the hypothesis of tectonic alteration of the planetary waves (Ruddiman *et al.* 1986*b*), which relies on the Manabe & Terpstra (1974) model results (figure 7). That experiment eliminated all mountains, not just those in regions of major Plio-Pleistocene uplift. It thus remains an open question whether observed uplift of the Tibetan plateau, the Himalayas, the Sierras, and the southern Colorado Plateau was sufficient to affect the planetary waves, and to do so in the specific way discussed here. An especially critical question is whether the relatively modest net Plio-Pleistocene uplift in portions of the western U.S.A. would be sufficient to affect the downstream wave over east-central North America. Alternatively, did the much larger but geographically more remote Tibetan–Himalayan uplift amplify wave trends over eastern North America by reinforcing the orographic effects of the already elevated Rockies? The location of the Himalayas and Tibet almost directly opposite the Rockies and Colorado Plateau (figure 8) might be particularly conducive to amplification of meanders on the other side of the world (planetary wave 2 and possibly wave 3). GCM modelling experiments to test these ideas are under way.

Critical to both theories is the need for geological evidence providing more detail about the timing of uplift during the past 3 Ma. Is there evidence that Tibetan–Himalayan and southwest U.S.A. uplift rates were greatly accelerated during the times of observed transition between major climatic régimes (3.15–2.4 and 0.3–0.45 Ma BP)? Or was uplift gradual during the entire late Pliocene and Pleistocene? In the latter case, it may be that key elevation thresholds were reached at which orographic control of the planetary waves suddenly became effective. This idea could be tested by sensitivity tests with a range of heights for the Tibetan Plateau.

One possible line of inquiry on the history of Tibetan–Himalayan uplift involves the Siberian high-pressure cell. Manabe & Terpstra (1974) note that, in their model, the wintertime strength of the Siberian High is markedly increased by the presence of the high-elevation Tibetan and Himalayan topography. Sancetta & Silvestri (1986) noted that at 2.4 Ma BP a diatom flora, interpreted to represent the first strong influence of cold dry winds from Siberian high-pressure cells, appeared in North Pacific sediments. Percentage variations in these species then intensified during the middle Pleistocene (1.0–0.35 Ma BP). These changes may be linked to intervals of rapid (or threshold-crossing) uplift in Tibet and the Himalayas. Similarly, increases in grain size and mass influx rates of aeolian material at 2.4 Ma BP in the western North Pacific are thought to reflect, in part, increased aridification of Asia (Janecek 1985). This, too, may reflect Tibetan–Himalayan uplift and isolation of Asia from moist subtropical air masses. Although these changes could of course simply be manifestations of the appearance and growth of Northern Hemisphere ice sheets, and their subsequent impacts on the climate system, they could also be indications that tectonic uplift altered climate.

Finally, it would be helpful to have more data on the transitions between the main Northern Hemisphere climatic régimes. Although it is quite clear that the Brunhes, Matuyama, and Gauss chrons all had distinctly different climatic responses, the sequence of change during the transitions between régimes is not yet well defined.

We thank Ann Esmay for assembling the figures. We particularly acknowledge Vivian Gornitz for her efforts in pursuing geological evidence of Plio-Pleistocene uplift in Asia and North America and for helpful general discussions about possible climate–tectonic connections. We also thank C. Sancetta and G. Kukla for critical reviews. This research was funded by proposals OCE82-19862 and OCE85-21514 from the Marine Geology and Geophysics Program in the Ocean Sciences Section of the National Science Foundation. This is LDGO contribution no. 4229.

References

Backman, J. 1979 Pliocene biostratigraphy of DSDP Sites 111 and 116 from the North Atlantic Ocean and the age of the Northern Hemisphere Glaciation. *Stockh. Contr. Geol.* **32**, 115–137.

Backman, J. & Pestiaux, P. 1986 Pliocene Discoaster abundance variations at Deep Sea Drilling Project Site 606: Biochronology and Paleoenvironmental implications. In *Init. Repts Deep-Sea Drilling Project*, vol. 94 (Ed. W. F. Ruddiman, R. B. Kidd, E. Thomas *et al.*), pp. 903–910. Washington, D.C.: U.S. Government Printing Office.

Barron, E. J. 1981 Paleogeography as a climatic forcing factor. *Geol. Rdsch.* **70**, 737–747.

Berger, A. L. 1984 Accuracy and frequency stability of the earth's orbital elements during the Quaternary. In *Milankovitch and climate* (ed. A. L. Berger, J. Imbrie, J. Hays, G. Kukla & B. Saltzman), pp. 3–39. Boston: D. Reidel.

Birchfield, G. E. & Weertman, J. 1983 Topography, albedo-temperature feedback, and climate sensitivity. *Science, Wash.* **219**, 284–285.

Birchfield, G. E., Weertman, J. & Lunde, A. T. 1982 A model study of the role of high-latitude topography in the climatic response to orbital insolation anomalies. *J. atm. Sci.* **39**, 71–87.

Boellestorf, J. 1978 North American Pleistocene Stages reconsidered in light of probable Pliocene–Pleistocene glaciation. *Science, Wash.* **202**, 305–307.

Bolin, B. 1950 On the influence of the earth's orography on the general character of the westerlies. *Tellus* **2**, 184–195.

Burbank, D. W. & Johnson, G. D. 1983 The late Cenozoic chronologic and stratigraphic development of the Kashmir intermontane basin, Northwestern Himalaya. *Palaeogeogr. Palaeoclim. Palaeoecol.* **43**, 205–235.

Charney, J. G. & Eliassen, A. 1949 A numerical method for predicting the perturbations of the middle-latitude westerlies. *Tellus* **1**, 38–54.

Curry, R. R. 1966 Glaciation about 3,000,000 years ago in the Sierra Nevada. *Science* **154**, 770–771.

Einarsson, T., Hopkins, D. M. & Doell, R. R. 1967 The stratigraphy of Tjornes, northern Iceland, and the history of the Bering Land Bridge. In *The Bering Land Bridge* (ed. D. M. Hopkins), pp. 312–325. California: Stanford University Press.

Eldholm, O., Thiede, J., Taylor, E. *et al.* 1987 *Init. Repts Ocean Drilling Project*, vol. 104A. Washington, D.C.: U.S. Government Printing Office.

Emiliani, C. & Geiss, J. 1958 On glaciations and their causes. *Geol. Rdsch.* **46**, 576–601.

Emiliani, C., Gartner, S. & Lidz, B. 1972 Neogene sedimentation on the Blake Plateau and the emergence of the Central American Isthmus. *Palaeogeogr. Palaeoclim. Palaeoecol.* **11**, 1–10.

Ewing, M. & Donn, W. L. 1956 A theory of Ice Ages. *Science, Wash.* **123**, 1061–1066.

Flint, R. F. 1957 *Glacial and Pleistocene geology*. New York: John Wiley.

Hartman, D. L. & Short, D. A. 1979 On the role of zonal asymmetries in climate change. *J. atm. Sci.* **36**, 519–528.

Hays, J. D., Imbrie, J. & Shackleton, N. J. 1976 Variations in the earth's orbit: pacemaker of the Ice Ages. *Science, Wash.* **194**, 1121–1132.

Hodell, D. A., Williams, D. F. & Kennett, J. P. 1985 Late Pliocene reorganization of deep vertical water mass structure in the western South Atlantic: faunal and isotopic evidence. *Bull. geol. Soc. Am.* **96**, 459–503.

Hsu, J. 1978 On the paleobotanical evidence for continental drift and Himalayan uplift. *Paleobotany* **25**, 131–142.

Imbrie, J. 1985 A theoretical framework for the Pleistocene Ice Ages. *J. geol. Soc. Lond.* **142**, 417–432.
Imbrie, J., Hays, J. D., Martinson, D. G., McIntyre, A., Mix, A. C., Morley, J. J., Pisias, N. G., Prell, W. L. & Shackleton, N. J. 1984 The orbital theory of Pleistocene climate: support from a revised chronology of the marine $\delta^{18}O$ record. In *Milankovitch and climate* (ed. A. L. Berger, J. Imbrie, J. Hays, G. Kukla & B. Saltzman), pp. 269–305. Boston: D. Reidel.
Izett, G. A. 1975 Late Cenozoic sedimentation and deformation in Northern Colorado and adjacent areas. *Mem. geol. Soc. Am.* **144**, 179–209.
Janecek, T. R. 1985 Eolian sedimentation in the Northwest Pacific Ocean: A preliminary examination of the data from Deep Sea Drilling Project Sites 576 and 578. In *Init. Repts Deep Sea Drilling Project*, vol. 85 (ed. G. R. Heath, L. H. Burckle *et al.*), pp. 589–603. Washington, D.C.: U.S. Government Printing Office.
Kasahara, A. & Washington, W. M. 1971 General circulation experiments with a six-layer NCAR model, including orography, cloudiness, and surface temperature calculations. *J. atm. Sci.* **28**, 657–701.
Keigwin, L. D. 1978 Pliocene closing of the Isthmus of Panama, based on biostratigraphic evidence from nearby Pacific Ocean and Caribbean Sea cores. *Geology* **6**, 630–634.
Keigwin, L. D. 1982 Isotopic paleoceanography of the Caribbean and east Pacific: role of Panama uplift in late Neogene time. *Science, Wash.* **217**, 350–353.
Keigwin, L. D. 1986 Pliocene stable-isotope record of Deep Sea Drilling Project Site 606: sequential events of ^{18}O enrichment beginning at 3.1 Ma. In *Init. Repts Deep-Sea Drilling Project*, vol. 94 (ed. W. F. Ruddiman, R. B. Kidd, E. Thomas *et al.*), pp. 911–920. Washington, D.C.: U.S. Government Printing Office.
Kennett, J. P. & Thunell, R. C. 1975 Global increase in Quaternary explosive vulcanism. *Science, Wash.* **187**, 497–503.
Liu, Dongsheng & Menglin, D. 1984 The characteristics and evolution of the paleoenvironment of China since the Late Tertiary. In *The evolution of the East Asian environment* (ed. R. O. Whyte, T.-N. Chiu, C.-K. Leung & C.-L. So), pp. 11–40. Hong Kong: University of Hong Kong Centre of Asian Studies.
Loubere, P. & Moss, K. 1986 Late Pliocene climatic change and the onset of Northern Hemisphere glaciation as recorded in the northeast Atlantic Ocean. *Bull. geol. Soc. Am.* **97**, 818–828.
Lucchiatta, I. 1979 Late Cenozoic uplift of the southwestern Colorado Plateau and adjacent lower Colorado River region. *Tectonophysics* **61**, 63–95.
Manabe, S. & Broccoli, A. J. 1985 The influence of continental ice sheets on the climate of an ice age. *J. geophys. Res.* **90**, 2167–2190.
Manabe, S. & Terpstra, T. B. 1974 The effects of mountains on the general circulation of the atmosphere as identified by numerical experiments. *J. atm. Sci.* **31**, 3–42.
McDougall, I. & Wensink, H. 1966 Paleomagnetism and geochronology of the Pliocene–Pleistocene lavas in Iceland. *Earth Planet. Sci. Lett.* **1**, 232–236.
McKee, E. D. & McKee, E. H. 1972 Pliocene uplift of the Grand Canyon region – Time of drainage adjustment. *Bull. geol. Soc. Am.* **83**, 1923–1932.
Mehta, P. K. 1980 Tectonic significance of the young mineral dates and the rates of cooling and uplift in the Himalaya. *Tectonophysics* **62**, 205–217.
Milankovitch, M. M. 1941 *Canon of insolation and the ice-age problem.* Koniglich Servische Akademie, Beograd. [English translation by the Israel Program for Scientific Translations; published by the U.S. Department of Commerce and the National Science Foundation, Washington, D.C.].
North, G. R., Mengel, J. G. & Short, D. A. 1983 Simple energy balance model resolving the seasons and continents: Application to the astronomical theory of the ice ages. *J. geophys. Res.* **88**, 6576–6586.
Opik, E. 1958 Climate and the changing sun. *Scient. Am.* **198**, 85–92.
Pisias, N. G. & Moore, T. C. Jr 1981 The evolution of Pleistocene climate: a time series approach. *Earth Planet. Sci. Lett.* **52**, 450–458.
Plass, G. N. 1956 The carbon dioxide theory of climatic change. *Tellus* **8**, 140–156.
Prell, W. L. 1982 Oxygen and carbon isotope stratigraphy for the Quaternary of Hole 502B: evidence for two modes of isotopic variability. In *Init. Repts Deep-Sea Drilling Project*, vol. 68 (ed. W. L. Prell, J. V. Gardner *et al.*), pp. 455–464. Washington, D.C.: U.S. Government Printing Office.
Prell, W. L. 1984 Covariance patterns of foraminiferal $\delta^{18}O$: an evaluation of Pliocene ice-volume changes near 3.2 million years ago. *Science, Wash.* **226**, 692–694.
Raymo, M. E., Ruddiman, W. F. & Clement, B. M. 1986 Pliocene–Pleistocene paleoceanography of the North Atlantic at Deep Sea Drilling Project Site 609. In *Init. Repts Deep-Sea Drilling Project*, vol. 94 (ed. W. F. Ruddiman, R. B. Kidd, E. Thomas *et al.*), pp. 895–902. Washington, D.C.: U.S. Government Printing Office.
Rogers, K. L., Repenning, C. A., Forester, R. M., Larson, E. E., Hall, S. A., Smith, G. R., Anderson, E. & Brown, T. J. 1985 Middle Pleistocene (Late Irvingtonian: Nebraskan) climatic changes in South-central Colorado. *Natn. geogr. Res.* **1**, 535–563.
Ruddiman, W. F. 1977 Late Quaternary deposition of ice-rafted sand in the subpolar North Atlantic (lat. 40° to 65°). *Bull. geol. Soc. Am.* **88**, 1813–1827.
Ruddiman, W. F. & McIntyre, A. 1981 Oceanic mechanisms for the amplification of the 23,000-year ice-volume cycle. *Science, Wash.* **212**, 617–627.

Ruddiman, W. F. & McIntyre, A. 1984 Ice-age thermal response and climatic role of the surface North Atlantic ocean, 40° to 63° N. *Bull. geol. Soc. Am.* **95**, 381–396.
Ruddiman, W. F., McIntyre, A. & Shackleton, N. J. 1986*a* North Atlantic sea-surface temperatures for the last 1.1 million years. In *North Atlantic paleoceanography* (ed. C. P. Summerhayes & N. J. Shackleton), (*Geol. Soc. Spec. Publ.* 21), pp. 155–173.
Ruddiman, W. F., Raymo, M. & McIntyre, A. 1986*b* Matuyama 41,000-year cycles: North Atlantic Ocean and Northern Hemisphere Ice Sheets. *Earth planet. Sci. Lett.* **80**, 117–129.
Ruddiman, W. F., McIntyre, A. & Raymo, M. 1986*c* Paleoenvironmental results from North Atlantic Sites 607 and 609. In *Init. Repts Deep-Sea Drilling Project*, vol. 94 (ed. W. F. Ruddiman, R. B. Kidd, E. Thomas *et al.*), pp. 855–878. Washington, D.C.: U.S. Government Printing Office.
Ruddiman, W. F., Kidd, R. B., Thomas, E. *et al.* 1986 *Init. Repts Deep-Sea Drilling Project*, vol. 94. Washington, D.C.: U.S. Government Printing Office.
Saini, H. S., Waraich, R. S., Malhotra, N. K., Nagpaul, K. K. & Sharma, K. K. 1979 Cooling and uplift rates and dating of thrusts from Kinnaur Himachal, Himalaya. *Himalayan Geol.* **9**, 549–567.
Sancetta, C. D. & Silvestri, S. 1986 Pliocene–Pleistocene evolution of the North Pacific ocean-atmosphere system, interpreted from fossil diatoms. *Paleoceanography* **1**, 163–180.
Schneider, D. A. & Kent, D. V. 1986 Influence of non-dipole field on determination of Plio-Pleistocene true polar wander. *Geophys. Res. Lett.* **13**, 471–474.
Seeber, L. & Armbruster, J. G. 1983 Continental subduction along the northwest and central portions of the Himilayan arc. *Boll. geof. Teoretica Appl.* **25**, 409–425.
Shackleton, N. J. & Opdyke, N. D. 1973 Oxygen isotope and paleomagnetic stratigraphy of equatorial Pacific core V28-238: oxygen isotope temperatures and ice volumes on a 10^5 and 10^6 year scale. *Quat. Res.* **3**, 39–55.
Shackleton, N. J. *et al.* 1984 Oxygen isotope calibration of the onset of ice-rafting and history of glaciation in the North Atlantic Region. *Nature, Lond.* **307**, 620–623.
Smagorinsky, J. 1953 The dynamical influence of large-scale heat sources and sinks on the quasi-stationary mean rotations of the atmosphere. *Q. Jl r. met. Soc.* **79**, 342–366.
Srivastava, S. P., Arthur, M., Clement, B. *et al.* 1987 *Init. Repts Deep-sea Drilling Project*, vol. 105A. Washington, D.C.: U.S. Government Printing Office.
Sutcliffe, R. C. 1951 Mean upper-air contour patterns of the Northern Hemisphere – the thermal-synoptic viewpoint. *Q. Jl r. met. Soc.* **77**, 435–440.
Trenberth, K. E. 1983 Interactions between orographically and thermally forced planetary waves. *J. atm. Sci.* **40**, 1126–1153.
Weissert, H. J., McKenzie, J. A., Wright, R. C., Clark, M., Oberhansi, H. & Casey, M. 1984 Paleoclimatic record of the late Pliocene at Deep Sea Drilling Project Sites 519, 521, 522, and 523 (central South Atlantic). In *Init. Repts Deep-Sea Drilling Project*, vol. 73 (ed. K. J. Hsu, J. L. LaBrecque *et al.*), pp. 701–715. Washington, D.C.: U.S. Government Printing Office.
West, R. M. 1984 Siwalik Faunas from Nepal: paleoecological and paleoclimatic interpretations. In *The evolution of the East Asian environment* (ed. R. O. Whyte, T.-N. Chiu, C.-K. Leung & C.-L. So), pp. 724–744. Hong Kong. University of Hong Kong Centre of Asian Studies.
Winograd, I. J., Szabo, B. J., Coplen, T. B., Riggs, A. C. & Holesar, P. T. 1985 Two-million year record of deuterium depletion in Great Basin ground waters. *Science, Wash.* **227**, 519–521.
Zimmerman, H. *et al.* 1985 History of Plio-Pleistocene climate in the Northeast Atlantic, Deep Sea Drilling Project Hole 552A. In *Init. Repts Deep-Sea Drilling Project*, vol. 81 (ed. D. Roberts, D. Schnitker *et al.*), pp. 861–875. Washington, D.C.: U.S. Government Printing Office.

Discussion

H. H. Lamb (*Climatic Research Unit, University of East Anglia, Norwich, U.K.*). There is surely no difficulty in supposing that a great stream of northwesterly winds from the north side of the Laurentide ice-sheet would produce a surface ocean drift and a great spread of sea ice towards mid Atlantic, if we interpret this feature of the modelled atmosphere in terms of very great persistence of the windstream (like the cold winds off the Antarctic ice-sheet today) rather than concentrating on the occasions of greatest strength, which would doubtless break up and blow away a good deal of ice.

But I am surprised that Dr Ruddiman lays so little stress on the drift of ice and polar water in glacial times from the East Greenland Current to mid ocean, which his own studies of ice-rafted material in the ocean-bed deposits first revealed. I am similarly surprised that he makes no mention at all of the (admittedly weaker) supply east of Iceland from the north, directly towards the British Isles. This can be substantiated as a feature of the recent Little Ice Age from the actual reports of drift ice between the late seventeenth century and the early part of this century. In the worst year, A.D. 1695, it seems (from ice and fisheries reports) that the polar water dominated the surface of the ocean across the whole width of the Norwegian Sea, extended south to the area of the Faeroe Islands and approached Shetland.

It is hard to believe that both these branches of the East Greenland cold current were not of some importance also in the major glaciation(s). Perhaps it is that they made less mark on the situation because of an inherent variability of the longitude position in which their main southward thrust occurred. This has to do with the anchoring of the east-Canadian–west-Atlantic cold trough in the upper westerlies, east of the great ice massif, and a greater variability of downstream cold troughs in the east Atlantic and European sectors.

W. F. RUDDIMAN. There is support for at least one of the drift patterns Professor Lamb mentioned. The general circulation model results of Manabe & Broccoli (1985) indeed show that the ice sheets intensified the northeasterly winds along the east coast of Greenland. As for the question of northerly winds producing a stronger southwards drift toward the British Isles east of Iceland, the evidence is more equivocal. Deposition patterns of ice-rafted sand published by Ruddiman (1977) suggest that Scandinavian icebergs did not move southwards along the European coast during glaciations; a depositional minimum located northwest of Britain argues against such a path. On the other hand, stronger north–south drift is more likely further out from the coast in the central North Atlantic, as well as in the southern Norwegian Sea near Iceland.

J. T. ANDREWS (*Department of Geological Sciences, University of Colorado, U.S.A.*). The suggested strong winds that are shown in various climatic models as moving along the northern perimeter of the ice sheet and then moving SE through Baffin Bay are a consequence of the height of the ice sheet. I still wonder whether strong, persistent offshore winds would not cause upwelling and open-pack conditions in northernmost Baffin Bay? If the wind régime were as predicted I would expect to find aeolian deposits and ventifacted surfaces in ice-free areas of Baffin Island (cf. Loken 1966) but such have not been reported.

Reference

Loken, O. H. 1966 Baffin Island refugia older than 54000 years. *Science, Wash.* **153**, 1378–1380.

Phil. Trans. R. Soc. Lond. B **318**, 431–449 (1988)
Printed in Great Britain

Palynological records from northwest African marine sediments: a general outline of the interpretation of the pollen signal

By H. Hooghiemstra†

Institute of Palynology and Quaternary Sciences, University of Göttingen, Wilhelm-Weber-Strasse 2, D-3400 Göttingen, F.R.G.

Pollen analysis of over 100 modern surface-sediment samples from the Atlantic off northwest Africa (between 35° and 4° N) has improved the understanding of the relation between modern pollen source areas, modern pollen transport and the modern distribution patterns of the pollen concerned in the marine sediments. Aeolian pollen transport is dominant in this region and the distribution patterns reflect the modern average atmospheric circulation.

Palaeoisopollen maps of the time slices of 9 ka BP and 18 ka BP monitor the atmospheric circulation during the last glacial–interglacial transition, providing evidence for the latitudinal position of the northeast trade winds and the African Easterly Jet, and the average northernmost and southernmost position of the intertropical convergence zone. In this way the number of major variables in time-series (continuous pollen records) was reduced to one; this has made it possible to interpret these records in terms of vegetational change, climatic humidity and changes in the intensity of the northeast trade winds.

1. Introduction

Pollen and pteridophyte spores are essentially products of the land vegetation and are therefore indicators of continental environmental conditions. The way in which pollen and spores are transported to the marine sediments may be complex. For a long time river and ocean current transport was considered as being generally dominant (see, for example, Muller (1959), Orinoco delta; Traverse & Ginsburg (1966), Great Bahama Bank; Cross *et al.* (1966), Gulf of California; Koreneva (1971), Mediterranean Sea). This may certainly be so in many areas, but it is not appropriate to generalize about methods of transport. Mudie (1982) and Melia (1984) showed that, in eastern Canada and northwest Africa, respectively, the aeolian factor may be important in the transport of pollen to the marine sediments.

A palynological study of modern marine surface-sediments off northwest Africa (between about 35° and 4° N) has provided distribution patterns (displayed as isopollen maps) of the most important northwest African and south European pollen producers. By relating these modern distribution patterns to the modern position of the source areas concerned, the transport mechanisms can be deduced. In this respect it is important to establish, for each pollen-producer, the period of main pollen release and to take into account the average atmospheric circulation over northwest Africa, which changes from month to month. In this way a model of modern pollen transport in the northwest African area can be established (Hooghiemstra & Agwu 1986; Hooghiemstra *et al.* 1986). This model of modern pollen

† Present address: University of Amsterdam, Dept. of Palynology and Palaeo/Actuo-Ecology, Hugo de Vries Laboratory, Kruislaan 318, 1098 SM, Amsterdam, The Netherlands.

transport was verified for selected time-slices of the past, namely, for 9 ka BP (this paper) and for 18 ka BP (Hooghiemstra *et al.* 1987). The model was then applied to studies of continuous pollen records (time-series) of deep-sea cores from the southern sector (*ca.* 9° N) (Hooghiemstra & Agwu 1988), from the central part of the area under study (*ca.* 21° N) (Hooghiemstra 1988*a*) and from the northern sector (*ca.* 29–37° N) (Hooghiemstra 1988*b*). The time intervals represented by these marine pollen records are 140–70 ka BP (southern sector), 20–5 ka BP (central part of the area under study) and (250) 140–3 ka BP (northern sector), including several glacial phases, interglacial phases and terminations.

The objective of this paper is to integrate the results of these studies, to present a general outline of the interpretation of pollen records of deep-sea cores from the northwest African area, and to emphasize the potential of marine palynology in palaeoclimatological and palaeoecological studies.

2. Setting of present vegetation and atmospheric circulation

The distinct climatic gradient between the arid Sahara and the humid climate in equatorial Africa is clearly reflected in a number of latitudinal vegetation zones (White 1983) (figure 1). These vegetation zones are characterized by the presence of certain taxa, which are palynologically mostly identified to the generic or family level.

Northwest Africa is further characterized by a pronounced atmospheric circulation system (figure 2) with surface winds (northeast trade winds, January trades, southerly trades) and a zonal wind belt at higher altitudes (African Easterly Jet or Saharan Air Layer).

The northeast trade winds reach as far south as *ca.* 23° N in July and August, and to *ca.* 4–6° N in January (January trades), depending on the main position of the intertropical convergence zone (ITCZ) in the course of the year (Leroux 1983). Although the January trades are most frequent in January and February, the associated sandstorms and dry hazes can also occur with high frequency as early as October and may persist until May (Maley 1982). The northeast trade winds today contribute only a limited amount of dust (including pollen), originating from the northwestern fringe of the Sahara and carried only a limited distance offshore. The trade winds obtain their maximum strength during late spring and blow basically parallel to the shoreline, and south of 18° N even slightly onshore towards the SSE (Sarnthein *et al.* 1982; Leroux 1983). From April to November the southwesterly trades account for a northeastward transport at the surface in the southern sector of the area under study.

The Saharan Air Layer (figure 2) is a zonal wind, which originates in the southern Sahara (corresponding today mainly to the source area of the Chenopodiaceae–Amaranthaceae pollen) and moves west above the trade-wind inversion between 8° and 23° N. A maximum concentration of transported dust (including pollen) is found at an altitude of *ca.* 3000 m between about 17° and 21° N. It passes around the upper-air high-pressure system found over the western Sahara, and finally forms a sickle-shaped course of trajectory over the eastern Atlantic (Tetzlaff & Wolter 1980; Sarnthein *et al.* 1982). Favourable transport conditions occur in the summer season between May and September, but most of the Saharan dust arrives at the Atlantic during July and August, when the Saharan Air Layer functions as a real jet wind between 17° and 21° N (African Easterly Jet).

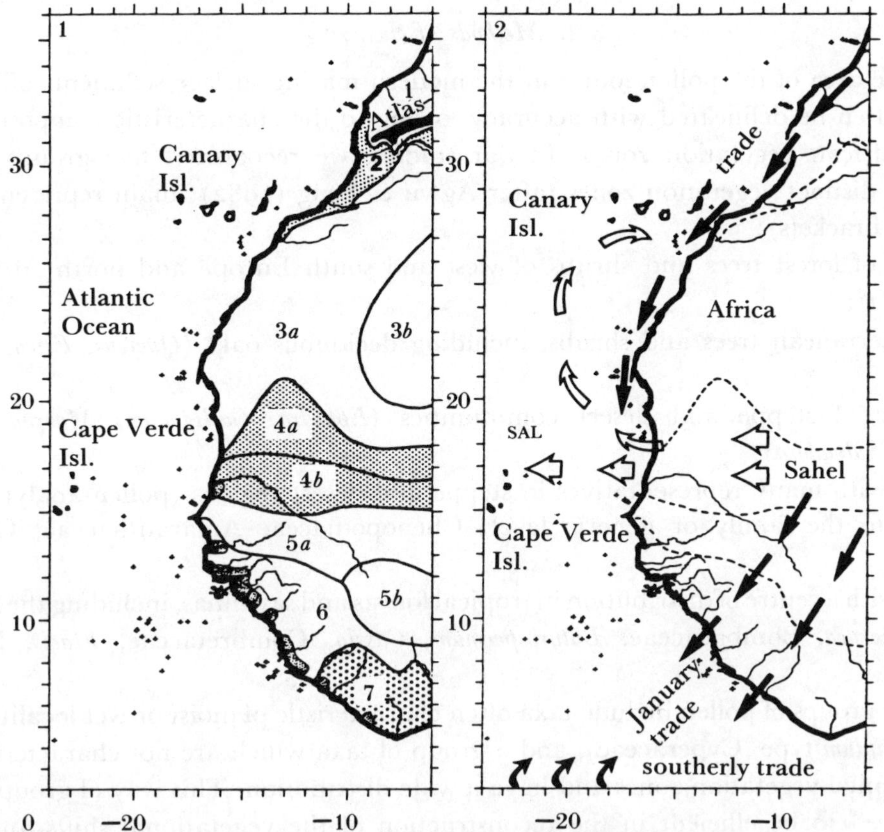

FIGURE 1. Vegetation of northwest Africa (after White 1983). From north to south, the numbers indicate: 1, Mediterranean vegetation zone; 2, steppes (semi-desert grassland and shrubs) of the western Atlas region; 3*a*, deserts and semi-deserts (regs, wadis, hamadas) of the Sahara; 3*b*, desert dunes without perennial vegetation, and absolute deserts; 4*a*, northern Sahel zone: semi-desert grassland and shrubs (dry thorn savannas); 4*b*, southern Sahel zone: *Acacia* wooded grassland and deciduous bush; 5*a*, Sudanian undifferentiated woodland (dry savannas); 5*b*, Sudanian woodland with abundant *Isoberlinia*; 6, Guinea savanna zone: mosaic of lowland rainforest and secondary grassland; 7, rainforest. The small stippled areas along the coast between 16° and 5° N indicate areas with mangroves (*Rhizophora*). As well as the climatic gradient, local orographic and edaphic factors determine the vegetational distribution and complicate the general pattern presented here. Crosses indicate the sample stations.

FIGURE 2. Major wind belts of northwest Africa. Solid arrows indicate surface winds (northeast trades, January trades, southerly trades); open arrows indicate zonal winds at higher altitudes (African Easterly Jet or Saharan Air Layer). Broken lines indicate the boundaries of the major vegetation belts; crosses, positions of sample stations.

3. POLLEN RECORDS AND METHODOLOGY

3.1. *Pollen records*

Intervals of different lengths from about 30 deep-sea cores from the Atlantic off northwest Africa have been palynologically analysed by several authors. A survey is presented by Hooghiemstra (1987). In most of the cores, time control is provided by oxygen-isotope chronology. The cores are located between about 37° and 9° N and mainly represent the time interval of 150–5 ka BP.

3.2. *Methods of clustering*

The source area of the pollen found in the modern marine surface sediments off northwest Africa can often be delineated with accuracy, owing to the characteristic composition of the northwest African vegetation zones. In our studies, we recognized five groups of pollen, representing distinct vegetation zones (after Agwu & Beug (1982); main representatives are indicated in brackets):

(i) pollen of forest trees and shrubs of west and south Europe and north Africa (*Pinus*, *Corylus*, *Betula*);

(ii) Mediterranean trees and shrubs, including deciduous oaks (*Quercus*, *Erica*, Oleaceae, *Rhus*);

(iii) plants of steppe and desert communities (*Ephedra*, *Calligonum*, *Maerua*, *Balanites*, *Heliotropium*, *Salvadora*);

(iv) taxa with many representatives in steppe and desert regions (pollen-analytically only identifiable to the family or generic level; Chenopodiaceae–Amaranthaceae, Gramineae, *Artemisia*);

(v) taxa with a centre of distribution in tropical forests and savannas, including the Sahel zone (*Acacia*, *Alchornea*, Bombacaceae, *Butyrospermum*, *Cassia*, Combretaceae, *Elaeis*, Meliaceae, *Rhizophora*).

Two other groups of pollen include taxa often characteristic of moist or wet localities (*Isoëtes*, *Typha–Sparganium* type, Cyperaceae), and a group of taxa which are not characteristic of the above mentioned vegetation zones and have a wide distribution. This way of grouping pollen types appeared to be efficient in the reconstruction of the vegetational shifts and climatic history of northwest Africa. Several authors, however, have grouped pollen taxa from northwest Africa (mostly land-based studies) in a more-or-less different way; see, for example, Bonnefille (1982), Caratine *et al.* (1979), Caratini & Cour (1980), Cour *et al.* (1973), Cour & Duzer (1976), Lezine (1987), Maley (1983), Rossignol-Strick & Duzer (1979) and Van Campo (1975). The most favourable way of grouping the northwest African pollen taxa depends heavily on the objective of the study concerned. The type of grouping given above is effective in monitoring large-scale latitudinal shifts of the northwest African vegetation zones. However, a different method of clustering has been used to monitor changes in the intensity of the northeast trade winds (see, for example, Hooghiemstra 1988*a*).

3.3. *Time-slices and time-series*

When studying changes of the northwest African vegetational and climatic conditions by means of pollen analysis of deep-sea cores we are dealing with two major variables. Firstly, the latitudinal position of the northwest African vegetation zones has changed and, as a consequence, the pollen source areas have changed positions. Secondly, the transporting mechanisms of the pollen concerned may have changed. In order to deal with these two variables we have studied three time-slices, namely, the modern situation, the time-slice of 18 ka BP (representing the situation during the last glacial maximum), and the time-slice of 9 ka BP (representing the phase of maximum northward expansion of the vegetation zones south of the Sahara).

The modern relation between the latitudinal position of the pollen source areas, the transporting mechanisms (wind, water currents) and the resulting distribution patterns of the pollen concerned in the marine surface-sediments can be investigated in detail, as the grid of surface-sediment samples is relatively dense. Comparing the modern situation with the 'snapshots' of the situation at *ca.* 9 ka BP and 18 ka BP (on the basis of a series of well-dated deep-sea cores), relevant changes in the transport of pollen during the last glacial–interglacial transition can be inferred. In this way the second variable mentioned above may be eliminated, this makes an interpretation possible of the time–series (continuous pollen records) into terms of changes in the northwest African vegetation.

4. Time-slices

4.1. *Modern situation*

The modern distribution of pollen and fern spores, originating from the northwest African vegetation, in the marine surface-sediments is based on pollen analysis of 109 sample stations, located between 35° and 4° N. (Full data and detailed discussion of the results are given by Hooghiemstra *et al.* (1986); concise results have been published by Hooghiemstra (1986) and Hooghiemstra & Agwu (1986).) The distribution of the mapped taxa (displayed as isopollen maps, figures 3–12) corresponds to the modern pattern of atmospheric circulation. Evidence for transport by water-currents is poor; this is not surprising in an area mainly characterized by climatic aridity. Most of the isopollen maps display a close relation between the geographical location of the source area and the distribution of the associated pollen and spores in the marine surface-sediments, taking into account for each taxon the main period of pollen release and the atmospheric circulation pattern characteristic of that part of the year. Thus the isopollen maps have the potential to record seasonal wind patterns. The above-mentioned close relation is most evident for *Quercus*, *Olea*, *Artemisia*, Chenopodiaceae–Amaranthaceae, Gramineae, Combretaceae, *Rhizophora*, *Alchornea*, *Elaeis* and the fern spores.

Studies on transportation and sedimentation of sediment fractions (Sarnthein *et al.* 1982) and microorganisms (Honjo 1976) demonstrated a process of lumping of dust particles into 'faecal pellets' in the water column near the sea surface, causing an accelerated sinking and preventing marked horizontal transport by ocean currents. Rapid sinking of pollen and spores to the ocean floor, after pollen and spores have passed the air–water boundary, can also be inferred from the clear-cut, not smeared, distribution patterns in the marine surface-sediments.

4.2. *Isopollen maps of* 9 *ka* BP *and* 18 *ka* BP

Palaeoisopollen maps of the time-slice of 18 ka BP are based on pollen analysis of core intervals of 14 well-dated deep-sea cores, located between 37° and 9° N (full data and detailed discussion of the results in Hooghiemstra *et al.* (1987). Eleven deep-sea cores, located between 37° and 15° N, were available with the time-slice of 9 ka BP and form the data set of the 9 ka BP palaeoisopollen maps (tables 1 and 2). The isopollen maps of 10 taxa and groups of taxa at these three time-slices are presented (figures 3–12).

The isopollen maps of the elements with a source area north of the Sahara (e.g. *Pinus*, Mediterranean elements, *Artemisia*) indicate that during all time-slices trade winds transported pollen from the Iberian Peninsula and the northern fringe of the Sahara in southern and southwestern directions. It may be concluded, therefore, that the belt with trade winds did not shift latitudinally during the last glacial–interglacial transition. This evidence is not compatible with the hypothesis assuming a zone of surface westerlies in the northern part of northwest

TABLE 1. MARINE POLLEN RECORDS OFF NORTHWEST AFRICA REPRESENTING THE TIME-SLICE 9 ka BP

core number	position: lat.	position: long.	water depth m	core interval (representing conditions at 9 ka BP) cm	dating method[a]	reference[b]
8057 B	37° 41′	10° 05′	2811	101–130	R_p/R_b index	3
M 15669-1	34° 53′	07° 49′	2030	6–17.5	$\delta^{18}O$	3
M80-17B	33° 37′	09° 25′	3016	50–80	^{14}C	1
M 16004-1	29° 59′	10° 39′	1512	9–25	$\delta^{18}O$	5, 3
M 15627-3	29° 10′	12° 05′	1024	0–14	$\delta^{18}O$	3
M123-92-1	25° 10′	16° 51′	2575	10–25	$\delta^{18}O$	1
M123-10-4	23° 30′	17° 43′	3080	10–35	$\delta^{18}O$	1
M 16017-2	21° 15′	17° 48′	800	50–72	$\delta^{18}O$, ^{14}C	2
M 13289-2	18° 05′	18° 01′	2490	35–65.5	^{14}C	5
M12347-2	15° 50′	17° 51′	2576	90	foram.	4
M12345-5	15° 29′	17° 22′	945	90–145	foram., $\delta^{18}O$	4

[a] R_p/R_b index: planktic/benthic foraminifera index.
[b] References: 1, Agwu & Beug (1982); 2, Hooghiemstra (1988*a*); 3, Hooghiemstra *et al.* (1988); 4, Rossignol-Strick & Duzer (1979); 5, this paper.

Africa (Rognon 1976; Rognon & Williams 1977; Nicholson & Flohn 1980). Changes in the intensity of the northeast trade winds during the last glacial–interglacial transition are amply demonstrated in the literature and can also clearly be inferred from our pollen data. The isopollen maps of 18 ka BP also suggest that, in the northern sector (around 30° N), the glacial trade winds had a stronger eastern component; this is in agreement with the sedimentological studies of Sarnthein & Walger (1974).

When the 18 ka BP and 9 ka BP isopollen maps are compared with the map of the modern situation it is evident that the latitudinal position of the Chenopodiaceae–Amaranthaceae maximum in the marine sediments is exactly at the same place (figure 8). The zone between *ca.* 19–22° N has apparently been continuously characterized by an abundant pollen supply from the east. It may be inferred, therefore, that the zonal belt with African Easterly Jet transport was stationary during the last glacial–interglacial transition and was continuously situated around 17–21° N. As the latitudinal position of the African Easterly Jet is related to the northernmost position of the ITCZ, it may be concluded that the latter was also stationary.

The stationary position of the belt with African Easterly Jet transport during the last glacial–interglacial transition is an important fact in the reconstruction of latitudinal shifts of the northwest African vegetation zones: Which pollen type is mainly transported by the African Easterly Jet to the Atlantic depends on the type of vegetation present between *ca.* 16 and 22° N.

Around 18 ka BP, when the Sahara had expanded maximally into northern and southern directions, this area was occupied by a chenopod-rich desert vegetation. This is reflected by high values (*ca.* 65%) of chenopod pollen in the marine sediments around 21° N. At about 9 ka BP the vegetation zones south of the Sahara had shifted to their northernmost position. It was estimated that the graminaceous-rich Sudanian and Sahelian vegetation zones extended at that time between about 16–17° and 23–24° N (Hooghiemstra 1988*a*; Lezine 1987). Thus

TABLE 2. POLLEN PERCENTAGES OF SELECTED TAXA FOR 9 ka BP DEEP-SEA CORE INTERVALS

(Mean values are given in the second column of each taxon.)

deep sea core	depth/cm	*Pinus*		Mediterranean elements		*Ephedra*		*Artemisia*		Chenopodiaceae–Amaranthaceae		Gramineae		tropical elements		Compositae Tub. + Lig.		fern spores		trade-wind indicators	
8057 B	101–110	60.0	61.4	14.3	18.2	0.2	0.1	0.2	0.3	1.6	1.1	1.2	1.9	0	0	2.1	2.0	1.7	1.2	62.6	63.8
	111–120	61.6	—	21.4	—	0.1	—	0.3	—	0.4	—	2.2	—	0	—	1.4	—	1.1	—	63.4	—
	121–130	62.5	—	19.0	—	0.1	—	0.4	—	1.2	—	2.2	—	0	—	2.4	—	0.9	—	65.4	—
M 15669-1	6–7.5	17.8	22.1	13.3	14.6	2.2	3.8	—	2.7	0	0	2.2	1.1	0	2.7	8.9	9.7	15.6	10.5	28.9	38.2
	16.0–17.5	26.3	—	15.8	—	5.3	—	5.3	—	0	—	0	—	5.3	—	10.5	—	5.3	—	47.4	—
M80-17B	50–60	18.4	20.4	34.6	32.0	4.7	5.4	2.1	3.4	14.3	15.7	6.1	4.7	0	0	0.9	1.7	0	0	26.1	31.0
	60–70	15.4	—	22.9	—	7.9	—	4.8	—	21.1	—	4.7	—	0	—	2.5	—	0	—	30.6	—
	70–80	27.3	—	38.5	—	3.7	—	3.4	—	11.8	—	3.4	—	0	—	1.7	—	0	—	36.2	—
M 16004-1	9–10	17.9	26.3	2.6	7.1	5.1	6.4	5.1	2.6	7.7	7.7	7.7	7.7	0	0	48.7	34.0	5.1	14.1	76.9	69.2
	24–25	34.6	—	11.5	—	7.7	—	—	—	7.7	—	7.7	—	0	—	19.2	—	23.1	—	61.5	—
M 15627-3	0–4	17.9	14.4	7.5	7.8	—	4.9	9.0	10.2	13.4	22.4	6.0	4.5	0	0	10.4	11.2	0	0.6	37.3	40.7
	10–14	10.9	—	8.0	—	9.7	—	11.4	—	31.4	—	2.9	—	0	—	12.0	—	1.1	—	44.0	—
M 123-92-1	10–15	57.1	58.0	3.4	3.0	2.6	2.9	2.2	2.7	18.4	17.5	2.2	2.1	0	0	3.0	2.7	0	0	64.9	66.2
	20–25	58.9	—	2.5	—	3.1	—	3.1	—	16.6	—	1.9	—	0	—	2.3	—	0	—	67.4	—
M 123-10-4	10–15	34.9	20.5	4.8	5.9	0.8	1.3	1.3	2.9	16.7	18.9	5.4	4.0	0.8	1.3	2.4	3.9	0	0	39.5	28.5
	30–35	6.1	—	7.0	—	1.8	—	4.4	—	21.1	—	2.6	—	1.8	—	5.3	—	0	—	17.5	—
M 16017-2	50–52	0.8	0.7	1.4	1.4	1.2	1.8	2.3	2.6	38.5	44.0	26.7	22.9	5.5	3.7	6.1	5.8	0.2	0.3	10.3	10.8
	60–62	0.5	—	1.8	—	1.8	—	2.3	—	55.2	—	13.3	—	3.1	—	5.1	—	0.3	—	9.7	—
	70–72	0.8	—	1.1	—	2.3	—	3.2	—	38.2	—	28.7	—	2.5	—	6.1	—	0.3	—	12.4	—
M 13289-2	35–36	5.8	3.1	1.4	0.6	5.8	4.7	1.4	1.0	14.5	12.7	24.6	24.8	20.3	25.0	2.9	4.0	2.9	2.7	15.9	12.8
	56–57	2.6	—	0.5	—	2.6	—	1.0	—	11.5	—	21.4	—	28.1	—	4.7	—	3.6	—	10.9	—
	64.5–65.5	1.0	—	0	—	5.6	—	0.7	—	12.2	—	28.4	—	26.7	—	4.3	—	1.7	—	11.6	—
M 12347-2	90	0	0	0	0	0	0	0	0	2.0	2.0	37.2	37.2	36.1	36.1	2.0	2.0	0.2	0.2	2.0	2.0
M 12345-5	90	0	0.1	0	0	0	0.2	0	0	2.1	2.8	37.3	35.1	35.5	40.9	0.3	0.4	0.5	0.3	0.3	0.6
	145	0.1	—	0	—	0.4	—	0	—	3.5	—	32.9	—	46.2	—	0.4	—	0	—	0.8	—

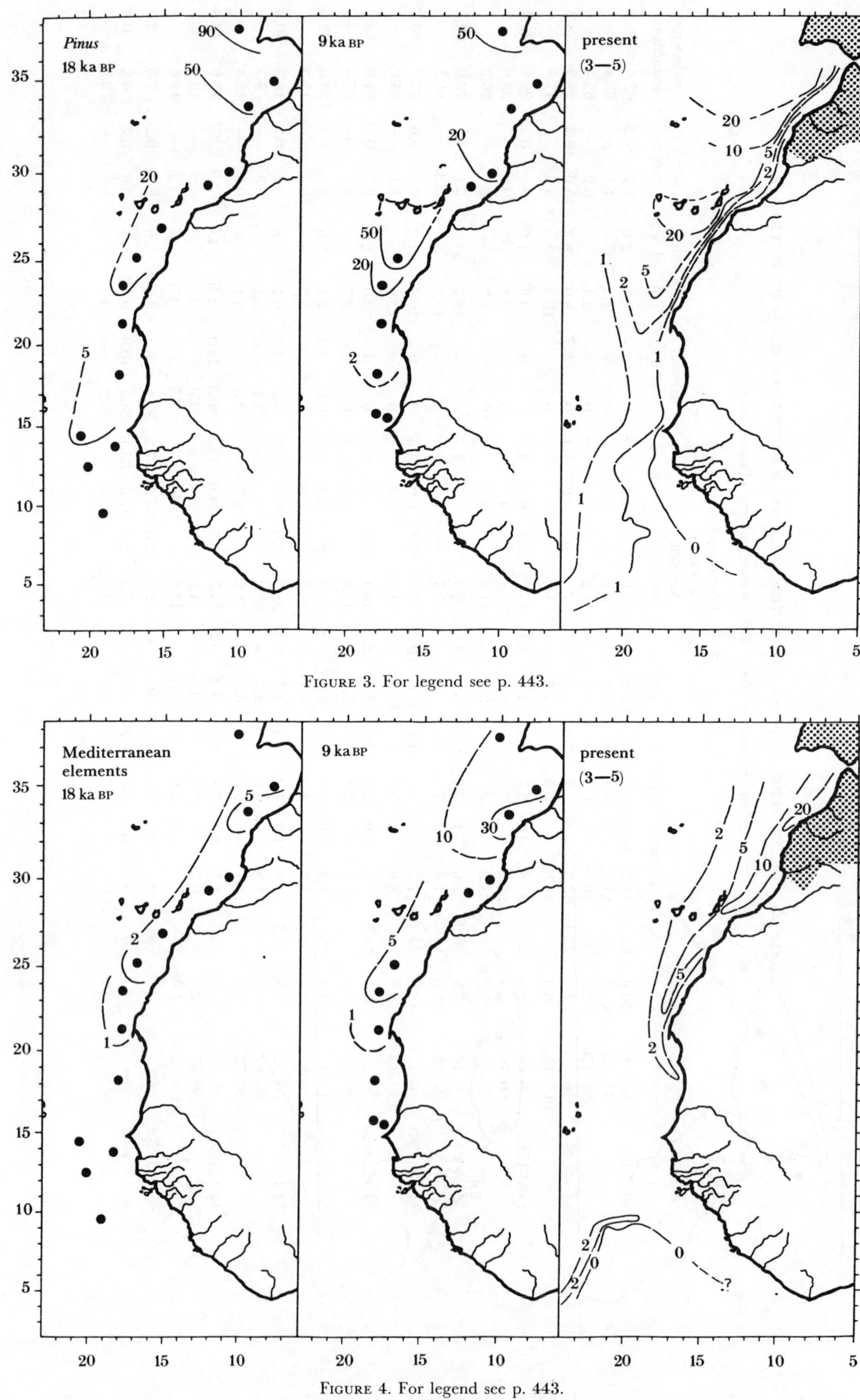

FIGURE 3. For legend see p. 443.

FIGURE 4. For legend see p. 443.

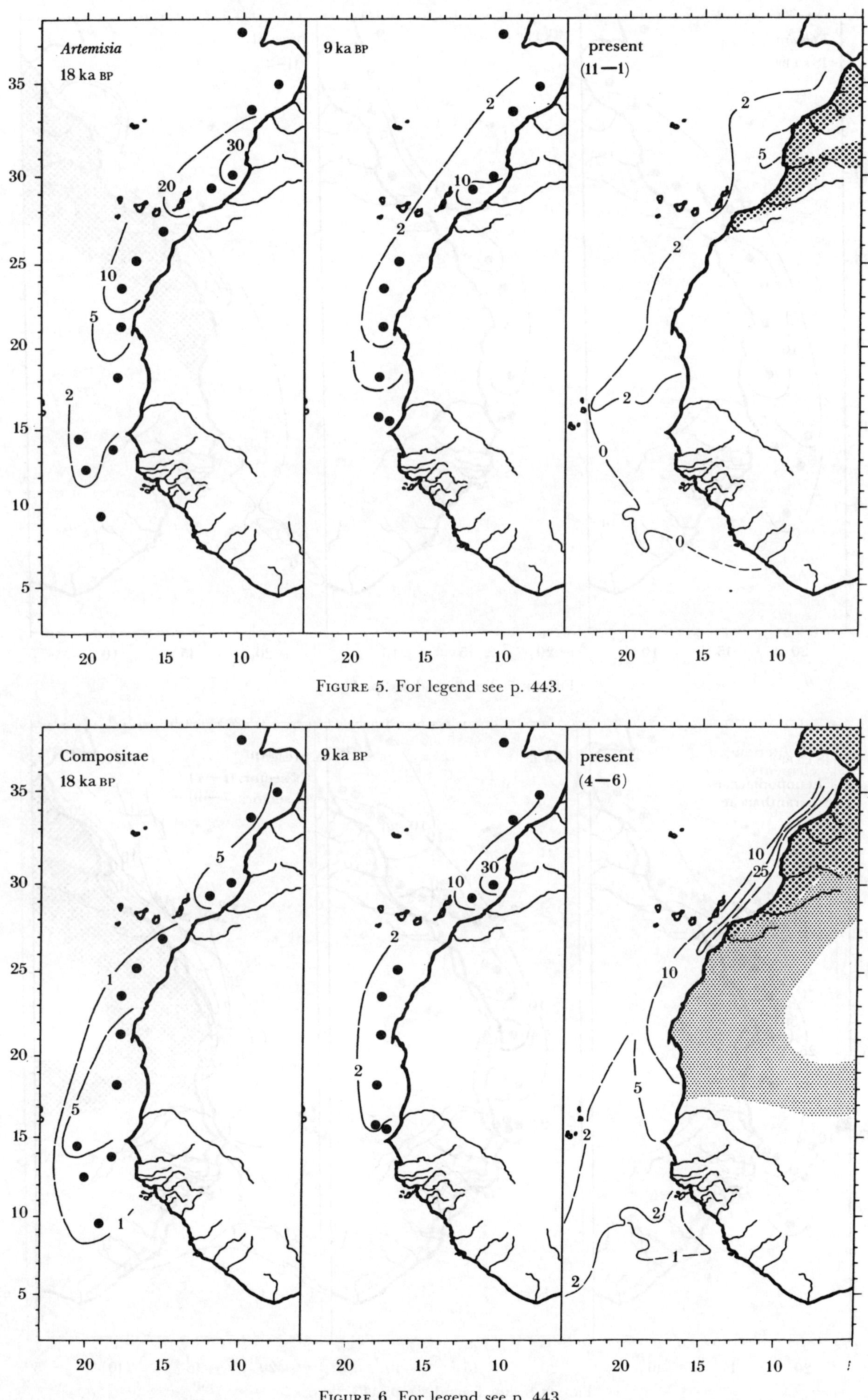

FIGURE 5. For legend see p. 443.

FIGURE 6. For legend see p. 443.

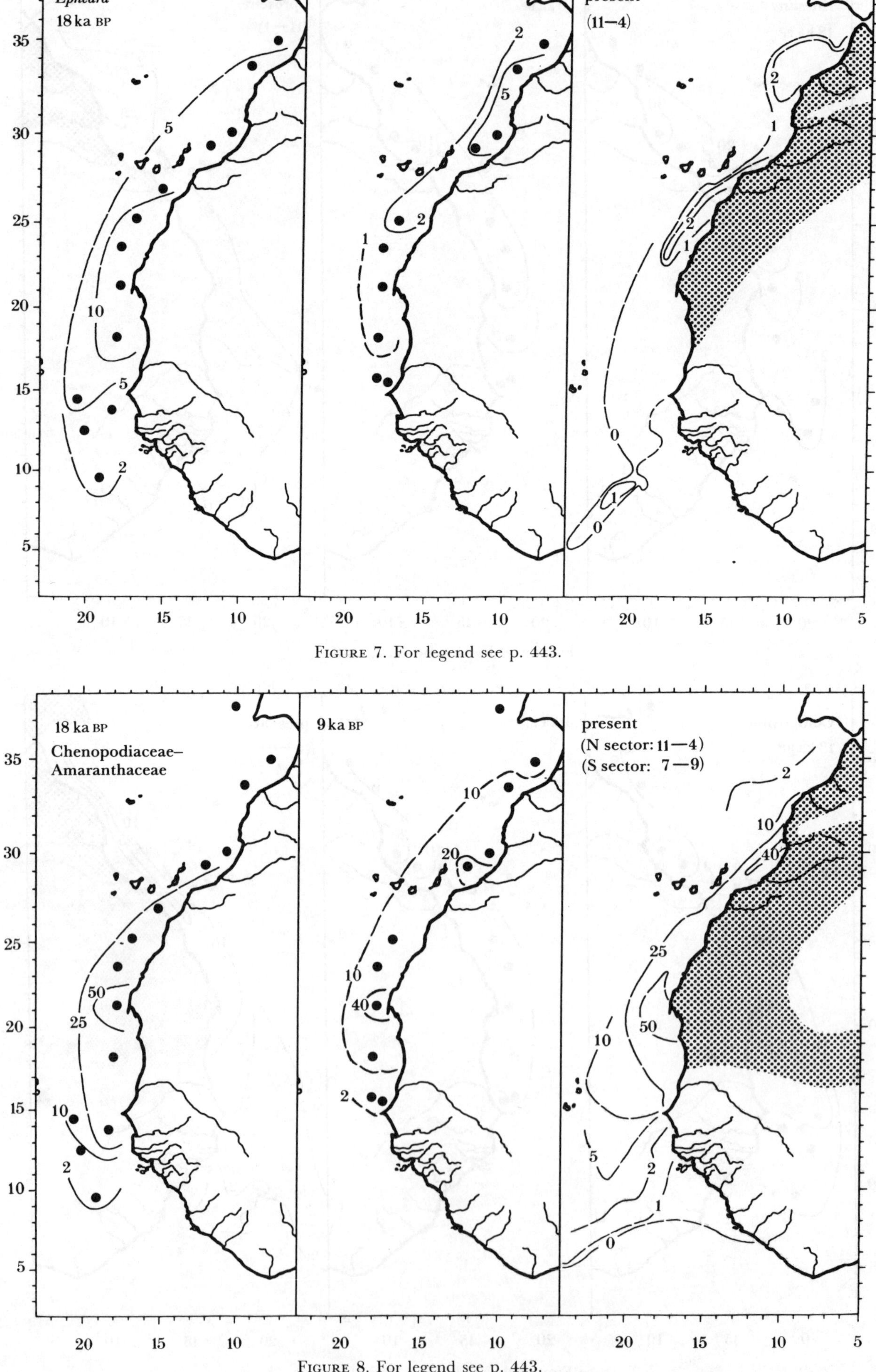

FIGURE 7. For legend see p. 443.

FIGURE 8. For legend see p. 443.

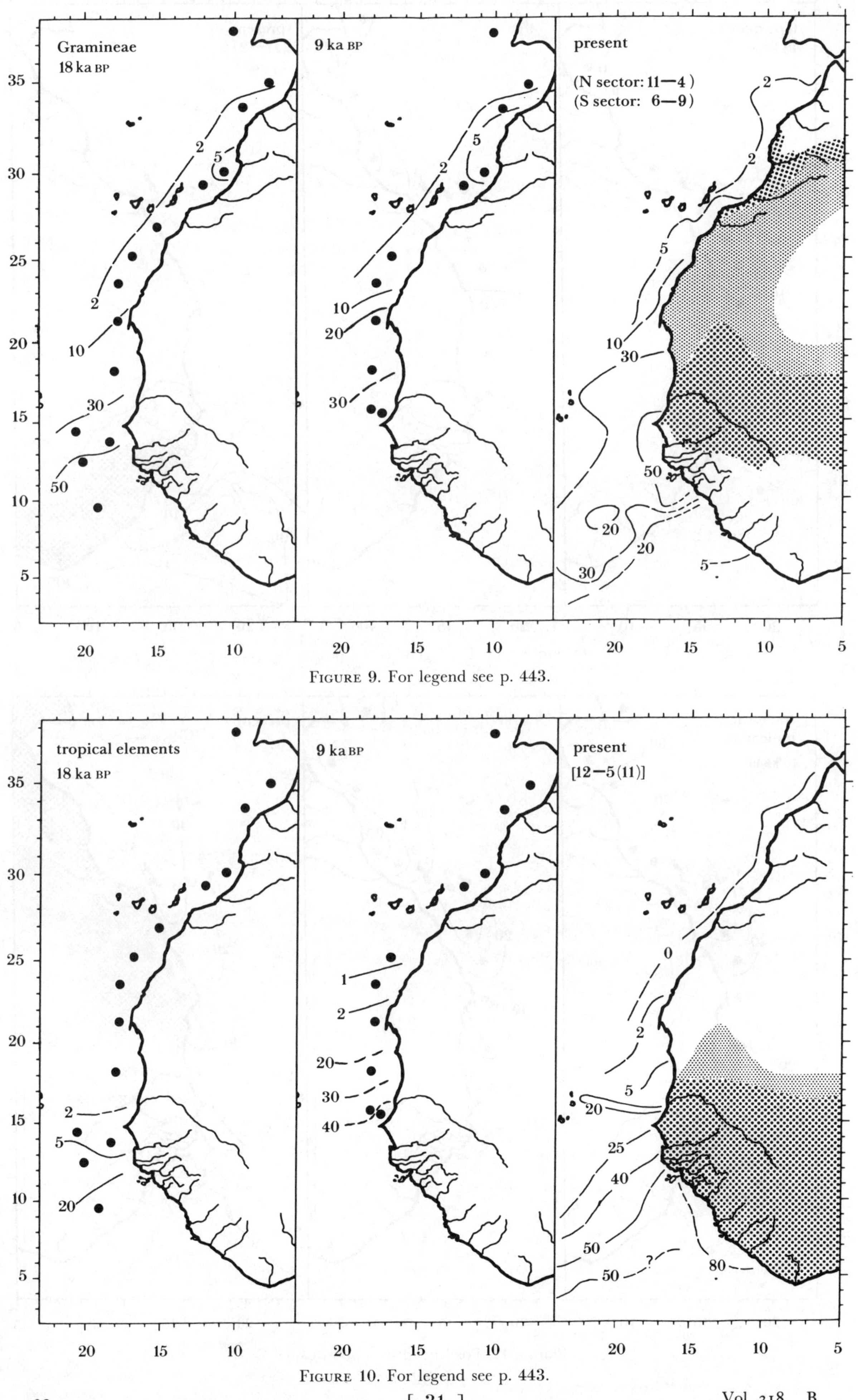

FIGURE 9. For legend see p. 443.

FIGURE 10. For legend see p. 443.

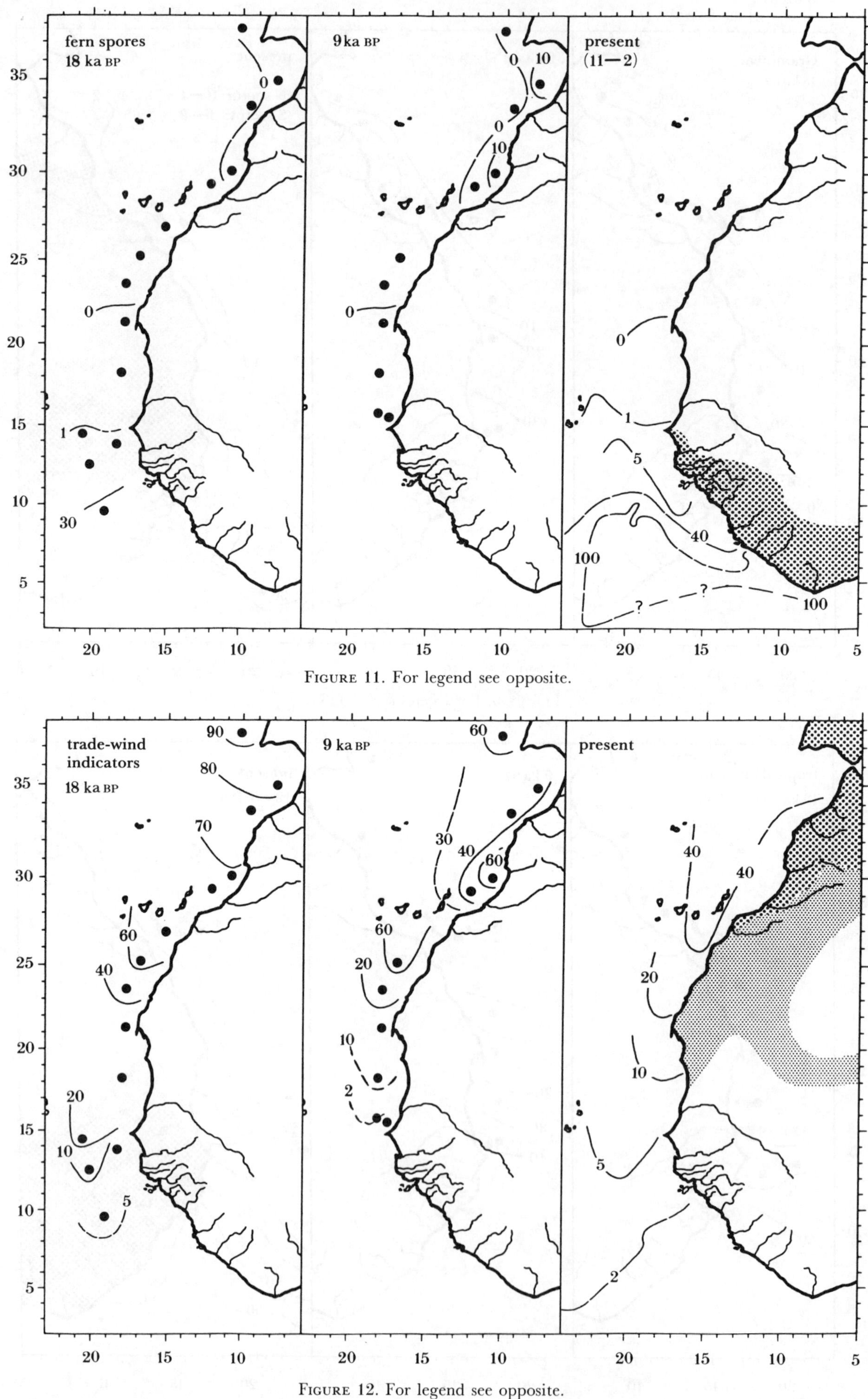

FIGURE 11. For legend see opposite.

FIGURE 12. For legend see opposite.

General description of figures 3–12

Isopollen maps of the Atlantic off northwest Africa, for various species at present and during the timeslices of 9 ka BP and 18 ka BP. Maps of the modern situation are based on 109 surface-sediment samples (locations indicated by crosses in figures 1 and 2); maps for 9 ka BP, based on 11 deep-sea core intervals (see tables 1 and 2); maps for 18 ka BP are based on 14 deep-sea core intervals. The modern isopollen maps and those of 18 ka BP were discussed in detail by Hooghiemstra *et al.* (1986 and 1987, respectively). Only a brief comment on the isopollen maps is provided in this paper.

Symbols: Bold stippled area, main source area; thin dotted area, source areas of secondary importance; dots, location of deep-sea cores with the concerning time-slice; figures in brackets, months of main pollen release (1–12, January–December). Changes in the river system and coastline of northwest Africa are not considered in the palaeoisopollen maps.

Figure 3. Isopollen maps of *Pinus*. Note: (1) pollen transport in southerly direction from the western Mediterranean area during all time-slices; (2) effective pollen transport far to the south *ca.* 18 ka BP, indicating strong trade winds; (3) the pine forests on the Canary Islands form a secondary source area for the trade winds.

Figure 4. Isopollen maps of the Mediterranean elements. Note: (1) A very poor representation *ca.* 18 ka BP, indicating the near-absence of the Mediterranean type of vegetation in the western Mediterranean area; the isopolls, however, reach far to the south, indicating strong trade winds. (2) around 9 ka BP this type of pollen is abundant, indicating a well-developed Mediterranean type of vegetation. The efficiency of pollen transport to the south has decreased markedly, indicating weak trade winds. (3) The modern isopollen map corresponds very well with the average flow pattern of the trade winds.

Figure 5. Isopollen maps of *Artemisia*. Note: (1) The very high representation *ca.* 18 ka BP, indicating abundant *Artemisia* vegetation at the northern fringe of the Sahara; (2) effective pollen transport to the south *ca.* 18 ka BP, indicating strong trade winds; (3) the source area of *Artemisia* is characteristic of the northern fringe of the Sahara during all time-slices.

Figure 6. Isopollen maps of the Compositae (subfamilies Tubuliflorae and Liguliflorae, *Artemisia* excluded). Note: (1) two main source areas *ca.* 18 ka BP at the northern and southern fringe of the Sahara; (2) a high representation *ca.* 9 ka BP and at present, indicating abundant Mediterranean vegetation in the western Mediterranean area; (3) the modern isopollen map corresponds very well with the average flow pattern of the trade winds; (4) the modern distribution of Compositae in northwest Africa is closely correlated with the distribution of the associated pollen in the marine sediments.

Figure 7. Isopollen maps of *Ephedra*. Note: (1) the high representation *ca.* 18 ka BP between 25° and 18° N, indicating extensive desert vegetation (expanded Sahara); (2) low efficiency of pollen transport to the south *ca.* 9 ka BP, indicating weak trade winds.

Figure 8. Isopollen maps of the Chenopodiaceae–Amaranthaceae. Note: (1) the latitudinally stationary area, with distinct pollen supply from the east by the African Easterly Jet, indicates that the African Easterly Jet did not shift latitudinally during the last glacial–interglacial transition. (2) The modern isopollen map corresponds between 35° and 5° N with the average flow pattern of the trade winds and between 17° and 22° N with the average course of trajectory of the African Easterly Jet. (3) In the modern isopollen map, the strong gradient in representation around 10° N reflects the southernmost position of the ITCZ; the accompanying rainbelt functions as a sharp cutoff for meridional aeolian pollen transport. (4) The northwards shift of the isopolls *ca.* 9 ka BP, in combination with a low representation, indicates a shrunken desert belt with a relatively northern geographical position.

Figure 9. Isopollen maps of the Gramineae. Note: (1) In the modern isopollen map, there is a close relation between the distribution of graminaceous-rich vegetation and the distribution of the associated pollen in the marine sediments. (2) In the modern isopollen map, there is a strong gradient in representation around 10° N, reflecting the southernmost position of the ITCZ; the accompanying rain belt functions as a sharp cutoff for meridional aeolian pollen transport. (3) A southwards shift of the isopolls *ca.* 18 ka BP, compared with the modern situation, indicates a southwards shift of the savanna zone. (4) The 9 ka BP isopollen map, especially, suffers from a shortage of core stations in the southern sector.

Figure 10. Isopollen maps of the tropical elements. Note: (1) The modern distribution of tropical forest is closely correlated with the distribution of the associated pollen in the marine sediments; (2) probable river-current transport of tropical forest pollen is evidenced by the Senegal river; (3) a southwards shift of the isopolls *ca.* 18 ka BP and a northwards shift of the isopolls *ca.* 9 ka BP (compared with the modern situation), indicates a shrunken and expanded area of tropical forest, respectively.

Figure 11. Isofrequency maps of fern spores. Note: (1) The modern isofrequency map shows a close correlation between the distribution of tropical forest and the distribution of fern spores in the marine sediments; (2) this potential for deducing the northernmost position of tropical forest could not be used for the timeslices of the past because of a lack of well-dated deep-sea cores in the southern sector; (3) an increase in the representation near the Mediterranean between 18 and 9 ka BP related to the development of the Mediterranean type of vegetation after about 10.5 ka BP; (4) in the modern surface-sediment samples of the northern sector, unfortunately, fern spores were not analysed: for this reason it is advisable not to use the map of the modern situation north of about 25° N.

Figure 12. Isopollen maps of the trade-wind indicators (*Pinus*, *Artemisia*, Compositae (Tubuliflorae + Liguliflorae, *Artemisia* excluded) and *Ephedra*). Note: (1) During all time-slices, pollen transport from the Mediterranean area to the south is demonstrated; this observation indicates a stationary belt with tradewind transport; (2) the position of the isopolls very far south *ca.* 18 ka BP indicates strong trade winds; (3) the main source areas *ca.* 9 ka BP are the Iberian Peninsula and the Canary Islands (contributing mainly *Pinus* pollen) and southern Morocco (contributing mainly *Artemisia* pollen); (4) low efficiency of southwards pollen transport *ca.* 9 ka BP indicates weak trade winds.

a great part of the Gramineae-rich vegetation zones was situated at that time in the belt with African Easterly Jet transport and, as a consequence, the offshore marine sediments contain chenopod pollen (*ca.* 44 %), as well as high percentages of graminaceous pollen (*ca.* 23–30 %, compared with about 10 % graminaceous pollen around 18 ka BP). It has to be noted that the offshore marine sediments between about 22 and 26° N receive African Easterly Jet-transported pollen, as well as trade-transported pollen. For this reason is cannot be expected that the representation of the Chenopodiaceae–Amaranthaceae decreased to a higher extend during that interval 18–9 ka BP. After the humid period of 9–7 ka BP climatic conditions turned more arid. The Sahara started to expand again in a southerly direction and the graminaceous-rich Sahelian and Sudanian vegetation zones shifted southwards and shrunk in north–south extension (see Hooghiemstra 1988*a*, figure 13). Thus the graminaceous-rich vegetation shifted out of the belt with African Easterly Jet transport, while the chenopod-rich desert vegetation became dominant again between 16 and 22° N. As a consequence, the representation of the Chenopodiaceae–Amaranthaceae in the marine sediments increased (to 50–60 % at present) and the representation of the Gramineae decreased (to 10–15 % at present).

5. Conclusions: interpretating the pollen signal in marine sediments off northwest Africa

On the basis of the modern isopollen maps and the palaeoisopollen maps of 18 ka BP and 9 ka BP the following conclusions may be drawn.

(i) In the marine sediments off northwest Africa, pollen is abundant in a relatively narrow offshore range. The method of grouping the pollen and pteridophyte spore taxa depends on the objective of the study.

(ii) The modern isopollen maps correspond to the modern flow pattern of atmospheric circulation, the northeast trade winds, the January trades and the African Easterly Jet being the major wind systems. Some water-current transport is indicated, but plays an unimportant part in the total pollen transport, owing to arid climatological conditions in a great part of northwest Africa and a pronounced atmospheric circulation.

(iii) The atmospheric circulation is driven by the intertropical convergence zone (ITCZ), which shifts over the continent between *ca.* 22° N (position during July and August) and *ca.* 4° N (position during December and January) in the course of the year. As many pollen-producers have a characteristic period of main pollen release, seasonal wind patterns can be recognized.

(iv) Washout of pollen from the atmosphere by the ITCZ-accompanying rainbelt functions as a sharp cutoff of meridional pollen transport. Pollen transport by long-distance transport (as in *Pinus*) shows, at the ITCZ, a slight increase in representation, whereas pollen from nearer source areas (e.g. Chenopodiaceae–Amaranthaceae, Gramineae) shows an abrupt decline in representation. The average southernmost position of the ITCZ can be inferred from these distribution patterns in the marine sediments.

(v) The modern isopollen maps show a close relation between the distribution of taxa in northwest Africa and the distribution of the associated pollen in the offshore marine surface-sediments. The average flow pattern of the major wind systems forms the link between both distribution patterns.

(vi) In studying the vegetational and climatic history of northwest Africa by means of pollen records of continuous deep-sea cores (time-series), we are dealing with two major variables: changing positions of the pollen source areas (i.e. vegetation zones) and changing transport systems. Palaeoisopollen maps ('snapshots' of the pollen distribution during time-slices of the past) of selected time-slices, on the basis of a number of well-dated deep-sea cores, provided qualitative evidence of changes in the atmospheric circulation during the last glacial–interglacial transition. In this way one major variable can be eliminated, making an interpretation of marine pollen records off northwest Africa possible in terms of vegetational changes.

(vii) Comparing the modern isopollen maps with those of 9 ka BP and 18 ka BP, we conclude that the trade winds did not shift latitudinally, but fluctuated only in intensity. Very effective pollen transport in the northern sector suggests that the last glacial trade winds intensified especially between about 36 and 24° N. The belt with maximum African Easterly Jet transport was situated, during the three time-slices, around 19–22° N; this indicates a stationary position of the African Easterly Jet during the last glacial–interglacial transition. This also implies a stationary northernmost position of the ITCZ.

(viii) The latitudinally stationary position of the belt with maximum zonal African Easterly Jet transport is an important datum for the reconstruction of latitudinal shifts of the northwest African vegetation zones: which pollen type is mainly transported by the African Easterly Jet to the Atlantic depends on the type of vegetation between about 16° and 22° N.

(ix) Climatic aridity may be the cause of a sparse vegetation cover and, as a consequence, a low pollen production in the area concerned. Relevant changes in the pollen production, displayed by pollen concentration and the pollen influx diagrams, may influence the quantitative record of the wind intensity. The quantitative pollen signal (pollen flux) in the marine sediments of an area with a very low pollen production probably has only a low correlation with the wind intensity (transport capacity).

(x) In the offshore sediments of northwest Africa, trade-transported pollen has source areas on the Iberian Peninsula, in Morocco and the northern fringe of the Sahara. In our studies we have used the taxa *Pinus*, Compositae (Tubuliflorae + Liguliflorae), *Artemisia* and *Ephedra* as trade-wind indicators.

(xi) The pollen influx record is the best proxy for evaluating changes in the trade-wind intensity. The pollen concentration record may be influenced by nonlinear accumulation processes. Knowledge of the regional vegetational history is necessary for a correct interpretation of the quantitative records.

(xii) A relative proxy for evaluating changes in the trade-wind intensity is the 'trade index', defined as the percentage of trade-transported pollen in the marine sediments. However, changes in the vegetational composition (e.g. leads and legs of the vegetational response to climatic change) may influence this relative signal to a considerable extent.

(xiii) Depending on the latitudinal position of the deep-sea cores studied, the pollen record has a different potential.

Marine pollen records from the northern sector (*ca.* 37–28° N) register mainly changes of the vegetation in the western Mediterranean area and the northern fringe of the Sahara. The trade wind record is influenced by vegetational changes to a considerable degree, as this area coincides with the source area of the trade-wind-indicating pollen.

Marine pollen records located in the area between *ca.* 22 and 16° N register changes of the vegetation from the western Mediterranean area to as far south as the Guinea zone, changes of fluvial runoff south of the Sahara (indicating climatic humidity) on the basis of the mangrove pollen record, changes in the intensity of the northeast trade winds, and African Easterly Jet transport.

Marine pollen records from the southern sector (*ca.* 16–7° N) record changes of the vegetation south of the Sahara, changes of the fluvial runoff south of the Sahara (indicating climatic humidity) on the basis of the mangrove record, seasonality in climatic humidity in the tropical forest area (climatic humidity all the year round is a prerequisite for tropical forest), and changes in the intensity of the northeast trade winds.

I thank U. Pflaumann, M. Sarnthein, K. Winn and R. Zahn (Geological Institute, Kiel University) for providing many samples and help with time control. I thank H.-J. Beug and E. Grüger (Institute of Palynology and Quaternary Sciences, Göttingen University), C. O. C. Agwu (Department of Botany, University of Nigeria, Nsukka), and G. Tetzlaff (Meteorological Institute, Hannover University) for many helpful discussions. C. O. C. Agwu, A. Bechler and H. Stalling carried out parts of the pollen analysis.

The opportunity to present parts of these studies at the INQUA–ASEQUA Conference (Dakar, April 1986) and at the 2nd International Conference on Paleoceanography (Woods Hole, U.S.A., September 1986) has contributed substantially to these studies. In this connection I also thank H. Faure (Marseille) and D. Rea (Washington).

This paper was based on the results of a number of previous studies, which were all financially supported by the German Federal Programme of Climate Research (Bundesministerium für Forschung und Technologie, Bonn: grant 200041 to the Institut für Palynologie und Quartärwissenschaften. Universität Göttingen). I thank the Netherlands Organization for the advancement of pure research (Z.W.O.) and the Hugo de Vries Laboratory, Department of Palynology (University of Amsterdam) for the opportunity to prepare this manuscript. A. M. Vink is acknowledged for improving the English text.

References

Agwu, C. O. C. & Beug, H.-J. 1982 Palynological studies of marine sediments off the West African coast. *Met. ForschErgebn.* **C 36**, 1–30.

Bonnefille, R., Rossignol-Strick, M. & Riollet, G. 1982 Organic matter and palynology of DSDP site 367 Pliocene–Pleistocene cores off West Africa. *Oceanologica Acta* **5**, 97–104.

Caratini, C., Bellet, J. & Tissot, C. 1979 Étude microscopique de la matière organique: palynologie et palynofaciès. In *Oregon III: Mauritanie, Sénégal, Isles du Cap-Vert* (ed. M. Arnould & R. Pellet), pp. 215–247. Paris: CNRS.

Caratini, C. & Cour, P. 1980 Aéropalynologie en Atlantique Oriental au large de la Mauritanie, du Sénégal et de la Gambie. *Pollen Spores* **22**, 245–256.

Cour, P., Guinet, P., Cohen, J. & Duzer, D. 1973 Reconnaissance des flux et retombées polliniques et de la sédimentation actuelle au Sahara nord occidental. In *Palynology in medicine* (*Proc. 3rd Int. palynol. Conf., Novosibirsk*), pp. 41–58. Moscow: Publishing House Nauka.

Cour, P. & Duzer, D. 1976 Persistance d'un climat hyperaride au Sahara central et méridional au cours de l'Holocène. *R. Géogr. phys. Géol. dyn.* **18**, 175–198.

Cross, A. T., Thompson, G. C. & Zaitzeff, J. B. 1966 Source and distribution of palynomorphs in bottom sediments, southern part of Gulf of California. *Mar. Geol.* **4**, 467–524.

Honjo, S. 1976 Coccoliths: production, transportation and sedimentation. *Mar. Micropaleont.* **1**, 65–79.

Hooghiemstra, H. 1986 Distribution patterns of pollen in marine sediments form a record for the seasonal wind patterns over NW Africa. INQUA–ASEQUA 1986 Dakar Symposium 'Changements globaux en Afrique'. *Trav. Documents* **197**, 191–194. Paris: ORSTOM.

Hooghiemstra, H. 1987 Survey of palynologically analysed deep-sea cores in the northeast Atlantic off northwest Africa. *Palaeoecol. Afr.* **18**, 47–53.

Hooghiemstra, H. 1988*a* Changes of major wind belts and vegetation zones in NW Africa 20,000–5000 yr B.P. as deduced from a marine pollen record near Cap Blanc. *Rev. Palaeobot. Palynol.* (In the press.)

Hooghiemstra, H. 1988*b* Variations of the NW African trade wind régime during the last 140,000 years: changes in pollen transport efficiency as evidenced by marine sediment records. NATO ASI series. (In preparation.)

Hooghiemstra, H. & Agwu, C. O. C. 1986 Distribution of palynomorphs in marine sediments: a record for seasonal wind patterns over NW Africa and adjacent Atlantic. *Geol. Rdsch.* **75**, 81–95.

Hooghiemstra, H., Agwu, C. O. C. & Beug, H.-J. 1986 Pollen and spore distribution in recent marine sediments: a record of NW-African seasonal wind patterns and vegetation belts. *Meteor ForschErgebn.* C **40**, 87–135.

Hooghiemstra, H., Belcher, A. & Beug, H.-J. 1987 Isopollen maps for 18,000 yr B.P. of the Atlantic offshore of northwest Africa: evidence for palaeo-wind circulation. *Paleoceanography* (In the press.)

Hooghiemstra, H. & Agwu, C. O. C. 1988 Changes in the vegetation and trade winds in west equatorial Africa 140,000–70,000 yr B.P. as deduced from two marine pollen records. *Palaeogeogr. Palaeoclim. Palaeoecol.* (In the press.)

Hooghiemstra, H., Stalling, H. & Agwu, C. O. C. 1988 Vegetational and climatic change at the NW fringe of the Sahara during the last 140,000 years: evidence from 4 new marine pollen records. (In preparation.)

Koreneva, E. V. 1971 Spores and pollen in Mediterranean bottom sediments. In *The micropaleontology of oceans* (ed. B. M. Funnell & W. R. Reidel), pp. 361–371. New York: Cambridge University Press.

Leroux, M. 1983 *The climate of tropical Africa.* Paris: Ed. Champion.

Lezine, A. M. 1987 Paleoenvironnements vegetaux d'Afrique nord-tropicale depuis 12,000 B.P. Analyse pollinique de series sedimentaires continentales (Sénégal-Mauritanie). Thèse, Université de Marseille (2 vols).

Maley, J. 1982 Dust, clouds, rain types, and climatic variations in tropical north Africa. *Quat. Res.* **18**, 1–16.

Maley, J. & Livingstone, D. A. 1983 Extension d'un élément montagnard dans de sud du Ghana (Afrique de l'Ouest) au Pleistocène supérieur et a l'Holocène inférieur: premières données polliniques. *C. R. Séanc.* Acad. Sci. Paris **296**, sér. 2, 1287–1292.

Melia, M. B. 1984 The distribution and relationship between palynomorphs in aerosols and deep-sea sediments off the coast of northwest Africa. *Mar. Geol.* **58**, 345–371.

Mudie, P. J. 1982 Pollen distribution in recent marine sediments, eastern Canada. *Can. J. Earth Sci.* **19**, 729–747.

Muller, J. 1959 Palynology of recent Orinoco delta and shelf sediments. *Micropaleontology* **5**, 1–32.

Nicholson, S. E. & Flohn, H. 1980 African environmental and climatic changes and the general atmospheric circulation in Late Pleistocene and Holocene. *Climatic Change* **2**, 313–348.

Rognon, P. 1976 Essai d'interprétation des variations climatiques au Sahara depuis 40,000 ans. *R. Géogr. phys. Géol. dyn.* **18**, 251–282.

Rognon, P. & Williams, M. A. J. 1977 Late Quaternary climatic changes in Australia and North Africa: a preliminary interpretation. *Palaeogeogr. Palaeoclim. Palaeoecol.* **21**, 285–327.

Rossignol-Strick, M. & Duzer, D. 1979 Late Quaternary pollen and dinoflagellate cysts in marine cores off West Africa. *Met. ForschErgebn.* C **30**, 1–14.

Sarnthein, M. & Walger, E. 1974 Der äeolische Sandstrom aus der West-Sahara zur Atlantikküste. *Geol. Rdsch.* **63**, 1065–1087.

Sarnthein, M., Thiede, J., Pflaumann, U., Erlenkeuser, H., Fütterer, D., Koopmann, B., Lange, H. & Seibold, E. 1982 Atmospheric and oceanic circulation patterns off northwest Africa during the past 25 million years. In *Geology of the northwest African continental margin* (ed. U. Von Rad *et al.*), pp. 545–602. Berlin: Springer-Verlag.

Tetzlaff, G. & Wolter, K. 1980 Meteorological patterns and the transport of mineral dust from the north African continent. *Palaeoecol. Afr.* **12**, 31–42.

Traverse, A. & Ginsburg, R. N. 1966 Palynology of the surface sediments of Great Bahama Bank, as related to water movement and sedimentation. *Mar. Geol.* **4**, 417–459.

Van Campo, M. 1975 Pollen analysis in the Sahara. In *Problems in prehistory, North Africa and the Levant* (ed. F. Wendorf & E. Marks), pp. 45–64. Dallas, Texas: S.M.U. Press.

White, F. 1983 *The vegetation of Africa. A descriptive memoir to accompany the UNESCO/AETFAT/UNSO vegetation map of Africa.* Paris: Unesco. (356 pages; with 3 maps.)

Discussion

JUDITH MAIZELS (*Department of Geography, University of Aberdeen, U.K.*). Dr Hooghiemstra has mentioned that evidence from the Senegal River suggested that a dramatic increase in discharge occurred around 11 ka BP. This statement raises two important points that may have some bearing on Dr Hooghiemstra's interpretation of the distribution pattern of pollen off the coast of northwest Africa. Firstly, it would be helpful if Dr Hooghiemstra would summarize the type of evidence available for estimating this change in river discharge. Is this estimate based on changes in sediment volume at the river mouth, or on changes in particular sediment and/or channel-form characteristics? In the former case, an increase in sediment volume may reflect an increase in aridity rather than increased humidity, as Dr Hooghiemstra suggests. Increased aridity could act to reduce the vegetation cover, thereby exposing greater volumes of sediment to erosion and fluvial transport during infrequent flood events. Changes in sediment characteristics associated with changes in channel morphology, by contrast, would provide a more reliable indicator of climatic change.

My second point is that the many large rivers, such as the Senegal River, that issue into the North Atlantic along this coast, are likely to input large amounts of pollen, as well as sediment. Much of this pollen may include reworked pollen from older sediments. Fluvial currents might also extend for some distance offshore, acting to redistribute bottom pollen deposits and modifying the original aeolian pollen distribution. Would Dr Hooghiemstra comment on the significance of these fluvial processes in affecting the distribution of pollen in the bottom deposits off the coast of northwest Africa?

H. Hooghiemstra. The estimation of changes in fluvial run-off of the Senegal river was based on fluctuations of the pollen record of *Rhizophora* (mangrove). Mangroves occur abundantly in river deltas and in narrow ranges along the coast. High fluvial runoff may cause extensive delta areas with a high mangrove pollen production. In the modern day arid parts of the Senegal River delta, extensive Early-Holocene buried mangrove peats show these fluctuations in the stands of mangroves. A high representation and influx of mangrove pollen in the offshore marine sediments was assumed to indicate high fluvial runoff. This pollen evidence is corroborated by geomorphological evidence (Michel 1984). Some river current transport of pollen was evidenced by the modern isopollen maps of *Rhizophora* and Cyperaceae. Ocean floor topography may also apparently influence distribution patterns of pollen in marine sediments, as suggested by the modern isofrequency maps of *Elaeis* and the fern spores. There is no evidence for a relevent contribution of reworked pollen. In the Gulf of Guinea it is difficult to estimate the contribution of fluvial pollen transport, compared with aeolian pollen transport by the January trades. It is estimated as to be low. In general, in the literature, evidence for aeolian pollen transport in tropical forest areas is increasing.

Reference

Michel, P. 1984 *Bull. Soc. Langued. Geogr.* **18**, 125–138.

R. G. W. Ward. (*Environmental and Geographical Studies, Roehampton Institute, Southlands College, London, U.K.*) Dr Hooghiemstra showed us isopollen maps of the present day and 18 ka BP, and inferred from their similarity that the main elements of the atmospheric circulation (the position of the northeast trades, the easterly wind from the Sahel, and the location of the ITCZ) had remained largely unchanged throughout this period. However, the final determinant of where pollen comes to rest in sediments must be the oceanic circulation, particularly if pollen has a long residence time within the zone of moving waters. What we may be seeing, therefore, is stability within the pattern of ocean currents, rather than stability in the wind pattern. More recent studies about sedimentation processes in the ocean describe an aggregation of suspended material into larger particles (inorganic aggregation (flocculation), as well as organic aggregation, forming faecal pellets reaching several millimetres in diameter), which cause higher settling velocities. This mechanism may explain why horizontal transport of pollen by ocean currents is less relevant than expected. Water current transport of pollen in the northwest African area is not excluded and evidenced by the isopollen maps of *Rhizophora* and Cyperaceae. Aeolian pollen transport, however, is apparently dominant to a large extent; this observation is corroborated by the clear-cut, not smeared distribution patterns in the modern surface sediments.

Phil. Trans. R. Soc. Lond. B **318**, 451–485 (1988)
Printed in Great Britain

Vegetational evidence for late Quaternary climatic changes in southwest Europe in relation to the influence of the North Atlantic Ocean

By C. Turner[1] and G. E. Hannon[2]

[1] *Department of Earth Sciences, The Open University, Walton Hall, Milton Keynes, Bucks MK7 6AA, U.K.; Subdepartment of Quaternary Research, Botany School, University of Cambridge, Downing Street, Cambridge CB2 3EA, U.K.*

[2] *School of Botany, Trinity College, University of Dublin, Dublin 2, Eire*

During the period 20–8 ka BP, movements of the polar front in the North Atlantic Ocean between the latitudes of Iceland and the Iberian peninsula greatly affected the climate of western Europe. During the Lateglacial, sea-surface temperature changes were particularly marked in the Bay of Biscay. Such migrations of the polar front, which have been shown to be time-transgressive, have been used to explain Lateglacial climatic events in northwestern Europe.

A comparative study of Lateglacial and early Holocene records from lacustrine sites in northern and northwestern Spain and the Pyrenees confirms that the Lateglacial climatic amelioration was time-transgressive along the seaboard of western Europe, beginning 500–1000 years earlier in northwestern Spain than in the British Isles. This time-lag is further exaggerated in the vegetational response by migrational lags and edaphic factors. There are marked differences in the nature and chronology of Lateglacial plant successions, not only between southwest and northwest Europe, but particularly between sites in northwestern Spain, the coastal lowlands of the Pays Basque and the Pyrenees.

Sites in northwestern Spain, including that of Sanabria Marsh, here published in detail for the first time, show the moderating climatic influence of the Atlantic Ocean throughout the Lateglacial. There, the climatic amelioration began early, perhaps before 14 ka BP. Deciduous oak forest had already begun to develop during Lateglacial times; this observation suggests that the perglacial refugia for these trees lay close to the maritime Atlantic coasts of Spain and Portugal, and not in the Pyrenees as some authors have proposed.

After the onset of the Lateglacial climatic amelioration, pine and birch forest became widespread in the Pyrenees but oaks were very sparse or absent. Oak forest only developed there after 10 ka BP in the early Holocene. The Younger Dryas episode of cooling can be detected, but only by a small expansion of herbaceous plant communities in some areas and with almost no lowering of the treeline.

In contrast, Lateglacial conditions in the Pays Basque appear to have been cold and bleak. Even birch and pine forest was poorly developed and may have disappeared with the onset of the Younger Dryas cooling. Acid heathland with *Empetrum* and ericaceous plants then developed, to be replaced by oak–hazel forest in the early Holocene. Here, clearly, the influence of cold polar water conditions in the Bay of Biscay was very strong.

Pollen diagrams from marine cores in the Bay of Biscay are also reviewed, but low sedimentation rates, bioturbation and differential transport and preservation of pollen make comparison with continental pollen diagrams difficult and correlation only possible in broad terms. Accurate vegetational interpretations are impossible.

Palynologists working on archaeological cave and rock shelter sequences in southwest France and northern Spain have claimed to recognize, between 32 and 14 ka BP, a series of interstadial intervals with expansions of temperate trees. Careful consideration of pollen diagrams covering the purported Laugerie and Lascaux interstadials, said to occur between 16 and 20 ka BP (conventionally the maximum period of glacial advance of the last glacial stage), suggests that temperate pollen has percolated down through overlying deposits and been preserved in certain sedimentologically favourable beds. Although widely accepted by archaeologists, these interstadials appear to have no reality and must be rejected. There is no trace of them in the long lacustrine records of Les Echets (Beaulieu & Reille 1984) and Grande Pile (Woillard 1975, 1978). There is thus no good palynological record for 30–16 ka BP from south-west Europe, other than the long pollen sequence from Padul in southern Spain (Pons & Reille 1986).

1. Introduction

This paper examines the evidence for vegetational changes in the northern and northwestern parts of the Iberian peninsula and in southwest France and their relation to conditions in the North Atlantic Ocean between approximately 30 and 7 ka BP, from the period just before the glacial maximum to the completion of the major warming of the early Holocene. In a series of papers that changed the whole background of European Lateglacial palaeoclimatic studies, Ruddiman & McIntyre described the migrations of the polar front in the North Atlantic Ocean during the late Quaternary and, in particular, its advances and retreats over the past 20 ka in the eastern sector, adjacent to the coastline of western Europe (Ruddiman & McIntyre 1973, 1981; Ruddiman *et al.* 1977). They showed that the maximum southward extension of polar water in the North Atlantic during the last interglacial–glacial cycle occurred between 20 and 16 ka BP at the period when the Laurentide and Scandinavian ice sheets reached their greatest size. At that time polar water, defined as water sedimenting low-carbonate sediment with only a single species of polar foraminiferan *Globigerina pachyderma* (s.), extended down to about 40 ° N, approximately the latitude of Lisbon. A later sharp readvance, to the latitude of northwest Spain, brought a return to almost full-glacial conditions in the Bay of Biscay between 11 and 10 ka BP (Duplessy *et al.* 1981). This readvance clearly corresponds to the Younger Dryas cooling in the continental Quaternary record of northwest Europe.

Numerous papers have discussed the climatic, vegetational and faunal history of northwest Europe during the Lateglacial period and its relation to changes in the North Atlantic (see particularly Lowe *et al.* 1980; Atkinson *et al.* 1987). The southwestern seaboard of Europe is also in a very critical geographical position with regard to these oceanic events, but the implications of studies there have received much less attention.

In this paper the term Lateglacial is used informally to cover the period between approximately 14 and 10 ka BP. There is an unresolved problem of nomenclature when discussing Quaternary stratigraphy with respect to southwestern Europe. As yet, no local stage-names have been defined for any glacial or interglacial stages. Archaeologists, particularly in France, have long used the scheme, developed by Bordes (1954), which subdivides the last glacial period into four substages, Würm I–IV, separated by three interstadials, Würm I–II, II–III and III–IV. This scheme has overtones that we very much wish to avoid. In any case, the stage name Würmian has now been formally defined, with a proper stratotype and in its original sense, as a local stage-name for the Alpine region (Chaline & Jerz 1983) and is really not relevant to the area under discussion. When unavoidable, we reluctantly use the more general term Weichselian to refer to the last glacial stage.

This area of southwestern Europe is also critical to Quaternary biogeographical studies, because it has long been assumed that the Iberian peninsula has been one of the major refuge areas in which plants and animals characteristic of interglacial stages in northern Europe survived the severe conditions of the glacial stages. Palynological evidence to support this is less well developed than, for example, for the Balkan area of southeast Europe (Huntley & Birks 1983).

2. The Pre-Weichselian vegetational record

Ideally, we would have reviewed in this paper vegetational development within this region throughout the Quaternary; however, unlike the situation in northwestern Europe, the number of sites in Spain and Portugal that have yielded palaeobotanical data for any part of the Quaternary before the last glacial stage is very small indeed. Briefly, there are a handful of Early and Middle Pleistocene sites in Catalonia (Elhai 1966; Deckker *et al.* 1979; Suc 1980) and three sites in central Spain, all associated with Palaeolithic industries of Acheulean type: terrace deposits of the river Manzanares at Villaverde (Madrid) and lacustrine deposits at Torralba and Ambrona (Soria) (Menendez Amor & Florschutz 1959; C. Turner, unpublished results). In the north, fragmentary coastal interglacial sites have just described in Galicia (Nonn 1966; Mary *et al.* 1975), and just across the border in the French Pays Basque, Oldfield (1968) has described a series of Early and Middle Pleistocene interglacial deposits at Bidart and Marbella, the latter of particular floristic interest, because of the indications of an oceanic vegetation with such taxa as *Arbutus*, *Daboecia* and *Rhododendron*.

Potentially the most important of all Spanish Quaternary sites is Padul (Granada), where a 50 m core indicated a succession of several cold and temperate stages (Menendez Amor & Florschutz 1964; Florschutz *et al.* 1971). Unfortunately, the sampling interval for this core, 50 cm, was far too large to give an adequate palynological record, as compared to the other long continental Quaternary pollen sequences from Grande Pile (Woillard 1978) and Les Echets (Beaulieu & Reille 1984) in France. The deposits at Padul are currently being reinvestigated and should yield a continuous pollen sequence which is directly comparable with the deep-ocean oxygen-isotope record, but so far only the results back to 30 ka BP have been released (Pons & Reille 1986).

At present, it must be concluded, these older Quaternary sites are simply too few in number, too scattered and too poorly dated to give a detailed and consistent picture of vegetational development, over this wide geographical area, that could be related accurately to the Atlantic record of climatic change. We considered it far better, therefore, to concentrate on the vegetational record since 30 ka BP, for which many more sites are available, and which is directly comparable with the best-studied part of the oceanic record. Of great importance too is the fact that this timespan lies within the range of radiocarbon dating, so that comparisons of vegetational development at different sites can be made with reference to an absolute timescale, a procedure still impossible for older Quaternary vegetational records.

3. Sources of evidence

Pollen analysis provides the only method for reconstructing detailed sequential changes in local and regional vegetation. The concomitant study of plant macrofossils can, however, be of the greatest help in the interpretation of pollen diagrams. The calculation of pollen concentrations or, where sequences have adequate radiocarbon dating control, of pollen influx rates, can also shed insights on vegetational dynamics, undetectable in relative percentage pollen diagrams. Unfortunately neither plant macrofossil records nor absolute pollen counts are available for most of the sites discussed.

For this study, palynological information is available from three very different sedimentary environments. Firstly there are pollen analytical studies carried out in freshwater aquatic environments on lacustrine and bog deposits. In the area under consideration, the available studies have largely been undertaken in the mountains of northern Spain, Portugal and the Pyrenees. The sites are mostly lake basins of glacial origin, some now overgrown by bog vegetation. They contain only sediments laid down since the melting of ice formed during the last glacial maximum. Consequently these records generally cover only the period since 16 ka BP. Exceptionally, lowland bogs have been studied, but these too go back no earlier into the last glacial period. We describe in detail the results of a new investigation into a particularly important site in northwestern Spain, Sanabria Marsh; this study provides a pollen and vegetational record as a cornerstone for the discussion and comparison of similar records from other sites. A second source of palynological evidence comes from two deep-sea boreholes from the Bay of Biscay, where cores have also been subjected to pollen analysis.

A third and more controversial source of palynological information is the quite extensive series of pollen-analytical studies that have been carried out on the sediments of caves and rock-shelters in association with archaeological excavations. Geographically these studies relate to the Dordogne in southwest France, the Pyrenees, and the coastal region of northern Spain. In time, they extend back from the Holocene through the Lateglacial to cover much of the last glacial period. The nature of cave sedimentation, however, means that individual records are fragmented rather than being continuous in time. Nevertheless, they are deemed to cover an important period of time, 30–15 ka BP, which is not (or barely) represented by lacustrine sites in the area under consideration. To compare these records with equivalents from lacustrine sites it is necessary to extend our view to the French sites, Grande Pile in the Vosges (Woillard 1978) and Les Echets near Lyon (Beauliu & Reille 1984) and to Padul in southern Spain (Pons & Reille 1986).

It must be stressed that palynological results from these three sedimentary environments are not directly comparable. There are important differences in the taphonomy of pollen assemblages and in their mode of incorporation into the sediments. Numerous studies have been made on the relations between pollen assemblages deposited in lake basins and the surrounding regional and local source vegetation, and also on the effects of such complications as redeposition and mixing of pollen assemblages because of sediment focusing on lake floors and the incorporation of pollen from eroded and inwashed soils. The taphonomy of marine pollen assemblages is not so well understood, but there tend to be recognizable patterns of differential transportation, degradation and deposition, such as the overrepresentation of conifer pollen, fern spores and certain types of Compositae pollen (Heusser & Florer 1973; Turon 1980). Another problem with marine sediments stems from their slow sedimentation rates and rather intense bioturbation and mixing. This generally precludes fine resolution, not only in pollen analysis but also in radiocarbon dating. By contrast, sedimentation in lakes may be an order of magnitude faster and bioturbation appears to be of lesser significance.

The taphonomy of pollen assemblages in cave and rock-shelter deposits is very much more complex. General comments have been made by Leroi-Gourhan & Renault-Miskovsky (1977), but a much more critical approach to the origin and interpretation of pollen from cave deposits has been taken by Couteaux (1977) and Turner (1985). Such problems will be discussed more fully later.

4. Present-day climatic and vegetational factors in southwestern Europe

The Iberian peninsula is the largest of the three Mediterranean peninsulas of southern Europe. It is also one of the most mountainous areas of Europe, with a mean altitude of more than 500 m above sea level. Indeed, more than 34 % of Spain lies between 800 and 2400 m in altitude. The central plateau of Spain forms a major climatic divide between the Atlantic coastal areas of the west and north of the peninsula and the Mediterranean coastlands. The former are characterized by mild, almost frost-free winters and cool summers, with rainfall fairly evenly spread through the cooler months of the year and particularly heavy precipitation in the mountains. The Mediterranean coasts also have mild winters but warm, dry summers with most rainfall in the spring and autumn. Because of its altitude, the plateau itself has very cold winters but hot, dry summers. The Pyrenees form a substantial physical barrier between the Iberian peninsula and France, and indeed a few small glaciers survive at altitudes above 3000 m (Hollermann 1968). However, from west to east the range itself forms a transitional area between the Atlantic climatic zone and that of the Mediterranean. A similar zone of transition occurs across the lower-lying areas of southwest France, from the Atlantic coasts of Gascony and Guyenne to the Mediterranean of Languedoc.

These climatic divisions are reflected in the modern vegetation patterns and particularly in the distribution of many individual taxa. The Cantabrian mountains in northern Spain form an important phytogeographical boundary. The natural forest vegetation of the coastal regions of northern and northwestern Spain contains many tree taxa common to the woodlands of northwestern Europe, particularly western France, such as *Quercus robur*, *Q. petraea*, *Q. pyrenaica*, *Ulmus glabra*, *Corylus avellana* and *Salix caprea*. Several of these taxa reach their southwestern limit of distribution in this area. *Carpinus betulus*, however, does not occur in Spain; its distribution in this part of Europe is apparently limited by the Pyrenees. *Fagus sylvatica* is an important tree in the Cantabrian forests but does not occur further south in the Iberian peninsula, although it is widespread in other parts of southern Europe. Many typical Mediterranean tree species fail to occur in the Cantabrian region, although they are found in southern and eastern Spain and in some cases in the Pyrenees and even on the central plateau. These include *Quercus coccifera*, *Pinus pinea*, *P. halepensis*, *Ephedra major* and *Celtis australis*. Mediterranean species that do occur in northern Spain and also along the Atlantic coasts of south-west France include *Quercus ilex*, *Q. suber* and *Pinus pinaster*. The distribution of *Q. ilex* has been of particular interest to Quaternary biogeographers. Widespread in the central and eastern Mediterranean, its range extends northwards along the Atlantic coast of France into southern Brittany. On the basis of pollen evidence, Van Campo & Elhai (1956) have suggested that it temporarily expanded into Normandy during the Middle Holocene. In Spain, although here as elsewhere it is much planted, it occurs as a native species largely along the eastern, Mediterranean coast and the northern, Atlantic coast. It has been suggested that the species has spread westwards into Spain from the central Mediterranean area during the Holocene (Huntley & Birks 1983). Over much of the Iberian peninsula *Q. ilex* is apparently replaced by the closely related *Q. rotundifolia*, a widely distributed species endemic to Spain and Portugal.

5. The North Atlantic climatic record 20–9 ka BP

As mentioned earlier, Ruddiman & McIntyre (1981) placed the position of the polar front in the North Atlantic at about 40 ° N between 20 and 16 ka BP. They suggest that, north of the polar front, ocean-surface conditions with only moderate meltwater influx and winter sea-ice cover favoured enough winter moisture to provide sufficient precipitation, falling as snow, to fuel the final southward advance of the major ice sheets. From 16 to 13 ka BP they conclude that decay of continental ice sheets led to the entire surface of the North Atlantic, down to at least 60 ° N, being flooded in summer by meltwater and icebergs, and consequently covered by sea ice in winter, thus starving the continents of moisture and creating rather arid conditions there. Van Campo (1984) has examined the vegetational evidence for southern Europe during this interval and strongly supports this conclusion. By 13 ka BP the polar front had withdrawn northwards and westwards, as far as Iceland, allowing warmer waters and much more temperate climatic conditions to reach the whole Atlantic coastline of western Europe. Between 13 and 11 ka BP Duplessy *et al.* (1981) estimate that summer sea-surface temperatures in the Bay of Biscay were even warmer that at present, particularly in the earlier part of this interval. A sharp readvance of the polar front in the eastern Atlantic to the latitude of northwest Spain brought a return to almost full-glacial cold conditions in the Bay of Biscay between 11 and 10 ka BP. The cause of this readvance is somewhat speculative, but Ruddiman & McIntyre (1981) consider that it may relate to the break-up of large ice-shelves in the Arctic Ocean, as suggested by Mercer (1969). From 10 to 9 ka BP there was a very rapid northwesterly retreat of the polar front and a warming to nearly full postglacial temperatures throughout most of the North Atlantic Ocean. The dating of these events is imprecise, partly because of the problems of bioturbation and slow sedimentation rates, but also because, as stressed by Ruddiman & McIntyre (1981), they were by their very nature time-transgressive.

6. The northwest European continental record 14–9 ka BP

This framework for events in the eastern North Atlantic not only integrates reasonably well with Van Campo's account of the climate and vegetation of much of Europe before 13 ka BP (Van Campo 1984), but is also in broad agreement with climatic interpretations of the much more complicated Lateglacial stratigraphic sequences of northwest Europe. These are based on many well-dated sites to which the study of palynology and also of fossil Coleoptera have made the most important contributions (Lowe *et al.* 1980). Indeed, analysis of fossil coleopteran data has recently provided an even more detailed and refined pictue of Lateglacial climatic change for the British Isles (Atkinson *et al.* 1987). In the oceanic record, however, it has not been possible to identify minor, short-lived climatic oscillations, such as the Older Dryas, which have been detected on land.

The northwest European Lateglacial continental record has virtually been used as a model by some authors, and particularly by archaeologists and their collaborators, in the interpretation of Lateglacial sequences not only in northern Europe but also in our study area in the southwest. For this reason we list Iversen's now classic sequence of Lateglacial pollen zones (Iversen 1954), together with the chronozones proposed by Mangerud *et al.* (1974) to formalize this sequence (table 1).

The use of this sequence and its nomenclature as a model for the interpretation of Lateglacial sequences at southwest European sites raises a number of questions. The first of these is the

TABLE 1. LATEGLACIAL POLLEN ZONES AND CHRONOZONES

(From Iversen (1954) and Mangerud *et al.* (1974).)

pollen zones (Denmark)		chronozones (Scandinavia)	
		chronozone	age/ka BP
		Preboreal	10–9
zone III	Younger Dryas period	Younger Dryas	11–10
zone II	Allerød period	Allerød	11.8–11
zone Ic	Older Dryas period	Older Dryas	12–11.8
zone Ib	Bølling period	Bølling	13–12
zone Ia	Daniglacial tundra period ('Oldest Dryas')		

often implicit assumption that the major episodes of climatic change during the Lateglacial were synchronous along the Atlantic seaboard. Ruddiman & McIntyre (1981) have shown that the movements of the polar front across the North Atlantic were clearly time-transgressive. The second point is that the criteria used to infer climatic change in most pollen diagrams have been the appearance and expansion of tree pollen taxa. Coope (1977) has pointed out repeatedly that Lateglacial beetle faunas demonstrate that climatic amelioration begins well before the expansion of tree birches in the British Isles. The spread of trees appears to be controlled by distance from refugia and by soil nutrient conditions (Van Geel *et al.* 1984). It must be assumed that refugia for temperate tree species lay much closer to sites in southwest Europe than to those in the north.

The third matter relates to the resolution of climatic events during the Lateglacial. Watts (1980) has reviewed the evidence for the recognition of Bølling and Younger Dryas climatic events in different areas of northern and central Europe and concluded that both of these are most marked in their expression in the British Isles. Evidence for a definable Bølling interval becomes more difficult to recognize with certainty eastwards beyond the Netherlands and certain sites in Germany. The Younger Dryas becomes less distinct in southern Germany and the Alps. Pine and birch remain present there during Younger Dryas time, but indications of climatic change may be limited to a decrease in birch, small rises in juniper and *Artemisia* pollen curves or inwash of inorganic sediment. Clearly the intensity, and consequently the expression, of Lateglacial climatic oscillations may vary geographically even over relatively short distances.

7. POLLEN-ANALYTICAL INVESTIGATIONS OF LACUSTRINE SEDIMENTS IN NORTHERN AND NORTHWESTERN SPAIN AND PORTUGAL

A relatively small number of sites has been investigated in the mountains of this region of the Iberian peninsula. Here we discuss only those which relate to the Lateglacial period and to which radiocarbon dating has been applied. The location of these sites is given in figure 1. The

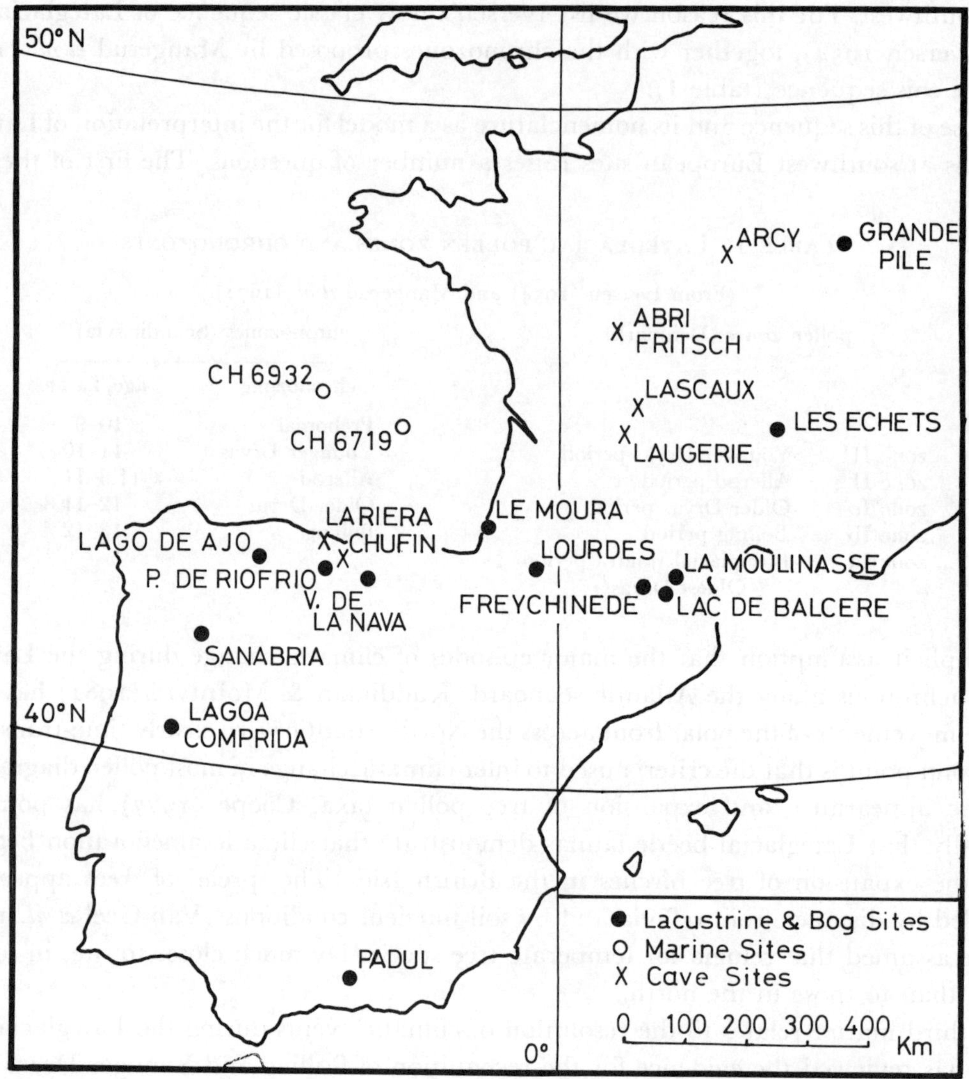

FIGURE 1. Sites of late Quaternary palynological investigations in southwest Europe and adjacent areas.

most important are Sanabria Marsh (Hannon 1984; Watts 1986), which is virtually the same locality as Laguna de las Sanguijuelas (Menendez Amor & Florschutz 1961), and Lago de Ajo, another recently investigated site (McKeever 1984; Watts 1986). Two sites further east near Santander, Puertos de Riofrio (Florschutz & Menendez Amor 1962) and Valle de la Nava (Menendez Amor 1968), and also Lagoa Comprida in the Serra da Estrela of northern Portugal (Janssen & Woldringh 1981; Van den Brink & Janssen 1985) are also discussed briefly.

(*a*) *Sanabria Marsh*

It was decided to reinvestigate the site of the Laguna de las Sanguijuelas examined by Menendez Amor & Florschutz (1961). This lies at an altitude of *ca.* 1050 m above sea level in a mountainous area of northwestern Spain about 20 km north of the frontier with Portugal (42° 06′ N, 06° 44′ W). The new investigations, hitherto largely unpublished and therefore described here in some detail, were carried out by G. E. Hannon at a site here referred to as

Sanabria Marsh. This lies about 700 m from the original site of Menendez Amor & Florschutz (1961) and is an overgrown small lake (100 m × 200 m) occupying what is probably an ice-block hollow in a moraine near a large natural lake, Lago de Sanabria.

There is geomorphological evidence for a recent glaciation in the area, and Schmidt-Thomé (1983) describes a young lowland glaciation, probably Weichselian, in the neighbouring part of Portugal (Minho district) very close to the study area. The mountains are of Cambrian and Silurian rocks with much granite. Much of the upland area around the site is above 1500 m, rising to over 2000 m in the Sierra de la Cabrera range to the north. The modern vegetation in the valley is woodland dominated by Pyrenean oak, *Quercus pyrenaica*, with virtually no other tree species present. The surrounding hillslopes are covered by multi-stemmed *Q. pyrenaica*–Leguminosae scrub, much of which has been burnt in the past. The mountain tops are dominated by ericaceous plant communities, with much *Calluna*. The area where coring took place is now covered with poor fen vegetation.

A Livingstone corer was used to recover sediment cores to a depth of 10.55 m. The sediments consisted of fibrous peat, fen peat, gyttja (lake mud) and clay (figure 2) and were polleniferous to a depth of 9.86 m. They also contained a rich macrofossil flora (figure 3). Three radiocarbon dates were obtained at important biostratigraphical horizons. The investigation of the upper part of the sequence relating to the later Holocene is not described here.

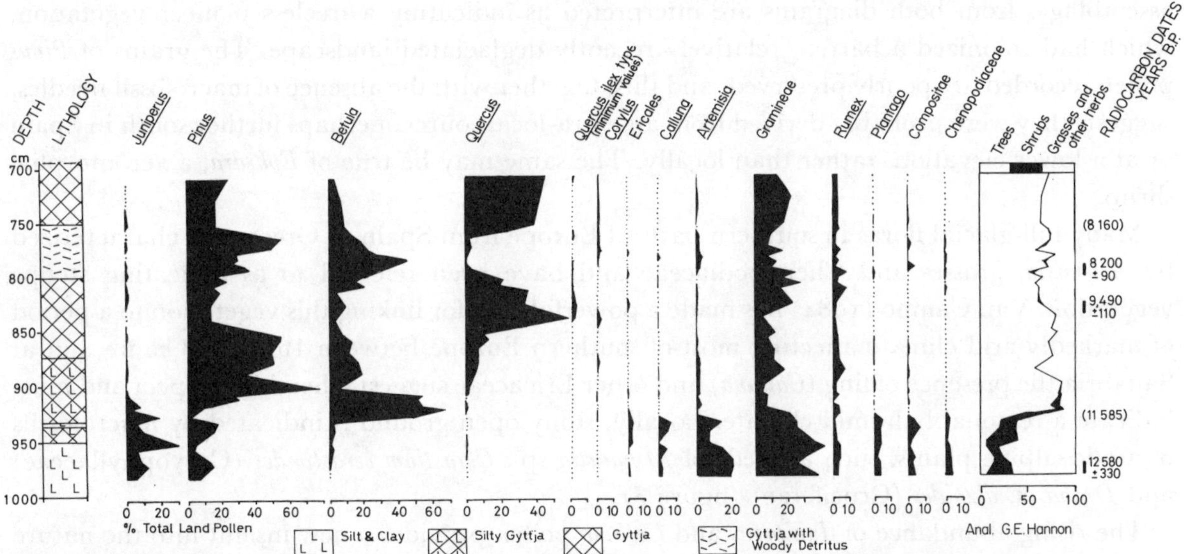

FIGURE 2. Pollen diagram (selected taxa only) from Sanabria Marsh, northwestern Spain. In the summary diagram *Corylus* is included with the tree pollen, and Ericaceae pollen with that of shrubs. Radiocarbon dates in brackets relate to the closely similar pollen diagram from Laguna de las Sanguijuelas.

Although the sediment depth was greater than that recorded by Menendez Amor & Florschutz (1961), the new pollen sequence (figure 2) does not extend quite so far back in time. Nevertheless, there is a very close similarity between the major pollen curves in both studies although a large number of taxa, and in particular *Juniperus*, which were not recognized in earlier work, have now been recorded. Because the pollen diagrams are so similar, it is also possible to apply the radiocarbon dates obtained for the earlier study to the interpretation of the present diagram. The granitic soils in the neighbourhood of the site make it very suitable for radiocarbon dating because the problem of 'old carbon' and hard-water error does not need to be considered. Pollen concentrations were also measured. The pollen-concentration and influx diagrams are not reproduced here, as they resemble the relative pollen diagram (figure 2) in all major features, but they confirm that the vegetational changes described are real rather than statistical artefacts.

Menendez Amor & Florschutz (1961) recorded an early zone with high percentages of herbs, grasses, *Artemisia* and *Pinus*. The early Lateglacial age of this zone is confirmed by radiocarbon dates of 13700 ± 300 years BP (Gro 705) and 12830 ± 280 years BP (Gro 702) and we suggest that it ended just before 11585 ± 220 years BP (Gro 688). Minor fluctuations in the *Betula* and *Pinus* curves suggest a small expansion of trees at about 13700 years BP, followed by a return to open vegetational conditions. The vegetational record for the new studies (figure 2) commences shortly before 12580 ± 330 years BP (Beta-9162), so that unfortunately it does not cover the period of this early presence of tree pollen, for which confirmation is very desirable.

The earliest pollen assemblages from Sanabria Marsh closely resemble those from the Laguna de las Sanguijuelas in showing high percentages of grasses, *Artemisia* and herbs such as Chenopodiaceae, Compositae, *Plantago* and *Rumex*, but they also contain significant amounts of *Juniperus*. The Ericaceae pollen, rising to a peak at this horizon in both pollen diagrams, is here identified as referable largely to *Calluna*. These early grass and *Artemisia*-rich pollen assemblages from both diagrams are interpreted as indicating a treeless pioneer vegetation, which had colonized a barren, relatively recently deglaciated landscape. The grains of *Pinus* pollen recorded are poorly preserved, and this, together with the absence of macrofossil needles, suggests they were probably derived from an extra-local source, perhaps further south in Spain or at a lower elevation, rather than locally. The same may be true of *Ephedra*, a xeromorphic shrub.

Many full-glacial floras in southern parts of Europe from Spain to Greece are characterized by *Artemisia*, grasses and Chenopodiaceae and have been referred to as indicating steppe vegetation. Van Campo (1984) has made a powerful case for linking this vegetation to a period of markedly arid climate affecting most of southern Europe between 16 and 13 ka BP, but at Sanabria the presence of ling (*Calluna*) and other Ericaceae suggest a heathlike aspect and must indicate a reasonably humid climate. Locally, stony open ground is indicated by macrofossils of arctic–alpine plants, such as seeds of *Minuartia* sp., *Cerastium cerastioides* (Caryophyllaceae) and *Draba* cf. *aizoides* (Cruciferae) (figure 3).

The rising abundance of *Juniperus* and *Calluna* pollen provides a new insight into the nature of these pioneer vegetation communities. *Juniperus* pollen was unrecognized or overlooked in most earlier pollen studies from Spain. At Sanabria Marsh, values of 30% are observed, accompaned by frequent macrofossil needles. European fossil pollen maps (Huntley & Birks 1983) show that juniper was widespread throughout much of western Europe between 13 and 10 ka BP, although not in the southwest. These new data extend the range to northwestern Spain. *Calluna vulgaris* has not previously been recorded from Spain during this time period, and the pollen is accompanied by abundant macrofossil remains which include leaves, flower-heads and seeds; these remains confirm its local presence and provide the earliest unambiguous evidence for extensive heath communities in the deglaciating landscape of northwestern Spain. The area probably provided a refugium for this species during the last glacial period.

A white silt band, devoid of both pollen and macrofossils, occurs between 935 and 937 cm, after which many of the pioneer taxa except *Juniperus* and Gramineae disappear. This band records a brief event, probably a flash flood.

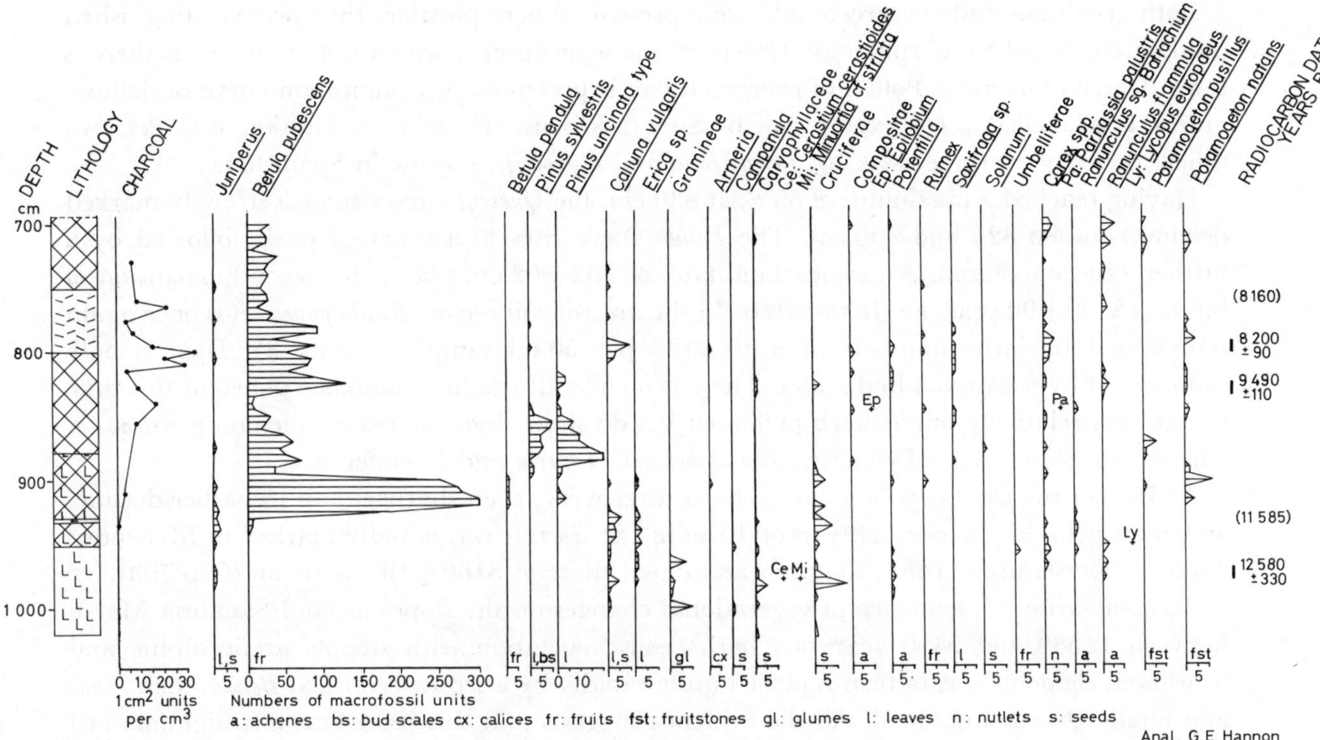

FIGURE 3. Plant macrofossil diagram (selected taxa only) from Sanabria Marsh, northwestern Spain. (Lithological symbols as for figure 2.)

Sedimentation of gyttja is resumed, whereupon a rapid rise in the *Betula* curve provides the first definite evidence of climatic change. This event has been dated to 11585±220 years BP (Gro 688) by Menendez Amor & Florschutz (1961). At the same time, herbaceous pollen types, mainly Gramineae with some *Rumex* and *Plantago*, fall to about 20%.

The pollen assemblages suggest a period when scrubby open woodland, with juniper and downy birch (*Betula pubescens*) colonized the slopes near the lake. Subsequently this developed into closed-canopy birch woodland; very large numbers of birch fruits (about 200 per 50 ml sample) and high pollen values (65%) are recorded. The macrofossils show that *Betula pubescens* was the predominant species, with some silver birch (*B. pendula*) also present.

Pinus pollen values rise gradually and replace *Betula* as the most abundant tree type above 910 cm. A low curve for *Corylus* begins at 915 cm but values for this tree are usually below 5% throughout the diagram. The *Quercus* pollen curve also starts to rise gradually. Macrofossil needles show that pine is first represented by a dwarf mountain pine, either *Pinus mugo* or *P. uncinata*, and at a higher level by Scots pine, *P. sylvestris* (figure 3). *Pinus uncinata* forms the natural treeline in the Pyrenees today, ascending to well over 2000 m (Gaussen 1948). *Pinus mugo* is common in the mountains of central Europe. The fossil needles of these two species are indistinguishable when studied in small fragments, but it seems more likely that the needles belonged to *P. uncinata*, given its modern distribution.

At 840 cm the *Pinus* pollen curve falls, after the gradual expansion of *Quercus*. A radiocarbon date of 9490±110 years BP (Beta-9161) was obtained between 827 and 837 cm; this result suggests that *Pinus sylvestris* woodland was being largely replaced by oak woodland at a time roughly corresponding to the onset of Holocene warming in northwestern Europe.

Both deciduous and evergreen oaks were present. Where possible, these were distinguished on the basis of pollen morphology. However, the separation is not entirely reliable, as there is some degree of overlap. Pollen of evergreen oak is presented as a minimum curve of definite grains. It is labelled as *Quercus ilex* type because (as discussed earlier) in addition to *Q. ilex*, two other species of evergreen oak, *Q. rotundifolia* and *Q. coccifera*, grow in Spain today.

Having reached a maximum of 55% at 830 cm, the *Quercus* curves show a strongly marked decline between 825 and 800 cm. The *Betula* curve rises to a renewed peak, followed by a further *Pinus* maximum. A radiocarbon date for 792–802 cm places this second expansion of birch at 8200 ± 90 years BP (Beta-9160). In the macrofossil record *Betula pubescens* is once again represented by large numbers of fruits (120 per 50 ml sample) (figure 3). *Pinus* is only represented by occasional bud scales. There is no overall rise in Gramineae pollen at this time but certain relatively minor herb pollen curves do show slight increases, including *Artemisia*, Chenopodiaceae, Caryophyllaceae, *Helianthemum*, *Armeria* and Cruciferae.

At 750 cm the *Quercus* pollen curve shows a recovery, even sharper than its earlier decline, and returns to dominance. This second rise in the *Quercus* was equally marked in Menendez Amor & Florschutz's (1961) diagram and dated there to 8160 ± 190 years BP (Gro 703).

To summarise the sequence of vegetational changes on the slopes around Sanabria Marsh between 12580 and 9490 years BP, herbaceous vegetation with steppe, arctic–alpine and heathland elements was in turn replaced quite rapidly by a sequence of first *Betula*, then *Pinus* and finally *Quercus* woodland. All the evidence suggests that this represents a straightforward ecological succession in response to ameliorating climatic and edaphic conditions. In particular, there is no record of any apparent climatic fluctuations within this timespan. After 8 ka BP there was a short interval when the oak population collapsed and *Betula pubescens* woodland developed very close to the site. The subsequent Pinus peak is not substantially supported by any macrofossil record, so the tree, although abundant, was probably not growing close to the site.

Menendez Amor & Florschutz (1961) in their study of the area referred the initial *Quercus* peak to the end of the Allerød and the second *Quercus* peak, which they had dated, to the onset of the postglacial *Quercus* rise. They therefore regarded the period of the second *Betula* maximum as the equivalent of the Younger Dryas. This correlation has been quoted in most subsequent papers on the Lateglacial period in southwestern Europe. However, their interpretation is in conflict with the new data from Sanabria, in particular with the radiocarbon dates for both the first *Quercus* rise (827–837 cm), 9490 ± 110 years BP, and the onset of the second *Betula* rise (792–802 cm), 8200 ± 90 years BP. It must be emphasized again that the two pollen diagrams show a remarkable degree of agreement, and clearly identify the same sequence of vegetational events. Likewise the two sets of radiocarbon dates combine to form a consistent and mutually supporting sequence.

The conclusions that can be drawn from this new study of the area are:

(i) There is no evidence for any marked cooling of climate at this site between 11 and 10 ka BP, equivalent to the Younger Dryas oscillation recorded elsewhere.

(ii) The *Quercus* decline identified by Menendez Amor & Florschutz (1961) as evidence for Younger Dryas cooling is clearly an early Holocene event, dated at about 8200 years BP.

(iii) Although the first *Quercus* maximum is dated to 9490 years BP, after the beginning of the Holocene, both *Quercus* and *Corylus* (hazel) appeared earlier and were present in low quantity in the area during the Lateglacial.

It will be shown that evidence for similar conclusions can be found in other sites in northwestern Spain and Portugal.

It is still necessary to explain the temporary fall in *Quercus* pollen at 8200 years BP and the subsequent reversion to a *Betula* phase. Although the dates bounding the event come from two separate diagrams, it seems to have occupied a relatively short period of time. It can be considered in two different contexts.

The first is that this represents a local forest catastrophe resulting in the destruction of the oak forest by fire or disease and the initiation of a seral woodland succession once again. In support of this hypothesis, the macrofossils and their residues are largely composed of charcoal between 820 and 800 cm, and the gytjja overlying this level from 800 to 760 cm shows an increase in its minerogenic component. Because this is an area of granitic bedrock, loss of nutrients as a result of a forest fire could have seriously affected the rate and species composition of forest regeneration. Macrofossils demonstrate that *Calluna* again invaded the site at this time.

There is, however, a certain amount of evidence to suggest that a more widespread episode of vegetational change may have taken place at this time, in which case a climatic explanation might be necessary. A fall in *Quercus* pollen accompanied by a rise in *Betula* is recorded at 8310 ± 160 years BP (GrN 9916) in the pollen diagrams from Lagoa Comprida in the Serra da Estrela (Janssen & Woldringh 1981; Van den Brink & Janssen 1985). In the new diagram from Padul there is a temporary fall in the curves for both deciduous and evergreen oak over a short interval between 9300 ± 90 years BP (Gif 6382) and 8200 ± 80 years BP (Gif 6383) (Pons & Reille 1986). Neither pine nor birch plays any significant role in the vegetation at Padul at that period. The decline in *Quercus* is associated with a rise in the Gramineae pollen curve and then with a great increase in monolete fern spores (included in the pollen sum in that diagram). Without pollen concentration measurements it is difficult to gauge whether or not there is a genuine decline of oak populations at Padul or alternatively whether the fall in *Quercus* pollen is largely a statistical artefact of the relative percentage diagram because of a highly localized input of fern spores. However, these observations indicate no common climatic trend, and only further observations will confirm or negate such an alternative explanation of the event at Sanabria Marsh.

(*b*) *Lago de Ajo*

Lago de Ajo is a lake occupying a large cirque in the north-west part of the Cantabrian mountains (43 ° N, 6° W). It lies about 135 km north-northeast of Sanabria Marsh but at a higher altitude (1570 m). The local bedrock, exposed in the steep walls of the cirque, consists of dolomitic limestone. The lake lies above the tree-limit and is surrounded by herbaceous alpine vegetation with juniper bushes. At a slightly lower level, beech (*Fagus sylvatica*) forest occurs, with some birch and hazel present. Pines and oaks are absent from the surrounding area, although deciduous oaks occur at lower altitudes.

A recently studied sediment core from this lake yielded a pollen diagram covering most of the Holocene and Lateglacial (McKeever 1984; Watts 1986). The principal features of the vegetational development at this site were as follows.

(i) The earliest part of this diagram suggests a vegetation dominated by *Artemisia*, grasses and other herbs. In terms of pollen percentages *Pinus* is quite abundant (30–35 %), but the low pollen concentrations suggest that vegetation was very sparse and treeless, much of the pollen, particularly that of *Pinus* and *Ephedra*, being carried in from lower altitudes elsewhere.

(ii) A sharp rise in the *Pinus* and *Betula* curves probably signals the return of trees to the area, although there are no supporting macrofossils at the peak of this event. This horizon has been dated to 14270±180 years BP (Beta-6740), but this determination may be somewhat too old; in this area of calcareous bedrock, early Lateglacial dates from predominantly inorganic sediments are likely to have been affected by hard-water error. There followed a brief resurgence of *Artemisia*, perhaps indicating a cooler oscillation, then, at 12610±90 years BP (Beta-9157), *Betula* again became abundant with macrofossils of *B. pubescens* indicating that trees were growing close to the site, which is now above the treeline.

(iii) This spread of birch was followed within a short time by a major expansion of oak. Nearly a metre of sediment in which *Quercus* pollen dominates was deposited below the level at which *Corylus* expanded, dated to 9780±80 years BP (Beta-6739). During this period of oak dominance, pine was present in the immediate vicinity of the lake, as indicated by finds of macrofossil needles.

To summarize, the lake basin was already deglaciated before 14 ka BP. The possibility of a minor episode of climatic warming, early in the Lateglacial, is suggested by a rise in *Betula* and *Pinus* pollen at the site, before the main expansion of *Betula* and other trees about 12610 years BP. If the radiocarbon date for this horizon, 14270 years BP, is too old then this oscillation may well represent the same event as the early *Betula* maximum in Menendez Amor & Florschutz's (1961) diagram from Laguna de Sanguijuelas, dated to 13700 years BP. As at Sanabria Marsh, expansion of oak takes place within the Lateglacial. Finally, the vegetational record at this site again shows no evidence of an identifiable Younger Dryas event.

(*c*) *Lagoa Comprida*

Investigations of this site at an altitude of 1600 m in the Serra da Estrela in northern Portugal relate exclusively to the Holocene (Janssen & Woldringh 1981; Van den Brink & Janssen 1985). Neither of the boreholes penetrated Lateglacial sediments, but it is significant that oak, as well as pine and birch, formed a major component of the forest at the base of the diagrams some time before 9200 BC. The fall in *Quercus* pollen at this site during the early Holocene, dated at 8310±160 years BP, has already been noted.

(*d*) *Valle de la Nava and Puertos de Riofrio*

Menendez Amor (1968) made three pollen diagrams from an area of peat bog in the Valle de la Nava near the boundary between the provinces of Burgos and Santander, well to the east of the sites previously discussed (figure 1). The altitude of the site is not stated, but it is less than 1000 m. A second site lies high up at an altitude of 1700 m in the Picos de Europa, called Puertos de Riofrio (Florschutz & Menendez Amor 1962). At both sites the diagrams for the early Holocene are generally similar and show a predominance of *Pinus* and *Quercus* with lesser amounts of *Betula*. The onset of the rise of *Quercus* is dated to 10210±115 years BP at Rifrio and to 10000±200 years BP at Valle de la Nava. Before this the vegetation at Riofrio, the higher site, appears to have been largely open with non-tree pollen curves, Gramineae, *Artemisia* and Chenopodiaceae dominating the lowest assemblages, together with 20–30% *Pinus* pollen. At Valle de la Nava, about 1000 m lower, pine forest clearly dominated the end of Lateglacial times; *Pinus* pollen values are high (70–80%) with 20–30% Cyperaceae and Gramineae pollen and only traces of other trees.

8. Lacustrine and bog sites in the Pyrenees and Pays Basque

Here the most important palynological studies come from three altitudinally distinct regions (figure 1). There is a cluster of high-level lake and bog sites in the eastern Pyrenees at altitudes between 1300 and 1800 m above sea level; a second cluster of sites in the neighbourhood of Lourdes in the central Pyrenees, lying at about 400–420 m above sea level; and a valley bog, Le Moura, in the Pays Basque near Biarritz, at only 40 m above sea level.

The three principal high-level sites, Lac de Balcère (Van Campo & Jalut 1969), La Moulinasse (Jalut 1973) and Freychinède (Jalut *et al.* 1982), have all yielded pollen records covering the Lateglacial, and their basins of deposition relate geomorphologically to the melting of the last major glaciers of the area. Lac de Balcère lies at an altitude of 1764 m and is surrounded by *Pinus uncinata* forest. La Moulinasse and Freychinède, lying at altitudes of 1330 m and 1350 m respectively, are now both peat bogs occupying basins where morainic barriers blocked drainage. Forests close to La Moulinasse consist primarily of *Pinus uncinata* and *Abies alba*, but the area around Freychinède has been extensively deforested. La Moulinasse lies in an area of schists and granites, but at Freychinède calcareous rocks outcrop, which may be important when considering radiocarbon dates.

Altitudinally and topographically these sites have much in common with those of the mountains of northwest Spain. The pollen records, initially similar, develop differently.

The oldest pollen assemblages suggest once again a treeless vegetation with low pollen productivity, dominated by grasses, *Artemisia* and other herbs, with *Pinus* pollen carried in by long-distance transport. A series of radiocarbon dates for this pollen zone at Freychinède range from 14600 ± 770 years BP (Gif 5017) to 21300 ± 760 years BP (Gif 4957) and older than 22000 years BP (Gif 5015), but older dates are interstratified with younger ones.

The first signs of climatic amelioration are provided by the spread of *Juniperus*, dated at Freychinède at 14700 ± 800 years BP (Gif 5018) and at La Moulinasse to before 13600 years BP (Gif 1775). As at Lago de Ajo, and in contrast to Sanabria Marsh, Ericaceae played no significant role.

This early phase of open Lateglacial vegetation was brought to an end by the invasion of *Pinus uncinata* and *Betula*. Subalpine dwarf pine forest, characteristic of the modern tree limit, must have spread to virtually its present altitudinal limit at that time. At this altitude *Betula* played a much less important role, except at Freychinède, where it appears to have arrived before *Pinus* at *ca.* 13150 ± 300 years BP (Gif 4958), and *Pinus* peaks at Freychinède and Lac de Balcere with dates of 11200 ± 250 years BP (Gif 4959) and 11240 years BP (Gif 792) respectively. *Pinus uncinata* continues to dominate the pollen diagrams at all three sites throughout the rest of the Lateglacial, until the expansion of *Quercus* and *Corylus* curves in the early Holocene. Curves for *Quercus* and *Corylus*, of course, represent the expansion of these trees at lower altitudes. They have played no role in the vegetation of these higher mountain areas at any time in the Holocene.

Evidence for a Younger Dryas oscillation is only weakly expressed. At the highest site, Lac de Balcère, there is a small rise in the Gramineae and *Artemisia* curves and a corresponding temporary fall in the *Pinus* curve from 60 to 50%. At La Moulinasse low sedimentation rates give a very condensed record. A single sample shows a Gramineae peak and a relatively low *Pinus* value, followed by a single-sample *Artemisia* peak corresponding with the recovery of *Pinus*. At Freychinède there is no appreciable vegetational change detectable.

It must be concluded that at high altitudes in the eastern Pyrenees there was no significant movement of the treeline corresponding to a Younger Dryas cooling; at most a slight opening up of the tree cover took place, allowing a small increase in herbaceous pollen.

With respect to the dating of these pollen diagrams, it must be concluded that the early dates for the Gramineae–*Artemisia* zone at Freychinède should be disregarded. It seems probable that many of the dates from this site are affected by hard-water error and are too old. Jalut and his co-authors reject such a possibility but still fail to explain the inconsistent sequence of dates. They comment that, because the lowest sediments are less calcareous than those above, much of the carbonate must have been in solution. These are precisely the conditions under which 'old carbon' is taken up by aquatic plants. The comparatively early date for the expansion of the *Quercus* pollen curve at Freychinède, 10850 ± 120 years BP (Gif 5524), serves to support the suggestion that some dates at this site are too old, because the *Quercus* expansion at La Moulinasse nearby is dated to shortly before 9150 years BP (Gif 1776) and at Lac de Balcère to shortly before 9250 years BP (Gif 791).

Although the sites in the neighbourhood of Lourdes are at a much lower altitude, they too relate to lakes formed behind morainic barriers after the melting of a glacier at the end of the last major glacial event in the Pyrenees. The investigations of Kolstrup (1980) of a core taken close to the still extant Lac de Lourdes cover only the older part of the Lateglacial sequence. A fuller record has been published by Mardones & Jalut (1983) for Biscaye, an adjacent lake basin, now totally infilled and overgrown by woodland swamp and bog vegetation. The latter basin also seems to include the site investigated by Florschutz and Menendez Amor under the name Poueyferré, for which only an outline pollen diagram has been published (De Vries *et al.* 1960; Alimen *et al.* 1964).

The oldest deposits occur at Biscaye, where organic peats and lake clays overlie at least 5 m of inorganic laminated clays. The basal 3.5 m contain many scattered cobbles and thin beds of gravel with striated clasts. These must represent the deposits of a proglacial lake into which ice was actually carrying rock debris. Mardones & Jalut (1983) nevertheless applied pollen analysis to these sediments and obtained low concentrations of pollen, particularly *Pinus* (70–90 %) together with small amounts of pollen of grasses, other herbs and even thermophilous trees. These assemblages must be interpreted as consisting overwhelmingly of reworked pollen from older peats and soils destroyed by glacial and periglacial processes. Such an interpretation is supported by radiocarbon analysis of small amounts of organic material incorporated in these otherwise inorganic sediments. The dates obtained are much older than any others for the region: 29500 ± 1200 years BP (Gif 5683), 31900 ± 2000 years BP (Gif 5684) and $38400 \pm ^{2000}_{1800}$ years BP (Gif 5685). Extraordinarily, Mardones & Jalut (1983) accept these as true dates for the age of the deposits and even propose an interstadial at that time on the basis of the high *Pinus* pollen percentages. Such an interpretation is quite inconsistent with either the actual lithology of the sediments or the generally accepted glacial history both of the Pyrenees in this region and of western Europe in general.

The earliest reliable pollen assemblages from all three pollen records again show the now familiar pattern of early Lateglacial vegetational development, namely a treeless open vegetation dominated by grasses and *Artemisia*, followed by the spread of juniper and then of trees, particularly tree birches and shortly afterwards pine. The *Betula–Pinus* maximum at these sites is accompanied by a low but consistent pollen curve for *Quercus* with values of 1–2 %. This suggests the possibility that oaks may have been present somewhere in the region in small quantity but not in the immediate neighbourhood of the basins of deposition. The major expansion of oak, closely followed by that of hazel, was clearly a Holocene event in this region.

Radiocarbon dates at Biscaye suggest that the *Juniperus* expansion took place after 14820±240 years BP (Gif 5682) and that the expansion of *Betula* there began at 13250±120 years BP (Gif 5693). These dates are in good agreement with that from the Poueyferré diagram of 13600±110 years BP (Gro 1679) for the upper part of the Gramineae–*Artemisia* zone (*Juniperus* was not counted in this diagram). At Lac de Lourdes the expansion of *Betula* is dated to 13480±140 years BP (GrN-8675). Older dates for the Gramineae–*Artemisia* zone from Lac de Lourdes, 18950±400 years BP (GrN-8512) and 19300±600 years BP (GrN-8682), and from Poueyferré, 18790±225 years BP (Gro 1890), interstratified between dates of 13600±110 years BP (Gro 1679) and 15800±120 years BP (Gro 1671), should be viewed with caution. At Biscaye the older lake sediments are characterized by the presence of the carbonate-secreting alga *Chara*, a good indicator of bicarbonate-rich waters. These rather old dates are likely to have been affected by hard-water error or possibly a small admixture of reworked organic matter. At Poueyferré, the *Pinus–Betula* maximum is dated to 12310±130 years BP (Gro 1681), compared with a date of 12760±200 years BP (Gif 5735) at Biscaye, and the major expansion of *Quercus* to just before 9260±100 years BP (Gro 1889).

At both Poueyferré and Biscaye there is evidence for vegetational revertence indicating some climatic deterioration at the time of the Younger Dryas event in the millenium before the onset of the Holocene. At Poueyferré the outline diagram shows a decline in the *Betula* and, to a lesser extent, the *Pinus* curves, but more significantly a temporary rise in the non-tree-pollen values from 25% to 60% of the pollen sum. At Biscaye at *ca.* 10860 years BP birch declined, and a very brief and small resurgence of *Juniperus* and *Artemisia* took place, correlated with a short-term fall in pollen concentration. Unfortunately this is followed by a sharp break in the pollen diagram at a critical horizon or lithological change, so there may be a hiatus here concealing the full nature of the event.

The site of Le Moura (Oldfield 1964) is at a much lower altitude (40 m above sea level) than previous sites discussed and remote from the direct effects of glaciation. It lies within 20 km of the Bay of Biscay and is a valley bog, surrounded by a region of heathland on poor sandy soils. The pollen diagrams from Le Moura, unlike others, were calculated on the basis of a pollen sum comprising only tree pollen. After recalculation, the pollen record at this site appears very different from the impression given by the published diagrams. Unfortunately the only relevant radiocarbon date for this site lies close to the onset of the Holocene: 9960±160 years BP (Q 616+617) but at a horizon in the pollen diagram where the *Quercus* and *Corylus* curves have attained moderately high values. The Lateglacial vegetational record itself is therefore undated, but comes from such a critical location close to the shores of the Bay of Biscay that it must be discussed.

In the principal borehole, five metres of peats and organic muds are underlain by one metre of inorganic silty clay. This basal deposit is represented by Oldfield's pollen zones F and L1 ('full-glacial' and 'Late-glacial 1'). Pollen assemblages for both zones are dominated overwhelmingly by Gramineae pollen with that of Compositae, *Polygonum* cf. *viviparum* and Caryophyllaceae frequent. Values for *Pinus* pollen are less than 1% in zone F and only 5–10% in zone L1. Towards the top of zone L1, *Juniperus* and *Artemisia* appear in small quantity. In zone L2, with the onset of organic sedimentation, the *Betula* and *Pinus* curves rise, but only to peaks of 6 % and 20 %; there is a slight increase in *Juniperus*.

In zone L3, tree pollen values fall again to less than 10 %. First the diagram shows high values for Gramineae and Cyperaceae pollen, then there is a massive increase in Ericaceae pollen, particularly *Calluna*, and in *Empetrum*. Towards the end of zone L3 the pollen curves for *Betula* and *Pinus* begin to rise again erratically, reaching peaks of over 30%. Ericaceae and *Empetrum* pollen curves remain high.

Zone P1 is marked by the expansion of first *Quercus* and then *Corylus*. The radiocarbon date was obtained from a separate borehole.

Zones F and L1 indicate a very open treeless environment with grassland and other herb communities that included a few arctic–alpine species such as the fern *Cryptogramma crispa* and possibly *Polygonum viviparum*. The so-called steppe elements such as *Artemisia* play almost no role in this vegetation. The virtual absence of even long-distance transport of *Pinus* pollen suggests that the prevailing wind direction was from the northwest, over the Bay of Biscay, perhaps as a result of anticyclonic conditions over an ocean covered with sea ice.

In zone L1, conditions ameliorated slowly; there was some long-distance transport of *Pinus* pollen, principally *P. uncinata*, and a small spread of *Juniperus*. In zone L2, the spread of *Betula* and *Pinus*, with *P. sylvestris* more frequent than *P. uncinata*, is restricted. It seems unlikely that there was woodland growing very close to the site, which was then an open-water habitat, but a notable increase in abundance of aquatic plants also suggests a climatic amelioration. In zone L3 the initial fall in tree pollen to very low values suggests a return, regionally, to open grassland conditions and the elimination of tree cover, but this is followed by the rapid spread of acid heathland, reminiscent of the Lateglacial heathlands of western Ireland, with crowberry (*Empetrum*), ling (*Calluna*) and other ericaceous plants. Since the actual site was undergoing a transition from reedswamp to bog, this development is particularly marked in the pollen record. The fluctuating record of birch and pine may be the result of individual trees that had actually colonised the bog surface at that time. The strong climatic amelioration in zone P1 is marked not only by the expansion of oak and hazel forest but the appearance of such thermophilous taxa as holly (*Ilex aquifolium*) and saw sedge (*Cladium mariscus*).

Oldfield broadly correlated his zones L1–L3 with Iversen's three Lateglacial pollen zones and P1 with the onset of the Holocene. The true age of the zones cannot be established without further investigation and radiocarbon dating. What does seem inescapable is that, for most of the Lateglacial, the area around Le Moura carried open grassland and then heathland vegetation with very little development of woodland. If zone L2 does represent the Lateglacial interstadial then at that time the spread of trees was particularly restrained, compared with other sites discussed, but in Ireland very oceanic conditions in the Lateglacial interstadial also seem not to have favoured the growth of birch. Zone L3 would represent a period of very dramatic vegetational change with grassland expanding and then being partially replaced by heathland with finally a local reinvasion by birch and pine. An alternative interpretation, that the main phase of Lateglacial climatic amelioration is represented by fluctuations of the tree pollen curve within zone L3 and that L2 represents a brief earlier climatic oscillation, cannot be excluded in the light of the comparatively early expansion of oak at this site, if the radiocarbon date is correct. In either case the vegetational development at Le Moura is very different indeed from that described from any other sites in southwest Europe. Such differences cannot be ascribed to altitude alone, but must also relate to the proximity of the Bay of Biscay with its rather exteme climatic conditions during the Lateglacial.

9. The marine palynological record from the Bay of Biscay

Pollen analysis has been applied by Turon to two deep-sea cores from the Bay of Biscay, where the marine planktonic fauna, particularly Foraminifera, has also been studied. Core CH 6932, taken from a site *ca.* 300 km from land (45° 24′ N, 05° 9′ W) covers approximately the past 20 ka (Pujol & Turon 1974; Turon 1974). Core CH 6719, taken 200 km from land (45° 45′ N, 3° 57′ W), a deeper sequence with a higher sedimentation rate, relates only to the Lateglacial and Holocene (Duplessy *et al.* 1981).

The diagrams bear a broad similarity to those from terrestrial sites, but are difficult to interpret in detail. The dominance of *Pinus* at every level, often 80–95 % of the total pollen, and the very low values for Gramineae, make it clear that the assemblages have undergone much differential sorting and degradation. Because of the high tree-pollen values, the diagram for core CH 6932 is calculated on the basis of a pollen sum based only on tree pollen, and core CH 6719 on a pollen sum excluding *Pinus*, by far the major constituent of the assemblages. This also makes comparison with other diagrams difficult. Neither of these cores has been subjected to radiocarbon dating; ages are inferred from faunal correlations with other dated cores. However, Ruddiman & McIntyre (1981) regarded Turon's dating of CH 6932 as 1500–2000 years too old.

The two issues raised by these diagrams are (i) whether they permit any interpretation of vegetation patterns and vegetational changes on the adjacent continent, and (ii) to what extent it is possible to correlate these pollen records with continental pollen diagrams. Because there is so much distortion of the pollen assemblages by processes of transport, sorting and degradation, the answer to the first question must be negative. With so many non-tree-pollen taxa under-represented or lost, cold-stage plant communities cannot be adequately recognized. For the Holocene, although both diagrams yield pollen curves for a variety of deciduous tree taxa, the abundance of pine gives the impression that western France and northern Spain were dominated by coniferous forest vegetation with a small admixture of deciduous trees. Continental diagrams show conclusively that this was not the case; nor, alternatively, can the marine assemblages be explained as representing simply a mixture of pollen from montane and lowland forest sources.

The possibilities for making detailed correlations between these marine and continental pollen diagrams are affected not only by problems of assemblage distortion but also by the effects of bioturbation and slow sedimentation rates, both of which obscure pollen-zone boundaries. In lacustrine pollen diagrams such boundaries, particularly for the Lateglacial, represent rapid vegetational changes of important climatic significance. However, broad correlations between the marine and continental pollen records are clear. The Holocene part of these marine diagrams is characterized, once the massive pine contribution is set aside, by relatively prominent *Quercus* and *Corylus* curves and by minor curves for other temperate trees such as *Ulmus* and *Tilia*, which are absent in the Lateglacial and earlier sections of the diagrams. Similarly the Lateglacial is marked by a prominent *Artemisia* curve, although grass values are only 5–10 %. It is very difficult to subdivide the Lateglacial part of the diagrams with any confidence. The different phases of the Lateglacial, as recognized by Turon, are based on only one or two samples. At these levels interpretation often depends on the behaviour of individual taxa. Information from other taxa may be contradictory. Thus in core CH 6719, the Younger Dryas cooling, although not well marked, is apparently recognized by a small decrease in *Quercus* and *Corylus* pollen which coincides with a fall in the *Artemisia* curve, whereas a rise in *Artemisia* might be expected on the evidence from continental pollen diagrams.

The lower part of the diagram from core CH 6932 shows assemblages where the highest values for non-tree pollen, including maxima for *Artemisia* and Gramineae, coincide with minor peaks for *Corylus*, *Betula* and *Quercus*. Turon suggests that this horizon represents the Würm III–IV interstadial, but elsewhere there is no evidence of such temperate conditions at this time, *ca.* 20 ka BP, either in the continental record or in other marine records. The peaks of temperate elements are over-emphasized because of the arboreal pollen sum, but the mixture of temperate and cold-climate elements in the pollen assemblages suggests some local reworking. It may be significant that the results from CH 6932 are not discussed at all by Duplessy *et al.* (1981) when presenting the pollen diagram from CH 6719.

The conclusion must be that these marine pollen diagrams can only be used to make very broad correlations with continental sequences. They are certainly not reliable indicators of short-lived events of less than 1000–2000 years' duration, such as the climatic changes of the Lateglacial. They may detect some features of these changes but the detailed record is erratic because of the unpredictable effects of bioturbation.

10. Pollen records from caves and rock shelters

For more than twenty years archaeological excavations of Palaeolithic caves and rock shelters in France, and recently in northern Spain, have included palynological investigations of the sedimentary infills. Because of the very low pollen concentrations, the sediment samples analysed generally have to be larger than those from lacustrine sites (25–50 g, as opposed to 1–2 g) and special techniques of preparation are used to extract the palynomorphs (Girard & Renault-Miskovsky 1969).

On the basis of these palynological studies, an increasingly detailed scheme, subdividing the latter part of the last glacial stage in France and adjacent areas into alternating stadial and interstadial intervals, has been proposed and developed by Leroi-Gourhan (1965; Leroi-Gourhan & Renault-Miskovsky 1977; Renault-Miskovsky & Leroi-Gourhan 1981). The scheme (figure 4) is closely linked to Upper and Middle Palaeolithic cultural sequences with a chronology based on radiocarbon dates from excavated bone and charcoal samples. A different stratigraphic classification of sedimentary sequences from caves and rock shelters in the Périgord (Dordogne) area of southwest France has been developed by Laville (Laville *et al.* 1980), largely on the basis of sedimentology but with strong climatic overtones, also with the support of palynological studies (Paquereau 1974–75, 1978). Laville's classification is essentially a detailed refinement of Würm I, II, III, IV subdivision proposed earlier by Bordes (1954).

After a period of competition these classifications now coexist with only relatively minor disagreements over the dating and possible overlap of some interstadial intervals (Laville *et al.* 1985). What is important is that both advocate a radically different interpretation of the sequence of climatic changes and vegetation patterns that took place in western, particularly southwestern, Europe between 35 and 10 ka BP from those suggested by non-archaeological palynologists. This alternative interpretation is widely accepted and taught in universities in Europe and North America and thus requires comment here.

As shown in figure 4, as well as the Allerød and Bølling interstadials, no less than six further

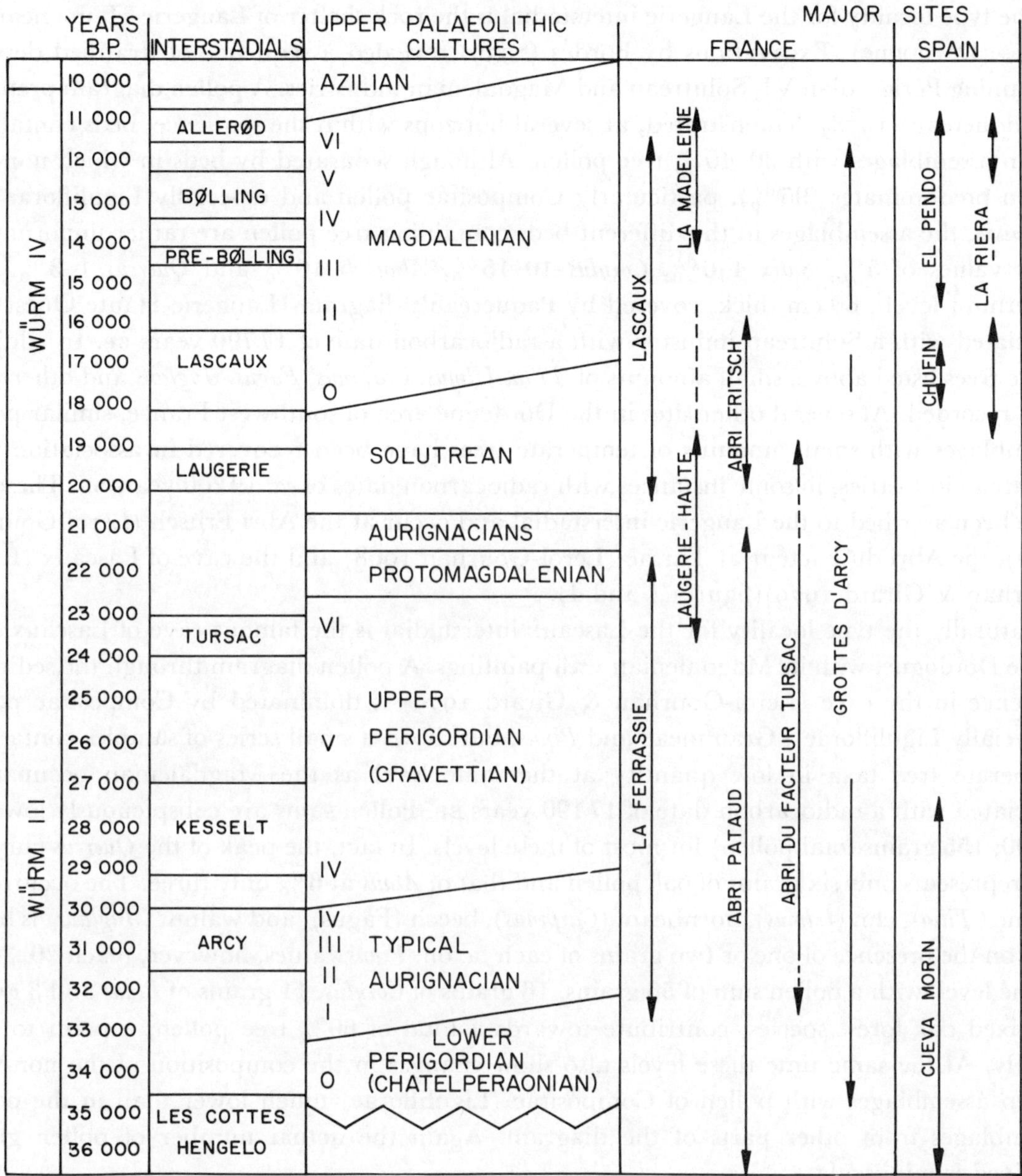

FIGURE 4. Climatic phases between 36 and 10 ka BP and their relation to major Palaeolithic sites in France and Spain, principally according to Renault-Miskovsky (1986).

interstadial intervals have been proposed between 32 and 10 ka BP. These interstadials are based on the occurrence of varying amounts of pine and relatively small percentages of temperate tree pollen at specific dated horizons in the cave sequences. As indicated in figure 4, these interstadials have been recognized at a range of sites in France and Spain.

The Laugerie and Lascaux interstadials are of particular importance because the radiocarbon-dated sequences suggest that these occurred between 19.5 and 18.5 ka BP and 18 and *ca.* 16.5 ka BP, respectively; these intervals are contemporary with those suggested by deep-ocean isotope and faunal records to have been very cold indeed just off the coasts of southwest France (Duplessy *et al.* 1981).

The type locality for the Laugerie interstadial is the rock shelter of Laugerie-Haute near Les Eyzies (Dordogne). Excavations by Bordes (1959) revealed a sequence of stratified deposits containing Perigordian VI, Solutrean and Magdalenian industries. A pollen diagram prepared by Paquereau (1978) demonstrated, at several horizons within the sequence, beds containing pollen assemblages with 30–40% tree pollen. Although separated by beds in which non-tree pollen predominates (95%), particularly Compositae pollen and especially Liguliflorae and *Artemisia*, the assemblages in the different beds containing tree pollen are rather uniform with *Pinus* values of 5%, *Salix* 4–6%, *Corylus* 10–15%, *Alnus* 5–10% and *Quercus* 1–3%. The uppermost levels, 60 cm thick, covered by Paquereau's diagram (Laugerie-Haute Ouest) are associated with a Solutrean industry with a radiocarbon date of 17790 years BP. In addition to the trees listed above, small amounts of *Tilia*, *Ulmus*, *Carpinus*, *Fagus*, *Juglans* and other trees were recorded. At several other sites in the Dordogne area of southwest France, similar pollen assemblages with small amounts of temperate trees have been recovered in association with Solutrean industries, in some instances with radiocarbon dates of *ca.* 19200 years BP. These too have been ascribed to the Laugerie interstadial and occur at the Abri Fritsch (Leroi-Gourhan 1967), the Abri du Facteur at Tursac (Leroi-Gourhan 1968) and the cave of Lascaux (Leroi-Gourhan & Girard 1979)(figures 1 and 4).

Naturally the type locality for the Lascaux interstadial is the famous cave of Lascaux, also in the Dordogne, with its Magdalenian wall paintings. A pollen diagram through the sediment sequence in the cave (Leroi-Gourhan & Girard 1979) is dominated by Compositae pollen (especially Liguliflorae), Gramineae and *Pinus*, but shows a small series of samples containing temperate tree taxa in low quantity at the same level as the Magdalenian occupation, associated with a radiocarbon date of 17190 years BP. Pollen sums are conspicuously low (56, 85, 90, 155 grains total pollen) for most of these levels. In fact, the peak of the *Quercus* curve at 7% represents only six grains of oak pollen and that of *Alnus* at 6% only three. The occurrence of lime (*Tilia*), elm (*Ulmus*), hornbeam (*Carpinus*), beech (Fagus), and walnut (*Juglans*) is based only on the presence of one or two grains of each taxon. *Pinus* values, however, reach 20–30%. At the level, with a pollen sum of 56 grains, 16 grains of *Corylus*, 11 grains of *Pinus* and 3 grains of mixed oak forest species, contribute towards a total of 60% tree pollen, a point to note shortly. At the same time these levels also show changes in the composition of the non-tree-pollen assemblages with pollen of Compositae (Liguliflorae) much lower than in the pollen assemblages from other parts of the diagram. Again the actual number of pollen grains involved is minuscule.

The Lascaux interstadial has been recognized at other sites (Leroi-Gourhan 1980*a*), particularly at the Abri Fritsch, where it stratigraphically overlies levels assigned to the Laugerie interstadial (Leroi-Gourhan 1967, 1980*b*), and at the caves of Chufin (Boyer-Klein 1980) and La Riera (Straus *et al.* 1981, 1983; Straus & Clark 1986) in northern Spain. Leroi-Gourhan's (1980*b*) outline pollen diagram for the site of Abri Fritsch (figure 5) shows features which she regards as important for the recognition and differentiation of the Lascaux and Laugerie interstadials: the fluctuations of the *Pinus* curves, including that for the southern European species *Pinus pinaster*, with a distribution just reaching south-west France today; the balance between tree and non-tree pollen; and the behaviour of the Gramineae and Compositae curves. The latter differ for the two interstadials, leading Leroi-Gourhan (1980*a*) to postulate that conditions during the Lascaux interstadial were more oceanic than during that of Laugerie.

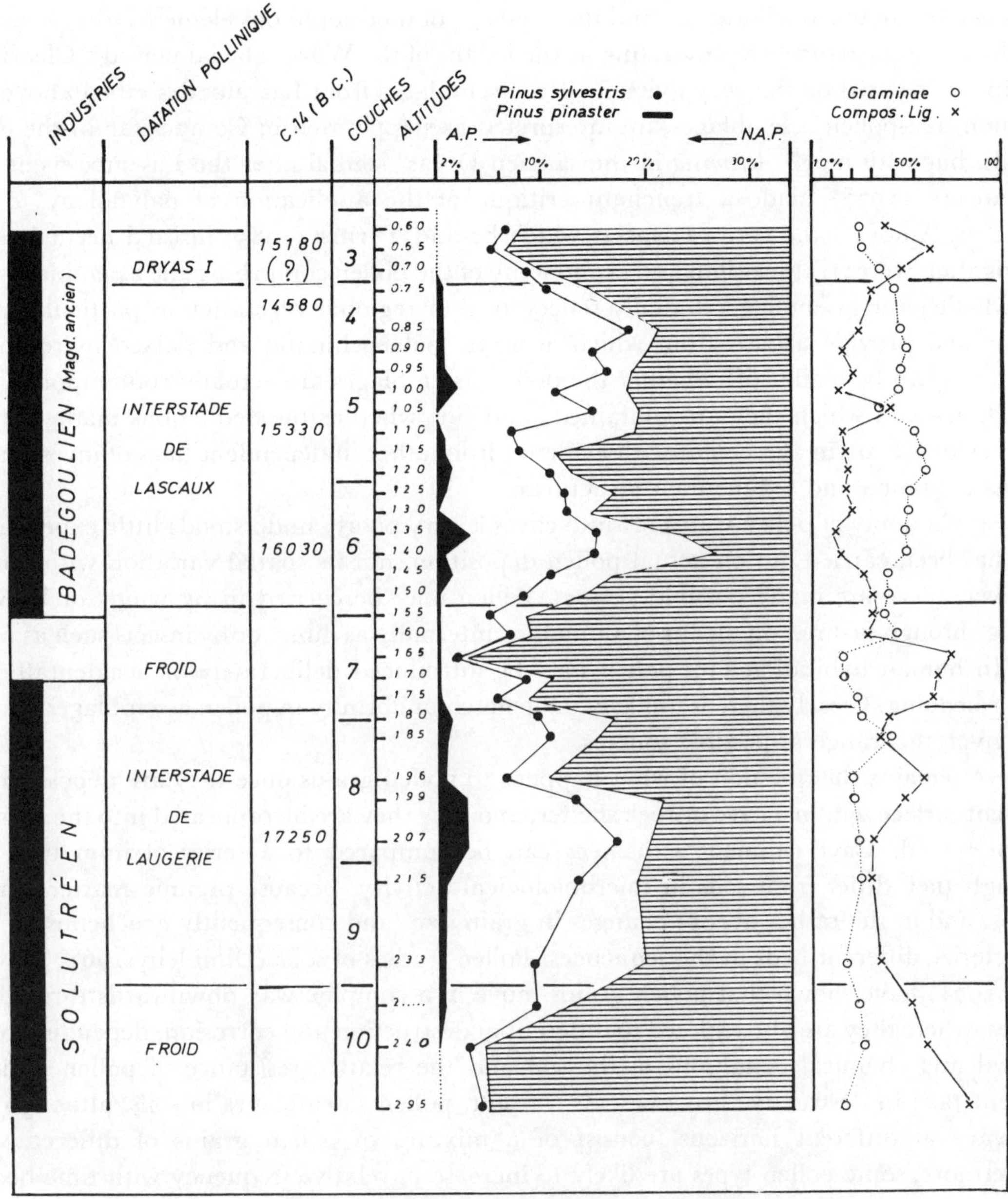

FIGURE 5. Pollen diagram from rock-shelter sediments of the Abri Fritsch (Indre), western France (Leroi-Gourhan 1980*b*).

The authors of these pollen diagrams have interpreted the temperate element in the assemblages in terms of the spread of either park woodland or of riverine forest in sheltered valleys. Implicit in both assumptions is the concept that these temperate trees had refugia close at hand within southwest France and northern Spain, for example on sheltered cliffs and gorges, where they survived the rigours of the colder intervals of the glacial period. The extremes of this interpretation are represented in a recently published volume (Renault-Miskovsky 1986). In translation: 'During this Lascaux interstadial, the principal features of the vegetation are characterized by: a significant forest cover (60% tree pollen), pines, but rather poorly represented, much hazel which declines with the advent of mixed oak forest at the maximum of the amelioration, and the presence of thermophilous elements, the occurrence of walnut being particularly interesting at the height of the Würm glacial period'. Clearly this description is based on the very sparse pollen assemblages from Lascaux described above. The high non-tree-pollen assemblages are interpreted as steppe, rich in Compositae in the earlier stadials, but with much *Artemisia* in the 'ancien Dryas' stadial after the Lascaux event.

Couteaux (1977) made a trenchant critique of the application of palynology to cave sediments. Amongst the various aspects which he and Turner (1985) insisted needed careful investigation were: (i) the origin and taphonomy of the pollen comprising the assemblages, and thus whether the assemblages actually reflect local or regional vegetation at particular points in time and provide a basis from which accurate palaeoclimatic and palaeoenvironmental deductions can be made; (ii) whether the pollen assemblages are actually contemporary with the sediments in which they are contained; and (iii) whether the deductions made from the pollen evidence are in agreement with evidence from other, independent lines of investigation, such as the fauna and sedimentary structures.

The taphonomy of pollen transport into caves is very poorly understood; little experimental work has been carried out on actual pollen deposition, nor its spatial variation within caves. However, there are many possible vectors. Pollen may be carried in by wind or by water seeping through fissures, on the fur of animals or internally as dung, or by insects such as mason bees. In human habitation sites pollen may be introduced deliberately or accidentally with food or bedding. Clearly there is unlikely to be much uniformity in pollen assemblages between sites, given this range of possible sources.

There remains the question of what happens to pollen grains once they are deposited on a sediment surface within a cave or rock shelter, and how they are incorporated into the sediment and preserved. Cave-sediment sequences can be compared to a series of immature soils, although they differ from soils in microbiological activity, because organic matter is much sparser, and in the rather abrupt changes in grain size (and consequently geochemistry) that characterize different beds in the sequences. Pollen studies of soils (Dimbleby 1961; Havinga 1968, 1971) have shown that pollen grains move in a complex way downwards through soil profiles, where they are also exposed to differential destruction and corrosion, depending on the physical and chemical conditions of the soil and the relative resistance of pollen grains of different taxa to destructive processes. As a result, pollen assemblages in soils, although they may vary at different horizons, consist of a mixture of pollen grains of different ages. Furthermore, some pollen types are likely to increase in relative frequency with time because they are resistant to destruction, whereas others become progressively diminished.

The pollen diagram from the Abri Fritsch (figure 5) has been redrawn in a more conventional way (figure 6), together with additional pollen curves published elsewhere (Leroi-Gourhan 1967, 1980*a*), so that it can be compared with diagrams from lacustrine sites. A full pollen diagram for this site has apparently never been published, so the data is necessarily incomplete. Pollen sums are also not published, but are generally above 250 grains. It is instructive to consider this diagram in the light of the discussion above. The first point to note is that, in this case, Holocene deposits actually overlie the older sequence. There is a discontinuous scatter of temperate pollen types, not only trees but also ivy (*Hedera*), throughout the diagram down to 185 cm, with only two levels lacking at least some representation.

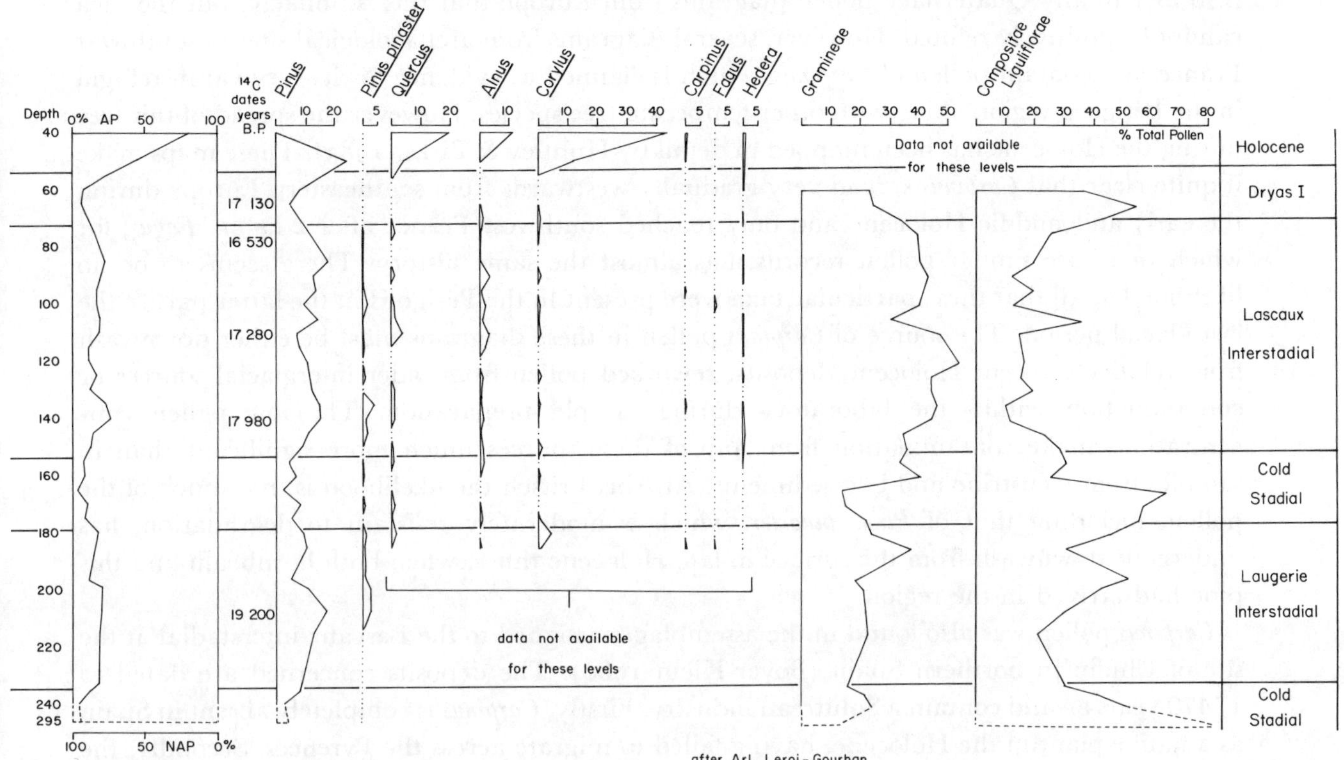

FIGURE 6. Pollen diagram from the Abri Fritsch (Indre), western France, redrawn with information from Leroi-Gourhan (1967, 1980*a*, *b*).

The interstadials and particularly their upper and lower boundaries are actually defined solely on the percentage curves of *Pinus* pollen; this is also largely true for the pollen diagram from Lascaux itself. When the diagram is drawn up in a conventional way, it becomes more obvious that the *Pinus* pollen curve shows comparatively low values. The fluctuations between samples can be seen to be grossly exaggerated by the graphic representation in figure 4. Equivalent values for *Pinus* pollen in diagrams from lacustrine sites are normally assumed to be referable to long-distance transport and show just as much variation.

There is also a critical ecological observation to be made. A well-established characteristic of pollen diagrams covering the climatic amelioration at the beginning of the Holocene and of interglacial stages is that they demonstrate how pioneer tree species immigrate and expand before the expansion of temperate taxa (Turner & West 1968). Interstadial vegetational successions show a similar but less developed succession. In virtually all pollen diagrams from caves and rock shelters, and specifically those from Laugerie-Haute, Lascaux and the Abri Fritsch, there is no evidence at all for vegetational succession. The temperate taxa simply appear 'unannounced'. In short, the pollen diagrams for the Lascaux and Laugerie, and likewise for other claimed interstadials, make no ecological sense.

Some of the less common thermophilous taxa can provide further evidence of the origin of the temperate pollen grains in these assemblages. Attention has been drawn to the occurrence of *Juglans* and likewise *Platanus* (plane) pollen in cave sites. Renault-Miskovsky *et al.* (1984) suggest that this provides evidence for their native status in western Europe. Pollen of these taxa is so rare in any Quaternary pollen diagrams from Europe that this is unlikely, but the idea cannot be positively refuted. However, several diagrams from archaeological sites in southwest France also contain pollen of *Carpinus*, which is claimed as evidence for its survival in refugia in the Périgord region along with other temperate tree species. However the spread of this tree during the Holocene has been mapped in detail by Huntley & Birks (1983). These maps make it quite clear that *Carpinus* spread very gradually westwards from southeastern Europe during the early and middle Holocene and only reached southwest France after 2 ka BP. *Fagus*, for which there are similar pollen records, has almost the same history. There seems to be no likelihood at all that these particular taxa were present in the Périgord in the latter part of the last glacial period. The source of *Carpinus* pollen in these diagrams must be either downwash from relatively recent Holocene deposits, reworked pollen from older interglacial sources or contamination within the laboratory during sample preparation. The low pollen concentrations make contamination from any of these sources much more significant than in samples from lacustrine and bog sediments. At Abri Fritsch the likelihood is that much of the pollen, including that of *Pinus pinaster*, which is moderately resistant to degradation, has undergone downwash from the surface in late Holocene times, when both hornbeam and this pine had arrived in the region.

Carpinus pollen was also found in the assemblages assigned to the Lascaux interstadial at the site of Chufin in northern Spain (Boyer-Klein 1980). The deposits concerned are dated to 17470 years BP and contain a Solutrean industry. Firstly, *Carpinus* is completely absent in Spain as a native plant in the Holocene, having failed to migrate across the Pyrenees. Secondly, the existence of a temperate period at that time in northern Spain (based on 50% absolute pollen (AP) values, but very low pollen sums in the Chufin diagram) seems very unlikely in view of climatic conditions inferred for the Cantabrian mountains and the Bay of Biscay at that time. Here, again the source of the pollen must be regarded as extraneous.

Finally, these cave pollen records must be compared with the evidence from other kinds of investigations. The archaeology is not particularly instructive in terms of palaeoclimatic reconstruction, but studies have generally been made on the sedimentology and vertebrate faunas from the same cave deposits. One of the reasons why the interpretation of pollen data from caves, as indicating relatively temperate conditions, has been so widely accepted is that it very often supports the climatic interpretations of the sedimentological studies (Laville 1980). Laville tends to make the assumption that deposition of fine-grained sediments takes place under mild, relatively humid conditions, whereas coarser rubbly deposits and breccias, which may or may not show evidence of solifluction and cryoturbation, are associated with colder climates. Other sedimentologists do not agree that this is always the case (Guillien & Lautridou 1974; Farrand 1985). However, fine-grained sediments certainly provide a more favourable matrix for the accumulation and protection of palynomorphs carried downwards over long periods of time, so that the relation between sediment type and pollen content may not, in fact, be independent. At Laugerie-Haute there seems to be a repeated correlation between beds of finer-grained sediment and the occurrence of less robust pollen types such as those of most temperate trees and also sedges, whereas assemblages from coarser grained beds, or those affected by weathering, tend to contain assemblages dominated by thick-walled or resistant pollen types such as Compositae, *Artemisia* and *Pinus*.

Turning to the vertebrate evidence, Delpech (1983) reports that the fauna from the interstadial horizon at Laugerie-Haute consists very largely of reindeer (*Rangifer tarandus*) with horse (*Equus caballus*) also frequent. This fauna shows, according to her, no trace of a climatic amelioration, not even an increase in humidity. With the exception of one site, excavated in the last century, the same conclusions are drawn from other sites of the same age in the Dordogne. Saiga antelope (*Saiga tatarica*) appears quite prominently in the faunal record shortly after this time, giving a strong impression of a cold, dry steppe environment. Likewise at Lascaux, despite the variety of animals painted on the cave walls, the actual faunal remains are predominantly those of reindeer. Remains of animals, such as roe deer (*Capreolus capreolus*) and wild boar (*Sus scrofa*), which today prefer wooded habitats, are rare.

The conclusions from this survey must be that these so-called interstadials based on pollen analyses of cave and rock shelter sediments do not, in fact, represent genuine climatic events or clearly demonstrable vegetational changes. This does not mean that all pollen in cave sediments is derived from younger deposits. At La Riera (Straus & Clark 1986) *Juniperus* pollen occurs in sediments dated to the early Lateglacial and *Artemisia* at lower horizons, but they form components of pollen assemblages which have at least in part moved downwards within the sediment column and also undergone differential destruction, hence the dominance of robust pollen grains, such as those of Compositae, identifiable even when partially degraded.

At Grande Pile and Les Echets in France (Woillard 1975, 1978; Beaulieu & Reille 1984) and at Padul in Spain (Pons & Reille 1986) there are now pollen diagrams from deep lacustrine sequences that cover the entire period of the last glacial stage. It is appropriate to complete this discussion by quoting the comments of Beaulieu & Reille (1984) on their sequence at Les Echets.

> The well established and dated sequence of Les Echets does not give evidence of the least climatic amelioration of any length between 25000 BP and the Lateglacial: none of the numerous interstadials that have been described and recognized in caves by non-botanist researchers and that encumber both the end of the Würm and the literature concerned can objectively be found; neither are they found in the diagrams from la Grande Pile published by Woillard (1975).

At first it seemed logical that in southwest Europe these purported interstadial intervals represented small oscillations in the position of the polar front, which brought about minor vegetational changes in coastal areas of Spain and France. These need not have had a widespread impact. Unfortunately it is clear that the palynological evidence itself must be rejected, on the grounds that the pollen assemblages that contain small quantities of tree pollen appear on balance of the evidence to be intrusive and not contemporary with the sediments in which they were contained.

11. Vegetational and climatic history of southwest Europe 30–10 ka BP: conclusions

30–22 ka *BP*

The only evidence for vegetational conditions before the glacial maximum comes from cave and rock-shelter sites. Proposed interstadial intervals, such as the Arcy, Kesselt and Tursac interstadials, are rejected as artefacts based on the presence of intrusive temperature pollen, although they have been claimed to be represented at a number of sites not only in the Pyrenees and Spain, as well as in other parts of France (Leroi-Gourhan 1959, 1971; Leroi-Gourhan & Leroi-Gourhan 1964; Bui-Thi-Mai & Girard 1984). Other levels in these sites indicate open vegetational conditions, although again the pollen may likewise not be contemporary with the enclosing sediments. These pollen assemblages, usually dominated by Compositae (Liguliflorae), are too distorted by differential destruction to permit a vegetational interpretation, but it does seem likely that at this period before the glacial maximum *Artemisia* was a very minor component of the vegetation, compared with its abundance after 16 ka BP.

22–16 ka *BP*

The Pyrenees and other mountain ranges in northern Spain and Portugal were quite extensively glaciated during the last glacial stage. It is concluded that, as elsewhere in western Europe, the glacial maximum occurred at approximately 18–20 ka BP, but it is virtually impossible to secure reliable absolute dates for such events. Dates suggesting that this episode of glaciation occurred much earlier (Mardones & Jalut 1983) or that the ice had melted substantially before 18 ka BP (Kolstrup 1980; Jalut *et al.* 1982) are rejected. Likewise rejected are the purported Laugerie and Lascaux interstadials of Paquereau (1978) and Leroi-Gourhan (1980*a*, *b*); Leroi-Gourhan & Girard (1979), not simply because they are in utter discord with other palaeoclimatic reconstructions for this interval, but primarily because, as discussed earlier, the palynological evidence on which they were based appears to be unsound from several lines of evidence.

16–*ca.* 14.5 ka *BP*

It can probably be assumed that by 16 ka BP melting of these mountain glaciers was well under way or complete and sediment was already accumulating in most of the lacustrine basins thereby created. In these mountain areas deglaciation was followed by the appearance of very open grass- and herb-dominated pioneer plant communities. In a short time there developed a vegetation dominated by grasses and *Artemisia*. Van Campo (1984), noting that vegetation of this kind became widespread over much of southern Europe at this time, has made a convincing case that this was essentially a steppe vegetation associated with an arid climate, where lack of rainfall restricted tree growth at least as effectively as low temperatures. This agrees with predictions by Ruddiman & McIntyre (1981) that sea ice cover between 16 and 13 ka BP would have led to a greatly reduced precipitation over the continent of Europe. Certainly pollen assemblages of this age are very uniform at Lago de Ajo, at the Pyrenean sites and even in Italy and Greece. However, at two sites in the area considered here, there is evidence for different conditions. In the undated basal levels at Le Moura, *Artemisia* is very sparse, but conditions seem to have been particularly chill and bleak there at low altitude on the very edge of the Bay of Biscay. By contrast at Sanabria, in the Laguna de las Sanguijuelas diagram, as later in time in the Sanabria Marsh diagram, evidence for extensive ericaceous plant communities from the earliest levels suggest a moderating oceanic influence on temperature and a much less arid climate. From no lacustrine site is there any indication of a major or even a minor climatic oscillation with any development of tree growth between deglaciation and the main Lateglacial spread of birch and pine.

ca. 14–10 ka *BP*

Clearly there is no uniform pattern of Lateglacial vegetational development in southwest Europe. Part of the variation between the sites discussed can be attributed to factors such as altitude, soil type and local hydroseral conditions. However, even given the uncertainties of dating suggested for some sites, there remain significant differences in the pattern and timing of immigration and expansion of particular taxa and in the impact and intensity of climatic changes as interpreted from the vegetational record.

In the Pyrenees, as in northern Europe, the earliest vegetational evidence for climatic amelioration during the Lateglacial seems to be an increase in *Juniperus* pollen productivity. At Biscaye, this took place shortly after 14820 years BP and at Freychinède, at 14700 ± 800 years BP, although neither of these dates is secure. At La Moulinasse, where hard-water dating error is unlikely, the *Juniperus* rise took place before 13600 years BP.

In the mountains of northwestern Spain at Sanabria, the climate was already mild enough for ericaceous plants to flourish, but information about the behaviour of juniper in this early period of the Lateglacial is sparse. Juniper was never abundant at Lago de Ajo, the record for Sanabria Marsh begins a thousand years later, and in the Laguna de las Sanguijuelas diagram *Juniperus* pollen was not counted. Nevertheless, it seems that at that site tree birches were already present by 13700 years BP. Thus, to the west, closer to the Atlantic seaboard, the expansion of trees, particularly birches, took place probably several hundred years earlier than in the Pyrenees and, of course, more than a thousand years before the expansion of birch in the British Isles; indeed, five hundred years before the earliest sign of climatic warming there postulated on the evidence of coleopteran faunas (Atkinson *et al.* 1987).

It is only from the two sites in northwestern Spain, Laguna de las Sanguijuelas and Lago de Ajo, that there is any evidence for two periods of expansion of birch and pine, separated in time by a minor re-expansion of open ground vegetation. Further investigations are needed to confirm that a minor climatic oscillation local to northwestern Spain took place and to date it more precisely. Note that evidence for climatic change is clearly time-transgressive along the seaboard of western Europe; this minor episode of warming and cooling is very unlikely, if confirmed, to correlate in time with the Bølling of northwestern Europe.

The concept of separate Bølling and Allerød interstadials has led other authors (e.g. Mardones & Jalut 1983) to try and identify, on very flimsy grounds, an interval of cooling in their pollen diagrams before the main Lateglacial expansion of birch and pine. There seems to be no good evidence for any such climatic oscillation from pollen diagrams in the Pyrenees. Indeed, it is doubtful whether those authors would have been tempted to identify such an event if they had not been influenced by the all too general use of the northwestern European evidence as a model.

Undoubtedly the most significant feature of the Lateglacial vegetational record for northwestern Spain is the relatively early appearance of oak, at least deciduous species, and the continuous record of their presence there into Holocene time. The Lateglacial occurrence of oak in the Pyrenees is rather tenuous; if established there, its occurrence must have been very sparse and local. It was certainly not present around Le Moura at that time.

As already noted, there is no evidence for any Younger Dryas cooling at either Sanabria Marsh or Lago de Ajo, nor, of course, from the Laguna de las Sanguijuelas pollen diagram in the light of new radiocarbon evidence. The mildness of conditions suggests that between 11 and 10 ka BP this part of Spain was still under the influence of weather patterns originating to the south of the polar front. Likewise there is very little evidence for climatic deterioration at high altitudes in the eastern Pyrenees, but to the west, closer to the Bay of Biscay, and with decreasing altitude some opening of the forest appears to have taken place at around 10860 years BP but again no real change in its composition. At Le Moura, where throughout the Lateglacial the vegetation had a rather bleak, open aspect, whether for edaphic or climatic reasons, there do seem to have been very substantial vegetational changes in the period immediately before 10 ka BP. These, if reinvestigated and properly dated, could well show the influence of the very cold conditions in the Bay of Biscay during Younger Dryas times postulated by Duplessy *et al.* (1981).

In the course of their reinvestigation of Padul, Pons & Reille (1986) have reported a marked vegetational event, which they relate to the Younger Dryas oscillation. Their pollen diagram shows a substantial rise in the *Artemisia*, Chenopodiaceae and *Pinus* pollen curves, accompanying a fall in those of Gramineae and *Quercus*, particularly evergreen oak type. These vegetational changes appear to represent a short period of aridity. The dating of this event is somewhat ambiguous, because it has a basal date of 9830±110 years BP (Gif 6006), but an older date of 10000±110 years BP (Gif 6212) marks the expansion of oak above. It seems unlikely that these vegetational changes are directly linked with the migration of the polar front in the North Atlantic. They may be more closely related to Lateglacial changes in aridity in North Africa.

The amelioration of climate at approximately 10 ka BP is well-marked in sites throughout the region. In northwestern Spain, where oak was already established, a substantial expansion of oak forest took place. Oak spread rapidly into other areas of northern Spain and to the lowlands around the Bay of Biscay. In the Pyrenees the amelioration is marked by an increase in birch and pine before the invasion of the lower and middle slopes of the mountains by oak. Everywhere open vegetational communities contracted with the spread of forest. Similarly the appearance and rapid expansion of hazel characterised this period, though the dating of the rise of the *Corylus* curve is not yet good enough to yield information on the migration routes and possible refugia for this species.

Finally, comparisons between these Lateglacial vegetational records do throw some new light on the probable refugia of a few important plant taxa. The claims by archaeological palynologists that their evidence shows that such refuges existed in the Dordogne and other areas have already been dismissed. If we turn to more reliable pollen diagrams, little can be said about tree birches and pines. These pioneer trees invaded many different areas soon after the amelioration of climate. Refugia were probably widespread and soil conditions, particularly enrichment in nutrients by blue–green algae, as well as climatic factors (Van Geel *et al.* 1984), may have acted as a control on the successful establishment of tree growth. It is clear that *Pinus uncinata* played an important role as a pioneer tree well beyond its present altitudinal and geographical limits.

More important is the new evidence that deciduous oak species were present during Lateglacial times in northwestern Spain and were abundant there, whereas they were sparse or absent in the Pyrenees. Jalut (1973) claimed in early publications that very small quantities of pollen of temperate trees, not only *Quercus* but also *Corylus*, *Alnus*, *Abies* and *Fagus*, detected in Lateglacial pollen sequences, indicated the local presence of refuges for these trees in the northern part of the Pyrenees. This pollen, if not related to contamination from use of Hiller corers, has almost certainly been reworked from older Quaternary peat and soil deposits. The new evidence from Sanabria and from Lago de Ajo suggests strongly that the refuges for these oaks lay in maritime areas of western Spain and Portugal. At *ca.* 18 ka BP, these were the only areas of the western European seaboard lying between the 12 °C and 18 °C August sea-surface temperature isotherms (McIntyre *et al.* 1976). Such isotherms bracket the northern and southern limits of deciduous and mixed coniferous–deciduous forests along both the eastern and western seaboards of the North Atlantic Ocean today (Van Campo 1984). As additional support, even in the early Holocene, there is evidence for increasingly later dates for the arrival and expansion of deciduous oaks eastwards from Cantabria along the Pyrenees.

The history of evergreen oak species is less clear, partly because of difficulties in recognizing and assessing the frequencies of their pollen. At Sanabria the record for evergreen oaks begins at *ca.* 9500 years BP and at Le Moura they were present at *ca.* 9960 years BP. In the south of Spain, at Padul, evergreen oaks were already abundant in the vegetation well before 12000 years BP. Because this pollen type is virtually absent from Lateglacial levels of pollen diagrams from northern and northwestern Spain, the refugia do not seem to have been nearby and were probably in the south or the maritime southwest of the Iberian peninsula, at least for species such as *Quercus rotundifolia* and *Q. coccifera*, although possibly not for *Q. ilex*.

There is good reason to suggest that *Calluna* and other ericaceous taxa, which today have an oceanic western European distribution, probably also survived the last glacial period in refuges in western Spain and Portugal.

12. Lateglacial vegetational changes and the climatic influence of the North Atlantic Ocean

A number of further conclusions may be drawn concerning the relation of Lateglacial climatic and oceanographic changes in the North Atlantic and vegetational changes in western Europe.

(i) Ruddiman & McIntyre (1981) have shown that movements of the polar front in the eastern North Atlantic during the period under consideration were time-transgressive. Clearly the vegetational response was also time-transgressive. In northwestern Spain climatic amelioration took place between 500 and 1000 years earlier than in northwestern Europe.

(ii) Vegetation patterns over most of western Europe were probably never in equilibrium with the rapidly changing climate, so that migrational factors increase still more the time differential between vegetational changes in different areas along the western seaboard of Europe.

(iii) The climate of the Atlantic Ocean not only influences the timing of vegetational changes but also affects some areas more intensely than others, either because they are ecotonal or because local factors magnify the climatic impact. Within southwest Europe the contrasts between Lateglacial vegetational development at Sanabria in the oceanic west, Le Moura, close to the Bay of Biscay and the Pyrenean sites are very considerable even if allowances are made for altitudinal and edaphic differences. Northwestern Spain remained under the influence of a moderating oceanic climate throughout the Lateglacial, so that even oak could survive when elsewhere conditions deteriorated during the Younger Dryas cooling. The shores of the Bay of Biscay, by contrast, seem to have been affected strongly by this cooling. Although Duplessy *et al.* (1981) thought that the interstadial conditions were, by contrast, very warm, there is no evidence from the Le Moura pollen record to confirm that idea. Lateglacial vegetational development in the Pyrenees, perhaps because of the altitude of the sites investigated, responded much less strongly to changing conditions in the North Atlantic than at sites farther west. During the Lateglacial interstadial there was little, if any, development of thermophilous trees in the area; the subsequent climatic deterioration of Younger Dryas times was likewise poorly marked.

(iv) Given the differences of timing and the intensity of vegetational change between sites, the traditional subdivision of the Lateglacial widely used in northern Europe (table 1), but already shown to have its areal limitations by Watts (1980), is clearly quite inapplicable farther south. Indeed, attempts to use this scheme in areas where neither the chronology nor the detailed climatic history apply have caused misinterpretations.

(v) The pattern of Lateglacial vegetational and climatic development along the western seaboard of Europe, although regionally diverse, clearly has a direct link to changes in atmospheric and oceanic circulation patterns in the eastern North Atlantic. It appears to be a complex, time-transgressive local event, and attempts to make long-distance correlations between Lateglacial sequences in distant extra-European areas with the classical tripartite sequence of the northwestern European Lateglacial should be regarded with scepticism.

We thank Professor W. A. Watts and Dr R. H. W. Bradshaw for helpful discussion and criticism; Professor R. G. West, F.R.S., for advice and support; and Mrs Sylvia Peglar for help in drafting the diagrams. Investigations at Sanabria Marsh were supported by the E.E.C. Climatology Programme.

References

Alimen, H., Florschütz, F. & Menendez Amor, J. 1964 Étude geologique et palynologique sur le Quaternaire des environs de Lourdes. *Actes 4e Congrès int. Études pyrénéennes*, pp. 7–26.

Atkinson, T. C., Briffa, K. R. & Coope, G. R. 1987 Seasonal temperatures in Britain during the past 22,000 years, reconstructed using bettle remains. *Nature, Lond.* **325**, 587–592.

de Beaulieu, J.-L., Pons, A. & Reille, M. 1985 Recherches pollenanalytiques sur l'histoire tardiglaciaire et holocène de la végétation des Monts d'Aubrac (Massif Central, France). *Rev. Palaeobot. Palynol.* **44**, 37–80.

de Beaulieu, J.-L. & Reille, M. 1984 A long Upper Pleistocene pollen record from Les Echets, near Lyon, France. *Boreas* **13**, 111–132.

Bordes, F. H. 1954 *Les limons Quaternaires du bassin de la Seine*. Paris: Archive de l'Institut de Paléontologie Humaine, Memoire 26.

Bordes, F. 1959 Laugerie Haute. *Gallia Préhist.* **2**, 156–167.

Boyer-Klein, A. 1980 Nouveaux résultats palynologiques de sites solutréens et magdaléniens cantabriques. *Bull. Soc. préhist. Fr.* **17**, 103–107.

Bui-Thi-Mai & Girard, M. 1984 L'analyse pollinique de la grotte de Saint-Jean-de-Verges (Ariège). *Bull. Soc. préhist. Ariège* **39**, 27–41.

Chaline, J. & Jerz, H. 1983 Proposition de création d'un étage würmien par la sous-commission de stratigraphie du Quaternaire européen de l'INQUA. *Bull. Ass. fr. Étude Quat.* (N.S.) **16**, 149–152.

Couteaux, M. 1977 A propos de l'interprétation des analyses polliniques de sédiments minéraux, principalement archéologiques. In *Approche écologique de l'homme fossil* (ed. H. Laville & J. Renault-Miskovsky), pp. 259–276. *Bull. Ass. Fr. Etude Quatern.*, **47** (suppl.).

Deckker, P. de, Geurts, M. A. & Julia, R. 1979 Seasonal rhythmites from a lower Pleistocene lake in northeastern Spain. *Palaeogeogr. Palaeoclimatol. Palaeoecol.* **26**, 43–71.

Delpech, F. 1983 *Les faunes du Paléolithique supérieure dans de Sud-Ouest de la France*. Cahiers du Quaternaire, no. 6. Paris: C.N.R.S.

De Vries, H., Florschütz, F. & Menendez Amor, J. 1960 Un diagramme pollinique simplifié d'une couche de 'gyttja', située a Poueyferré près de Lourdes (Pyrénées francaises centrales), daté par la méthode du radiocarbone. *Proc. ned. Akad. wet.* B **63**, 498–500.

Dimbleby, G. W. 1961 Soil pollen analysis. *J. Soil Sci.* **12**, 1–11.

Duplessy, J. C., Delibrias, G., Turon, J. L. & Duprat, J. C. 1981 Deglacial warming of the northeastern Atlantic Ocean: correlation with the palaeoclimatic evolution of the European continent. *Palaeogeogr. Palaeoclimatol. Palaeoecol.* **35**, 121–144.

Elhai, H. 1966 Deux gisements du quaternaire moyen: Bruges (SW de la France), Bañolas (Catalogne). *Bull. Ass. fr. Étude Quat.* **6**, 69–78.

Farrand, W. R. 1985 Rockshelter and cave sediments. In *Archaeological sediments in context* (ed. J. K. Stein & W. R. Farrand), pp. 21–39. (*Peopling of the Americas*: Edited volume series 1.) Orono: University of Maine.

Florschütz, F. & Menendez Amor, J. 1962 Beitrag zur Kenntnis der quartären Vegetationsgeschichte Nordspaniens. In *Festschrift Franz Firbas. Veröff. geobot. Inst., Zürich* **37**, 68–73.

Florschütz, F., Menendez Amor, J. & Wijmstra, T. A. 1971 Palynology of a thick Quaternary succession in southern Spain. *Palaeogeogr. Palaeoclimatol. Palaeoecol.* **10**, 233–264.

Gaussen, H. 1948 *Carte de la végétation de la France*. Perpignan: C.N.R.S.

Girard, M. & Renault-Miskovsky, J. 1969 Nouvelles techniques de préparation en palynologie appliquées à trois sédiments du Quaternaire final de l'Abri Cornille (Istres-Bouches-du-Rhone). *Bull. Ass. fr. Étude Quat.* **21**, 275–284.

Guillien, Y. & Lautridou, J. P. 1974 Conclusions des recherches de gélifraction expérimentale sur les calcaires des Charentes. *Bull. Cent. Géomorphol.* **19**, 25–34.

Hannon, G. E. 1984 Late Quaternary vegetation of Sanabria Marsh, northwest Spain. M.Sc. thesis, Trinity College, University of Dublin.

Havinga, A. J. 1968 Some remarks on the interpretation of a pollen diagram of a podsol profile. *Acta bot. neerl.* **17**, 1–4.

Havinga, A. J. 1971 An experimental investigation into the decay of pollen and spores in various soil types. In *Sporopollenin* (ed. J. Brooks, P. R. Grant, M. D. Muir, P. van Gijzel & G. Shaw), pp. 446–479. London: Academic Press.

Heusser, C. J. & Florer, L. E. 1973 Correlation of marine and continental Quaternary pollen records from the Northeast Pacific and Western Washington. *Quat. Res.* **3**, 661–670.

Höllermann, P. 1968 Die rezenten Gletscher der Pyrenäen. *Geographica helv.* **23**, 157–168.

Huntley, B. & Birks, H. J. B. 1983 *An atlas of past and present pollen maps for Europe: 0–13,000 years ago*. Cambridge University Press.

Iversen, J. 1954 The lateglacial flora of Denmark and its relation to climate and soil. *Danm. geol. Unders.* ser. 2, **80**, 87–119.

Jalut, G. 1973 Analyse pollinique de la Tourbière de la Moulinasse: nord oriental des Pyrénées. *Pollen Spores* **15**, 472–509.

Jalut, G., Delibrias, G., Dagnac, J., Mardones, M. & Bouhours, M. 1982 A palaeoecological approach to the last 21,000 years in the Pyrenees: the peat bog of Freychinède (alt. 1350 m, Ariège, South France). *Palaeogeogr. Palaeoclimatol. Palaeoecol.* **40**, 321–359.

Janssen, C. R. & Woldringh, R. E. 1981 A preliminary radiocarbon dated pollen sequence from the Serra da Estrela, Portugal. *Finisterra* **16**, 299–307.

Kolstrup, E. 1980 Climate and stratigraphy in northwestern Europe between 30,000 B.P. and 13,000 B.P. with special reference to the Netherlands. *Meded. Rijks geol. Dienst* **32**(15), 181–253.

Laville, H., Delpech, F. & Rigaud, J.-P. 1985 Sur la zonation du Pleistocène récent: les précisions du domaine Aquitain. In *Palynologie archéologique* (ed. J. Renault-Miskovsky, Bui-Thi-Mai & M. Girard), pp. 245–249. (Centre de Recherches Archéologiques; notes et monographies techniques no. 17.) Paris: C.N.R.S.

Laville, H., Rigaud, J.-P. & Sackett, J. 1980 *Rock shelters of the Perigord: geological stratigraphy and archaeological succession*. New York: Academic Press.

Leroi-Gourhan, A. 1959 Résultats de l'analyse pollinique de la grotte d'Isturitz. *Bull. Soc. préhist. Fr.* **56**, 619–624.

Leroi-Gourhan, A. 1965 Les analyses polliniques sur les sédiments des grottes. *Bull. Ass. fr. Étude Quat.* **7**, 145–152.

Leroi-Gourhan, A. 1967 Analyse pollinique des niveaux paléolithiques de l'abri Fritsch. *Rev. Palaeobot. Palynol.* **4**, 81–86.

Leroi-Gourhan, A. 1968 L'Abri du Facteur à Tursac (Dordogne): III Analyse pollinique. *Gallia Préhist.* **11**, 123–132.

Leroi-Gorhan, A. 1971 Analisis polinico de Cueva morin. In *Cueva Morin: Excavaciones 1966–1968* (ed. J. G. Echegaray & L. G. Freeman), pp. 359–365. Santander: Patronato de las Cuevas Prehistoricas.

Leroi-Gorhan, A. 1980*a* Interstades Würmiens: Laugerie et Lascaux. *Bull. Ass. fr. Étude Quat.* (N.S.) **3**, 95–100.

Leroi-Gourhan, A. 1980*b* Les interstades du Würm supérieur. In *Problemes de stratigraphie Quaternaire en France et dans les pays limitrophes* (ed. J. Chaline), pp. 192–194. (*Suppl. Bull. Ass. fr. Étude Quat.* (N.S.) **1**.)

Leroi-Gourhan A. & Leroi-Gourhan, A. 1964 Chronologie des grottes d'Arcy-sur-Curé. *Gallia Préhist.* **7**, 1–64.

Leroi-Gourhan, A. & Girard, M. 1979 Analyses polliniques de la grotte de Lascaux. In *Lascaux inconnu* (*Gallia Préhist.* Suppl. 12). pp. 75–80.

Leroi-Gourhan, A. & Renault-Miskovsky, J. 1977 La palynologie appliquée à l'archéologie: méthodes, limites, résultats. In *Approche écologique de l'homme fossile* (ed. H. Laville & J. Renault-Miskovsky), pp. 35–49. (*Bull. Ass. fr. Étude Quat.*, suppl. no. 47.)

Lowe, J. J., Gray, J. M. & Robinson, J. E. 1980 *Studies in the Lateglacial of North-west Europe*. Oxford: Pergamon Press.

McIntyre, A., Kipp, N., Be, A. W. H., Crowley, T., Gardner, J. V., Prell, W. L. & Ruddiman, W. F. 1976 Glacial North Atlantic 18,000 years ago: a CLIMAP reconstruction. In *Investigations of late Quaternary paleoceanography and paleoclimatology* (ed. R. M. Cline & J. D. Hays), pp. 43–76. *Mem. geol. Soc. Am.* 145.

McKeever, M. 1984 *Comparative palynological studies of two lake sites in western Ireland and northwestern Spain.* M.Sc. thesis, Trinity College, University of Dublin.

Mangerud, J., Andersen, S. T., Berglund, B. E. & Donner, J. J. 1974 Quaternary stratigraphy of Norden, a proposal for terminology and classification. *Boreas* **3**, 109–127.

Mardones, M. & Jalut, G. 1983 La Tourbière de Biscaye (Alt. 409 m, Hautes Pyrénées) : approche paléoécologique des 45,000 dernières années. *Pollen Spores* **25**, 163–212.

Mary, G., Médus, J. & Delibrias, G. 1975 Le quaternaire de la côte asturienne (Espagne). *Bull. Ass. fr. Étude Quat.* **42**, 13–24.

Menendez Amor, J. 1968 Estudio esporo-polinico de una turbera en el Valle de la Nava (prov. de Burgos). *Boln R. Soc. esp. Hist. nat.* (*Geol.*) **66**, 35–39.

Menendez Amor, J. & Florschütz, F. 1959 Algunas noticias sobre el ambiente en que vivio el hombre durante el gran interglaciar en dos zonas de ambas Castillas. *Estudios geol. Inst. invest. Geol. Lucas Mallada* **15**, 277–282.

Menendez Amor, J. & Florschütz, F. 1961 Contribucion al concimiento de la historia de la vegetacion en Espana durante el Cuaternario. *Estudios geol. Inst. invest. Geol. Lucas Mallada* **17**, 83–99.

Menendez Amor, J. & Florschütz, F. 1964 Results of the preliminary palynological investigation of samples from a 50 m boring in southern Spain. *Boln R. Soc. esp. Hist. nat.* (*Geol.*) **62**, 251–255.

Mercer, J. H. 1969 The Allerød oscillation: a European climatic anomaly? *Arct. alp. Res.* **1**, 227–234.

Nonn, H. 1966 *Les régions cotières de la Galice* (*Espagne*). Etude géomorphologique T. III. Faculté des Lettres de l'Université de Strasbourg, Fondation Baulig.

Oldfield, F. 1964 Late-Quaternary deposits at Le Moura, Biarritz, south-west France. *New Phytol.* **63**, 374–409.

Oldfield, F. 1968 The Quaternary vegetational history of the French Pays Basque: I Stratigraphy and pollen analysis. *New Phytol.* **67**, 677–731.

Paquereau, M. M. 1974–75 Le Wurm ancien en Périgord. Etude palynologique. Première partie: Les diagrammes palynologiques – la zonation climatique. *Quaternaria* **18**, 67–116. Deuxième partie: L'évolution des climats et des flores. *Quaternaria* **18**, 117–160.

Paquereau, M. M. 1978 Flores et climats du Wurm III dans le Sud-Ouest de la France. *Quaternaria* **20**, 123–164.

Pons, A. & Reille, M. 1986 Nouvelles recherches pollenanalytiques à Padul (Granada): La fin du dernier glaciaire et l'Holocène. In *Quaternary Climate in Western Mediterranean* (*Proceedings of the Symposium on Climatic Fluctuations during the Quaternary in the Western Mediterranean Regions, Madrid, June 16–21, 1986*) (ed. F. Lopez-Vera), pp. 405–420. Universidad Autonoma de Madrid.

Pujol, C. & Turon, J.-L. 1974 Paléoclimatologie et stratigraphie du Quaternaire terminal du Golfe de Gascogne déduites de l'analyse des Foraminifères planctoniques et des ensembles sporopolleniques des sédiments marins. *Boreas* **3**, 99–104.

Renault-Miskovsky, J. 1986 *L'Environnement au temps de la préhistoire: méthodes et modèles.* Paris: Masson.

Renault-Miskovsky, J., Bui-Thi-Mai, M. & Girard, M. 1984 A propos de l'indiginat ou de l'introduction de *Juglans* et *Platanus* dans l'ouest de l'Europe au Quaternaire. *Revue Paléobiol.*, volume spécial, 155–178.

Renault-Miskovsky, J. & Leroi-Gourhan, A. 1981 Palynologie et archéologie: nouveaux résultats du Paléolithique supérieur au Mésolithique. *Bull. Ass. fr. Étude Quat.* (N.S.) **7–8**, 121–128.

Ruddiman, W. F. & McIntyre, A. 1973 Time-transgressive deglacial retreat of Polar water from the North Atlantic. *Quat. Res.* **3**, 117–130.

Ruddiman, W. F. & McIntyre, A. 1981 The North Atlantic Ocean during the last deglaciation. *Palaeogeogr. Palaeoclimatol. Palaeoecol.* **35**, 145–214.

Ruddiman, W. M., Sancetta, C. D. & McIntyre, A. 1977 Glacial/Interglacial response rate of subpolar North Atlantic waters to climatic change: the record in oceanic sediments. *Phil. Trans. R. Soc. Lond.* B **280**, 119–142.

Schmidt-Thomé, P. 1983 Besonders niedrig gelegene Zeugen einer würmzeitlichen Vereisung in Nordwestspanien und Nord-Portugal. In *Late and post-glacial oscillations of glaciers: glacial and periglacial forms* (ed. A. A. Balkema), pp. 213–230. Rotterdam: Balkema.

Straus, L. G., Altuna, J., Clark, G. A., Gonzales Morales, M., Laville, H., Leroi-Gorhan, A., Menendez de la Hoz, M. & Ortea, J. A. 1981 Paleoecology at La Riera (Asturias, Spain). *Curr. Anthropol.* **22**, 655–682.

Straus, L. G. & Clark, G. A. 1986 *La Riera cave: stone age hunter-gatherer adaptations in northern Spain.* Arizona State University anthropological research paper no. 36.

Straus, L. G., Clark, G. A., Altuna, J. & Ortea, J. A. 1980 Ice-Age subsistance in Northern Spain. *Scient. Am.* **242**, 42–152.

Suc, J.-P. 1980 *Contribution à la connaissance du Pliocène et du Pleistocène inférieur des régions méditérranéennes d'Europe occidentale par l'analyse palynologique des dépôts du Languedoc-Roussillon* (*Sud de la France*) *et de la Catalogne* (*Nord-Est de l'Espagne*). Thèse, Université de Montpellier.

Turner, C. 1985 Problems and pitfalls in the application of palynology to Pleistocene archaeological sites in Western Europe. In *Palynologie archéologique* (ed. J. Renault-Miskovsky, Bui-Thi-Mai & M. Girard), pp. 347–373. (Centre de Recherches Archéologiques; notes et monographies techniques, no. 17.) Paris: C.N.R.S.

Turner, C. & West, R. G. 1968 The subdivision and zonation of interglacial periods. *Eiszeitalter Gegenw.* **19**, 93–101.

Turon, J.-L. 1974 Étude palynologique d'une carotte prélevée dans le Golfe de Gascogne. Intérêt paléoclimatique et stratigraphique. *Pollen Spores* **16**, 475–487.

Turon, J.-L. 1980 La sédimentation pollinique actuelle le long de la ride Reykjanes et dans la fracture de Gibbs. Intérêt dans la détermination des apports sédimentaires. *C. r. Acad. Sci. Paris* D**291**, 453–456.

Van Campo, M. 1984 Relations entre la végétation de l'Europe et les températures de surface océaniques après le dernier maximum glaciaire. *Pollen Spores* **26**, 497–518.

Van Campo, M. & Elhai, H. 1956 Étude comparative des pollens de quelques chênes: application a une toubière normande. *Bull. Soc. bot. Fr.* **103**, 254–260.

Van Campo, M. & Jalut, G. 1969 Analyse pollinique de sédiments des Pyrénées orientales: Lac de Balcère (1,764 m). *Pollen Spores* **11**, 117–126.

Van den Brink, L. M. & Janssen, C. R. 1985 The effect of human activities during cultural phases on the development of montane vegetation in the Sierra da Estrela, Portugal. *Rev. Palaeobot. Palynol.* **44**, 193–215.

Van Geel, B., De Lange, L. & Wiegers, J. 1984 Reconstruction and interpretation of the local vegetational succession of a Lateglacial deposit from Usselo (The Netherlands), based on the analysis of micro- and macrofossils. *Acta bot. neerl.* **33**, 535–546.

Watts, W. A. 1980 Regional variation in the response of vegetation to Lateglacial climatic events in Europe. In *Studies in the Lateglacial of northwest Europe* (ed. J. J. Lowe, J. M. Gray & J. E. Robinson), pp. 1–21. Oxford: Pergamon Press.

Watts, W. A. 1986 Stages of climatic change from Full Glacial to Holocene in northwest Spain, southern France and Italy: a comparison of the Atlantic coast and the Mediterranean basin. In *Current Issues in Climate Research* (*Proceedings EEC Climatology Programme, Sophia Antipolis, France, October 1984*) (ed. A. Ghazi & R. Fantechi), pp. 101–112. Dordrecht: D. Reidel.

Woillard, G. 1975 Recherches palynologiques sur le Pleistocène dans l'est de la Belgique et dans les Vosges Lorraines. *Acta geogr. lovaniensia* **14**, 114 pp.

Woillard, G. 1978 Grande Pile peat bog: a continuous pollen record for the last 140,000 years. *Quat. Res.* **9**, 1–21.

Turner, C. & West, R. G. 1968 The subdivision and zonation of interglacial periods. *Eiszeitalter Gegenw.* 19, 93–101.

[illegible] 1993 [illegible] de la Grande [illegible] et [illegible] 4–12.

[illegible] 1976 [illegible] dans la région des Vosges [illegible] *C. r. hebd. Séanc. Acad. Sci., Paris* D 283, 455–458.

Van Campo, M. 1969 Relations entre la végétation de l'Europe et les températures de surface océaniques après le dernier maximum glaciaire. *Pollen Spores* 11, 359–371.

Van Campo, M. & [illegible] 1966 [illegible] comparative des [illegible] de quelques [illegible]. Leur application à [illegible] *Pollen Spores* 8, 441–460.

Van [illegible] 1969 [illegible] des sédiments des [illegible] *Pollen Spores* 11, 171–200.

Van der [illegible] 1964 The effect of human impacts during cultural phases on the development of mountain vegetation in the Serra da Estrela, Portugal. *Rev. Palaeobot. Palynol.* 44, 179–211.

Van Zeist, W. [illegible] 1983 Reconstruction and interpretation of the [illegible] from a [illegible] in [illegible] based on [illegible] and macrofossils. *Acta bot. neerl.* 32, [illegible].

Watts, W. A. 1980 Regional variation in the response of vegetation to lateglacial climatic events in Europe. In *Studies in the lateglacial of north-west Europe* (ed. J. J. Lowe, J. M. Gray & J. E. Robinson), pp. 1–21. Oxford: Pergamon Press.

Watts, W. A. 1986 Stages of climatic change from Full Glacial to Holocene in northwest Spain, southern France and Italy: a comparison of the Atlantic coast and the [illegible] basin. In *Current issues in climate research* (ed. [illegible]), *Proceedings* [illegible] pp. [illegible]. Dordrecht: Reidel.

Woillard, G. 1978 [illegible] pollen [illegible] et [illegible] dans les Vosges [illegible] 16, 1–15.

Woillard, G. 1979 [illegible] 1–15.

Phil. Trans. R. Soc. Lond. B **318**, 487–504 (1988)
Printed in Great Britain

Global wind-induced change of deep-sea sediment budgets, new ocean production and CO_2 reservoirs *ca.* 3.3–2.35 Ma BP

By M. Sarnthein and J. Fenner

Geologisch-Palaeontologisches Institut, Universität Kiel, Olshausenstrasse 40, *D*-2300 *Kiel, F.R.G.*

The late Pliocene phase of large-scale climatic deterioration about 3.2–2.4 Ma BP is well documented in a number of (benthic) $\delta^{18}O$ records. To test the global implications of this event, we have mapped the distribution patterns of various sediment variables in the Pacific and Atlantic Oceans during two time slices, 3.4–3.18 and 2.43–2.33 Ma BP. The changes of bulk sedimentation and bulk sediment accumulation rates are largely explained by the variations of $CaCO_3$-accumulation rates (and the accumulation rates of the complementary siliciclastic sediment fraction near continents in higher latitudes).

During the late Pliocene, the $CaCO_3$-accumulation rate increased along the equatorial Pacific and Atlantic and in the northeastern Atlantic, but decreased elsewhere. The accumulation rate of organic carbon (C_{org}) and net palaeoproductivity also increased below the high-productivity belts along the equator and the eastern continental margins. From these patterns we may conclude that (trade-) wind-induced upwelling zones and upwelling productivity were much enhanced during that time. This change led to an increased transfer of CO_2 from the surface ocean to the ocean deep water and to a reduction of evaporation, which resulted in an aridification of the Saharan desert belt as depicted in the dust sediments off north-west Africa.

Introduction

Various lines of recently compiled evidence from sediments of the North Atlantic suggest that the Quaternary régime of pronounced climatic fluctuations and glaciations started at about 2.5 Ma BP, after a fairly short phase of massive climatic deterioration during the Late Pliocene, i.e. a major 'event' perhaps comparable with that of the Middle Miocene and that near the Eocene–Oligocene boundary.

The first and major evidence for this was based on the few detailed benthic $\delta^{18}O$ curves available from the depth range of North Atlantic Deep Water (NADW) between about 2800 and 4100 m (figure 1). These records of the Deep Sea Drilling Project (DSDP) Sites (141), 366, 397 and 552 reveal a Late Tertiary 'golden age' of stable climate, which was almost free of cold fluctuations and lasted until about 3.2 Ma BP. Subsequently, the fluctuations increased rapidly and had reached a first fully developed glacial amplitude by 2.43–2.33 Ma BP, with a magnitude comparable to those of the Early Quaternary before *ca.* 1 Ma BP.

Further evidence for major environmental changes during that time was obtained from site 397 (Stein 1984, 1985) offshore from the northwestern Saharan coast (27° N). At this site, the sediment composition after 3.0 Ma BP, and especially after 2.43 Ma BP, records a significant increase in the concentration of biogenic opal and in the accumulation rate of organic carbon, which both form reliable records of enhanced ocean carbon productivity (figure 2). The lower part of the section with enhanced palaeoproductivity was also marked by a substantial supply

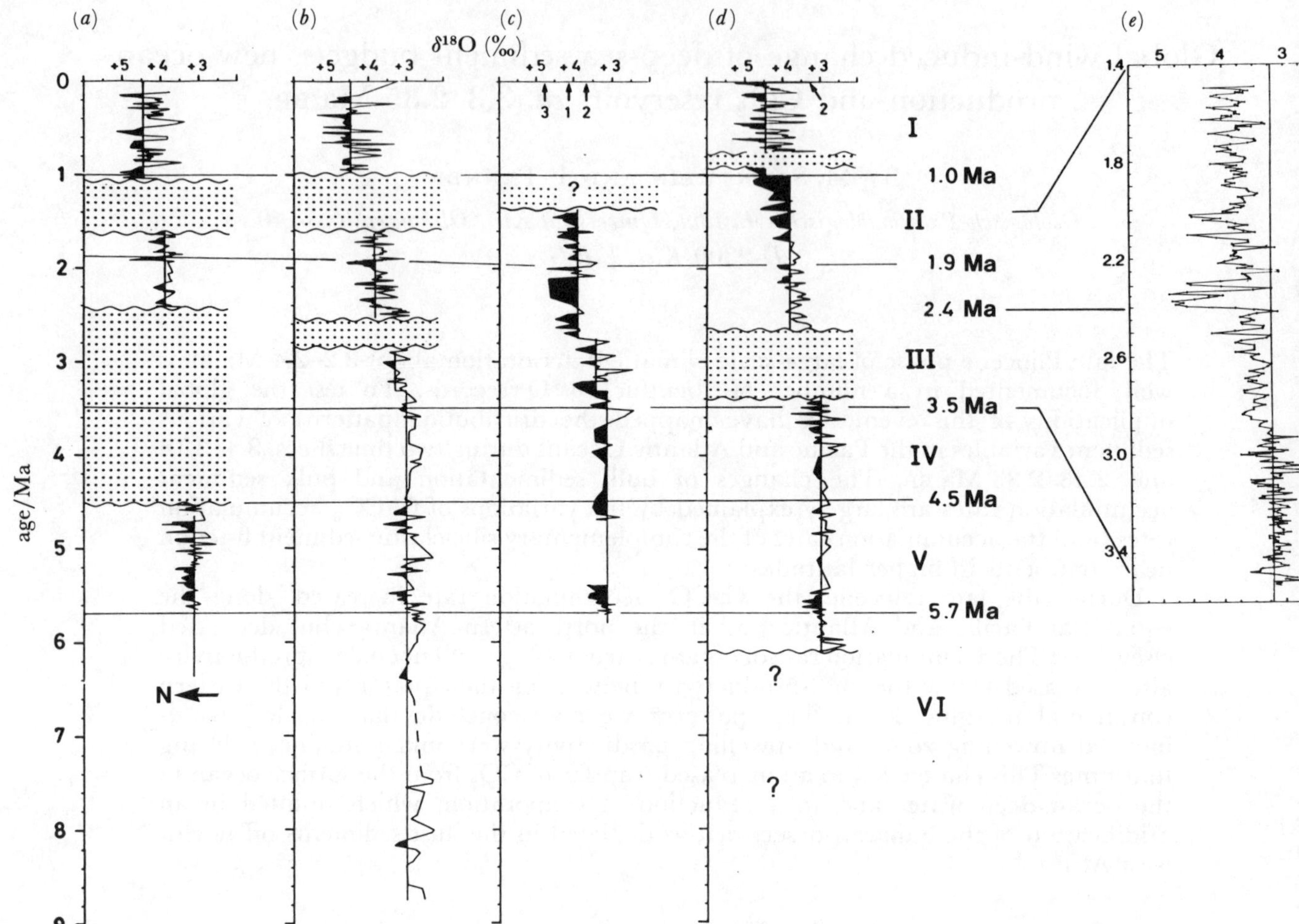

FIGURE 1. Benthos $\delta^{18}O$ records from DSDP sites 544B (*a*), 141 (*c*) and 366 (Stein 1984) (*d*), 397 (Shackleton & Cita 1979; supplemented by Stein 1984) (*b*) and 552 (Shackleton & Hall 1985) (*e*).

of fluvial mud. It was replaced by aeolian dust only after 2.38 Ma BP. This has two implications. One is the onset of strong aridity in the north Sahara during that time; the other is an ambiguity in interpreting the early phase of enhanced near-shore productivity. It may either be controlled by more intensive coastal upwelling or by stronger fertilization owing to fluvial discharge, as would happen, for example, if the sea level were lowered. Recent results (Ruddiman *et al.* 1987) from Ocean Drilling Project (ODP) site 658 off Cap Blanc (22° N) suggest a decrease in the fluvial sediment supply as early as *ca.* 3.0 Ma BP, i.e. they imply that upwelling is the main factor influencing primary productivity in that region.

A third, independent line of evidence for major changes in the ocean palaeoenvironment

FIGURE 2. Late Pliocene climatic deterioration as depicted by benthos $\delta^{18}O$ (*a*) and the variations of the sedimentary régime off northwest Africa (27° N) (data from DSDP site 397; modified and supplemented from Stein (1984, 1985)): concentrations (as percentage of carbonate-free sediment fraction over 6 μm) of biogenic opal (*b*); accumulation rates of organic carbon (*c*); accumulation rates of siliciclastic sediment fraction (grain size over 6 μm) (*d*); and palaeowind speeds (*e*) as deduced from aeolian dust-grain diameters via the term l/uz, where l is trajectory length, u is wind speed and z is the diameters thickness of the air mass. Arrows show samples with riverborne terrigenous sediment fraction.

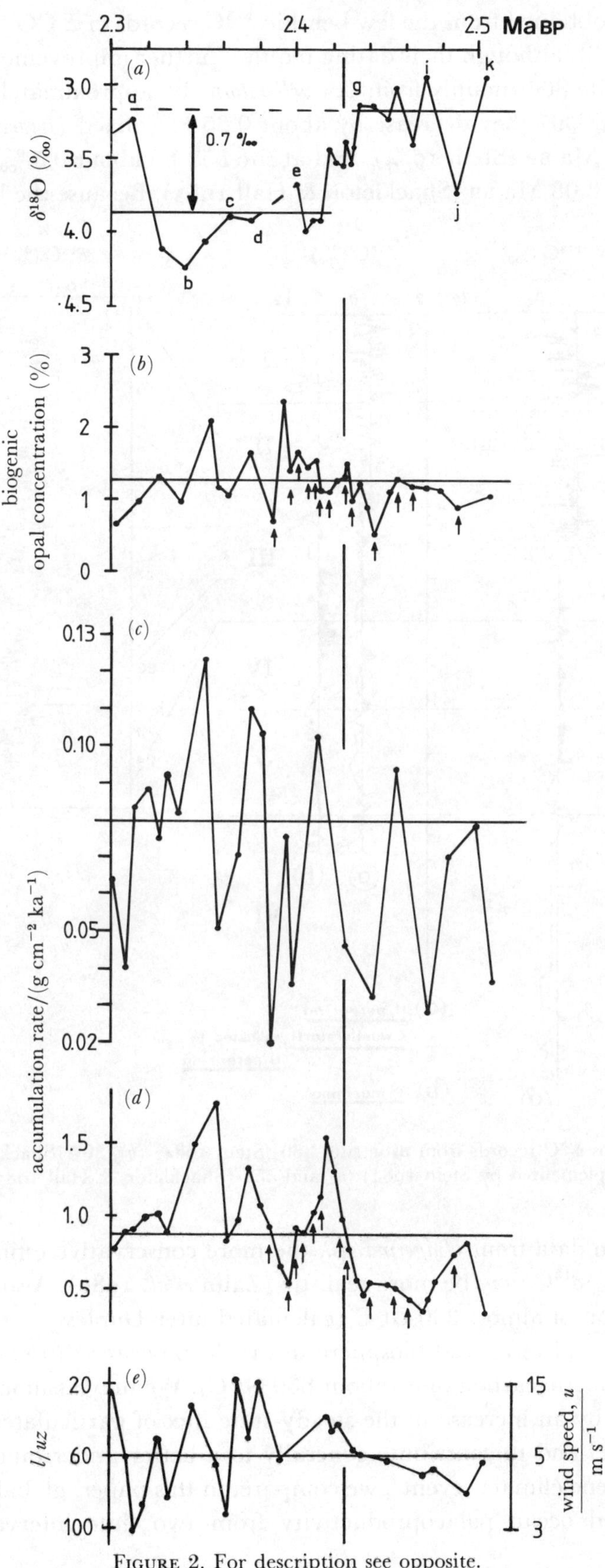

FIGURE 2. For description see opposite.

about 2.5 Ma BP is obtained from the few benthic $\delta^{13}C$ records of ΣCO_2 in the North Atlantic Deep Water (figure 3), although their dating requires further improvement. For example, $\delta^{13}C$ values decrease at site 366 (mainly from *C. wuellerstorfi*) by approximately 0.2‰ from *ca.* 2.25 to 2.0 Ma BP. At site 397 they decrease by about 0.25 ‰ (mixed *Uvigerina* and *C. wuellerstorfi* values) after *ca.* 2.5 Ma BP (Stein 1984), and at site 552 by about 0.5 ‰ from 2.4 to 2.05 Ma BP and also at 3.27–3.08 Ma BP (Shackleton & Hall 1985). Because the latter record is almost

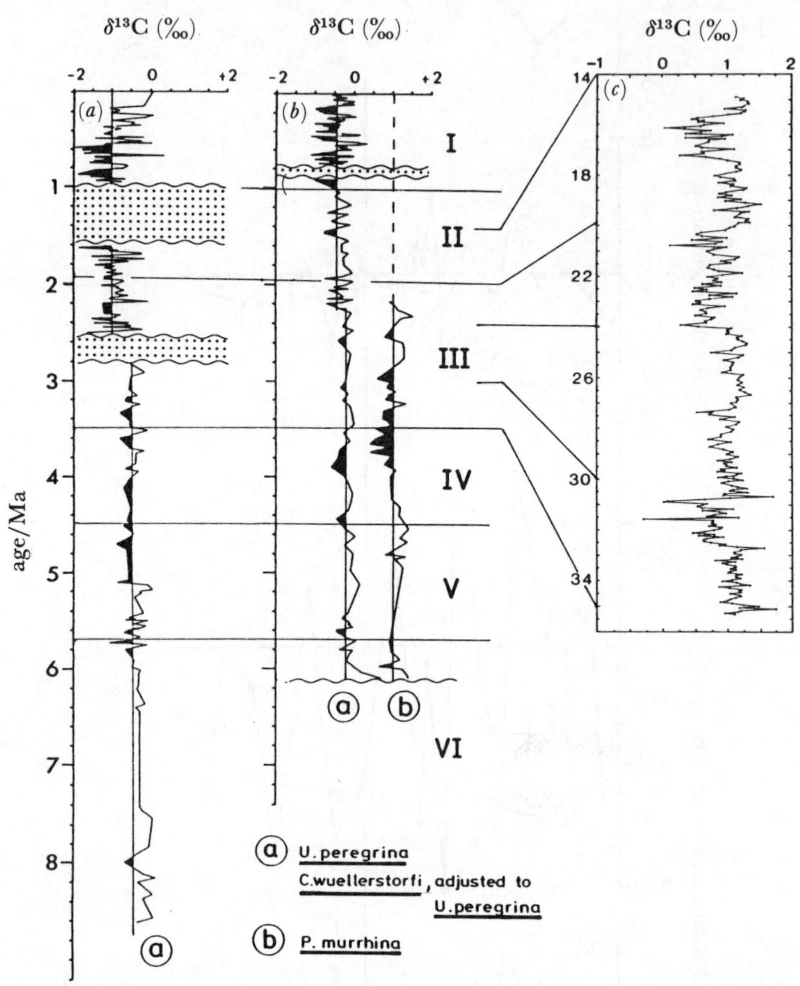

FIGURE 3. Benthos $\delta^{13}C$ records from DSDP sites 366 (Stein 1984) (*b*), 397 (Shackleton & Cita 1979; supplemented by Stein 1984) (*a*) and 552 (Shackleton & Hall 1985) (*c*).

exclusively based on data from *Uvigerina* sp., the more conservative estimate of the difference of about 0.2–0.25 ‰ $\delta^{13}C$ may be more realistic (Zahn *et al.* 1986). Also this estimate already records an extraction of almost 300 Gt C (calculated after Duplessy (1982) from the surface ocean, atmosphere, and terrestrial biosphere to the deep ocean; this value is almost one half of a modern atmospheric carbon unit (about 650 Gt C). We may assume that the transfer was possibly controlled by an increase in the steady-state flux of particulate organic matter.

To test this model and to contribute generally to a better understanding of the gradually emerging late Pliocene climate 'event', we compare, in this paper, global distribution maps of sediment fluxes and ocean palaeoproductivity from two time intervals, at 3.4–3.18 and

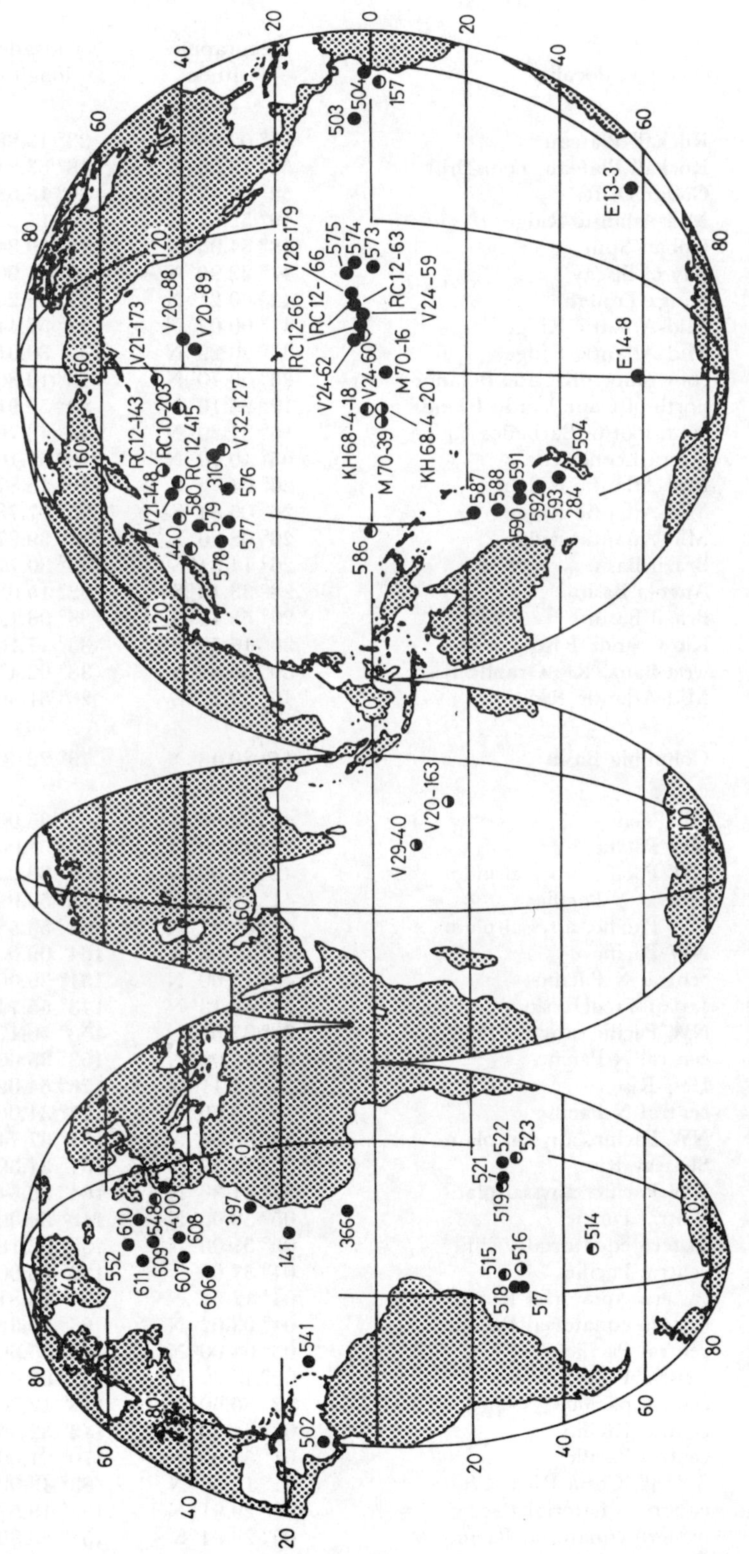

FIGURE 4. Locations of DSDP sites and coring stations used for reconstruction of time intervals described in this paper.

TABLE 1. LOCATION OF SITES

site	locality	geographic latitude	location longitude	water depth m
ATLANTIC				
DSDP 552A	Rockall Plateau	56° 02.56′ N	23° 13.88′ W	2301
610A	Rockall Plateau, Feni Drift	53° 13.30′ N	18° 53.21′ W	2417
611D	Gadar Drift	52° 50.47′ N	30° 18.58′ W	3203
609B	Mid-Atlantic Ridge	49° 52.67′ N	24° 14.29′ W	3884
548	Goban Spur	48° 54.95′ N	12° 09.84′ W	1256
400A	Bay of Biscay	47° 22.90′ N	09° 11.90′ W	4399
608	Kings Trough	42° 50.21′ N	23° 05.25′ W	3526
607	Mid-Atlantic Ridge	41° 00.07′ N	32° 57.44′ W	3427
606	Mid-Atlantic Ridge	37° 20.29′ N	35° 30.01′ W	3007
397	cont. slope off Cape Bojador	26° 50.70′ N	15° 10.80′ W	2900
141	north of Cape Verde Islands	19° 25.16′ N	23° 59.91′ W	4148
541	cont. foot of Barbados Ridge	15° 31.20′ N	58° 43.70′ W	4940
366A	Sierra Leone Rise	05° 40.70′ N	19° 51.10′ W	2853
521	Mid-Atlantic Ridge	26° 04.43′ S	10° 15.87′ W	4125
522	Mid-Atlantic Ridge	26° 06.84′ S	05° 07.78′ W	4441
519	Mid-Atlantic Ridge	26° 08.20′ S	11° 39.97′ W	3769
515A	Brazil Basin	26° 14.31′ S	36° 30.17′ W	4252
523	Angola Basin	28° 33.13′ S	02° 15.08′ W	4562
518	Brazil Basin	29° 58.42′ S	38° 08.12′ W	3944
516A	Rio Grande Rise	30° 16.59′ S	35° 17.10′ W	1313
517	west flank, Rio Grande Rise	30° 56.81′ S	38° 02.47′ W	2963
514	Mid-Atlantic Ridge	46° 02.77′ S	26° 51.30′ W	4318
CARIBBEAN				
DSDP 502C	Columbia Basin	11° 29.48′ N	79° 22.70′ W	3051
PACIFIC OCEAN				
V21-173	NE Pacific	44° 22.00′ N	163° 33.00′ W	5493
RC12-413	NW Pacific	43° 17.00′ N	166° 54.00′ E	5015
V21-148	NW Pacific, abyssal plain	42° 05.00′ N	160° 36.00′ E	5477
RC10-203	central N Pacific	41° 42.00′ N	171° 57.00′ W	5883
DSDP580	NW Pacific, abyssal plain	41° 37.47′ N	153° 58.58′ E	5375
RC12-415	NW Pacific	41° 17.00′ N	164° 09.00′ E	4872
V20-88	central N Pacific	40° 11.00′ N	151° 39.00′ W	—
DSDP440B	Japan Trench, slope terrace	39° 44.13′ N	143° 55.74′ E	4509
DSDP579A	NW Pacific, abyssal plain	38° 37.68′ N	153° 50.17′ E	5737
V20-89	central N Pacific	38° 12.00′ N	153° 35.00′ W	—
DSDP310	Hess Rise	36° 52.11′ N	176° 54.09′ E	3516
V32-127	central N Pacific	35° 28.00′ N	177° 34.00′ E	3927
DSDP578	NW Pacific, abyssal plain	33° 55.56′ N	151° 37.74′ E	6010
DSDP577A	Shatsky Rise	32° 26.53′ N	157° 34.39′ E	2678
DSDP576	NW Pacific, abyssal plain	32° 21.36′ N	164° 16.54′ E	6217
RC12-63	central Pacific	05° 58.00′ N	142° 39.00′ W	4949
DSDP575	eastern equatorial Pacific	05° 51.00′ N	135° 02.16′ W	4536
V28-179	central Pacific	04° 37.00′ N	139° 36.00′ W	4509
DSDP574	eastern equatorial Pacific	04° 12.52′ N	133° 19.81′ W	4561
DSDP503B	eastern equatorial Pacific	04° 03.02′ N	95° 38.32′ W	3672
V24-62	central Pacific	03° 04.00′ N	153° 35.00′ W	4834
V24-60	central Pacific	02° 48.00′ N	149° 00.00′ W	4859
RC12-66	central pacific	02° 36.60′ N	148° 12.80′ W	4755
V24-59	central Pacific	02° 34.00′ N	145° 32.00′ W	4662
KH68-4-18	central Pacific	01° 59.00′ N	170° 01.00′ W	5390
DSDP504	S flank, Costa Rica Rift	01° 13.58′ N	83° 43.93′ W	3460
DSDP573	eastern equatorial Pacific	00° 29.91′ N	133° 18.57′ W	4301
DSDP586B	western equatorial Pacific	00° 29.84′ S	158° 29.89′ E	2208
DSDP157	Carnegie Ridge	01° 45.70′ S	85° 54.17′ W	1591
M70-39	central Pacific	02° 27.00′ S	173° 20.00′ W	5412

TABLE 1. (*cont.*)

site	locality	geographic latitude	location longitude	water depth m
KH68-4-20	central Pacific	02° 28.00′ S	170° 00.00′ W	5535
M70-16	central Pacific	03° 13.00′ S	160° 15.00′ W	5543
DSDP587	Landsdowne Bank	21° 11.09′ S	161° 19.99′ E	1101
DSDP588	Lord Howe Rise	26° 06.70′ S	161° 13.60′ E	1533
DSDP590A	Lord Howe Rise	31° 10.02′ S	163° 21.51′ E	1299
DSDP591	Lord Howe Rise	31° 35.06′ S	164° 26.92′ E	2131
DSDP592	Lord Howe Rise	26° 28.40′ S	165° 26.53′ E	1088
DSDP593	Challenger Plateau	40° 30.47′ S	167° 40.47′ E	1068
DSDP284	Challenger Plateau	40° 30.48′ S	167° 40.81′ E	1066
DSDP594	Bounty Trough	45° 31.41′ S	174° 56.88′ E	1204
E13-3	Bellingshausen Basin	57° 00.00′ S	89° 29.00′ W	5090
E14-8	flank of Mid-Pacific Ridge	59° 40.00′ S	160° 17.00′ W	3875
INDIAN OCEAN				
V20-163	Mid-Indian Ridge	17° 12.00′ S	88° 41.00′ E	2706
V29-40	Mid-Indian Ridge	10° 29.00′ S	78° 03.00′ E	5325

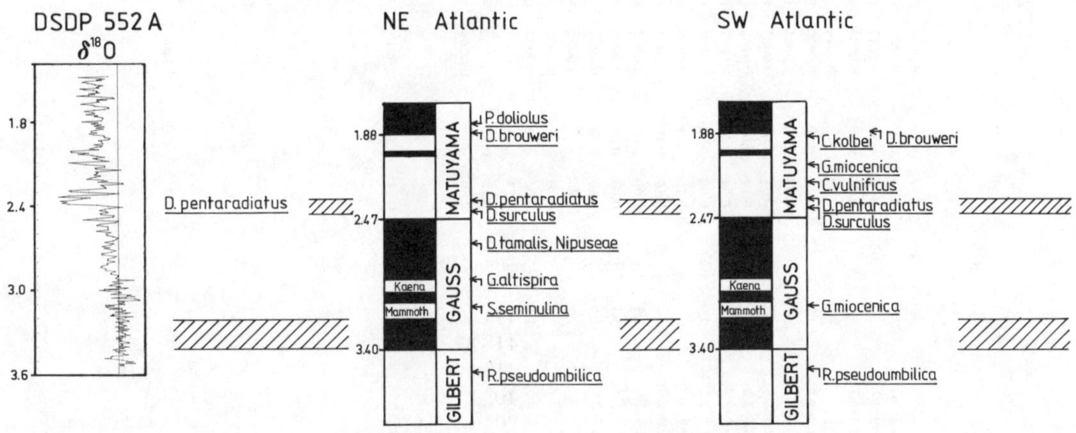

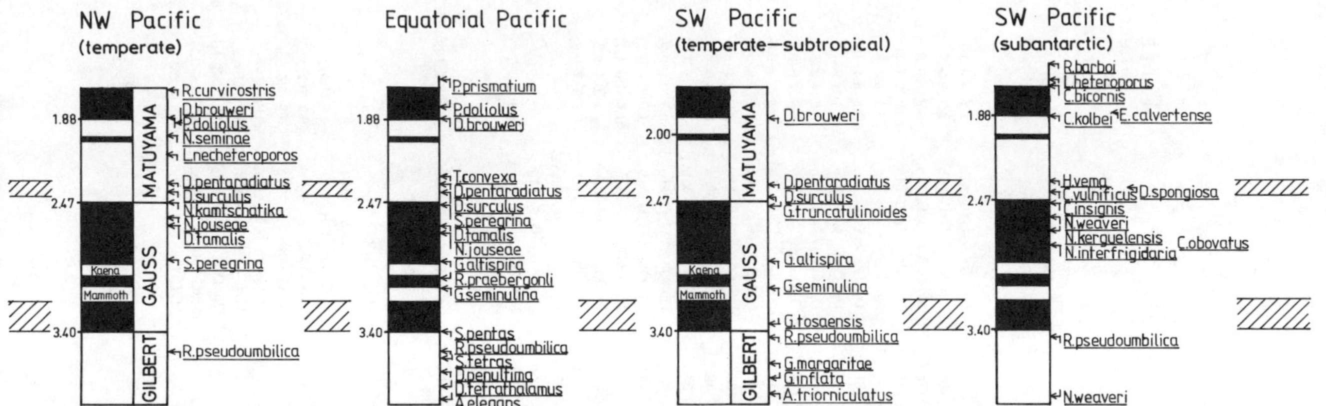

FIGURE 5. Standard stratigraphic sections showing the stratigraphic markers used in different parts of the ocean to define the two time intervals (hatched) described in this paper (table 2).

Table 2. Sediment data from DSDP sites and piston cores

site	average depth below sea floor (m) I	II	datums used for interval I	II	sedimentation rates (cm ka^{-1}) I	II	dry bulk density (g cm^{-3}) I	II	C_{org} (% by mass) I	II	C_{org} accum. rate (g cm^{-2} ka^{-1}) I	II	$CaCO_3$ (% by mass) I	II	$CaCO_3$ accum. (g cm^{-2} ka^{-1}) I	II	bulk accum. (g cm^{-2} ka^{-1}) I	II	PP_{new} (g cm^{-2} ka^{-1}) I	II
Atlantic																				
DSDP 552A	41.5	55.6	O–Rb	G/G–Mb	1.7 (1.6–1.8)	1.0 (1.0)	1.0	1.1	0.19	0.05	0.0033	0.0005	45.75	96.35	0.78	1.01	1.72	1.05	7.12	2.23
610A	123.5	165.8	Dp–Rb	G/G–Mb	5.3 (2.5–7.9)	2.9 (2.1–3.7)	—	—	0.18	0.08	0.0107	0.0025	72.07	88.18	4.27	2.73	5.93	3.10	14.68	6.29
611D	103.3	178.8	M/G–Rb	G/G–Mb	⟨8.9 (8.6–9.3)⟩	10.1 (9.4–10.7)	—	—	0.42	0.15	0.0317	0.0125	12.27	71.67	2.92	5.97	7.54	8.33	29.52	23.56
609B	165.2	241.8	Dp–Rb	G/G–Mb	10.2 (7.7–12.7)	5.0 (4.4–5.7)	—	—	0.24	0.06	0.0238	0.0033	58.70	90.01	5.81	4.91	9.91	5.46	35.05	9.43
548	130.7	178.7	Dp–Rb	—	5.7 (5.4–6.0)	—	1.3	—	0.36	—	0.0274	—	31.70	—	2.42	—	7.62	—	17.10	—
400A	123.1	161.0	Dp–Ob	G/G–Mb	3.8 (3.4–4.2)	3.2 (1.8–4.5)	1.3	—	0.49	0.20	0.0239	0.0067	30.20	55.50	1.48	1.87	4.89	3.37	42.21	21.58
608	80.9	108.5	Dp–Ob	G/G–Mb	3.1 (1.8–4.3)	0.9 (0.4–1.4)	—	—	0.03	0.03	0.0011	0.0003	95.00	97.31	3.36	0.99	3.54	1.02	4.69	2.16
607	105.6	141.8	Dp–Ob	G/G–Mb	5.2 (4.2–6.2)	3.0 (2.3–3.7)	—	—	0.04	0.04	0.0023	0.0014	88.00	94.18	5.04	3.39	5.73	3.60	7.45	5.48
606	76.3	113.8	O–Dp	G/G–Mb	8.5 (7.6–9.4)	7.3 (6.6–7.9)	—	—	0.11	0.04	0.0103	0.0035	94.55	97.52	8.89	8.56	9.41	8.78	17.76	8.51
397	179.5	272.2	O–Dp	G/G–Mb	9.2 (8.9–9.5)	4.9 (0.0–9.8)	—	—	0.94	0.44	0.0917	0.0315	44.75	65.00	4.37	4.66	9.76	7.17	70.92	34.96
141	22.6	41.2	O–Dp	G/G–O	4.1 (2.6–5.6)	⟨2.0 (1.9–2.1)⟩	1.0	1.1	0.05	0.08	0.0020	0.0018	79.90	82.62	3.18	1.83	3.97	2.22	8.20	7.67
541	106.3	138.0	Dp–Ob	G/G–Mb	2.5 (2.4–2.5)	0.9 (0.9)	—	—	0.13	0.14	0.0038	0.0015	43.00	45.00	1.27	0.48	2.95	1.08	14.23	6.00
366A	33.5	48.2	O–Dp	Rp–O	6.2 (6.0–6.3)	⟨2.3 (1.9–2.6)⟩	—	—	0.10	0.10	0.0022	0.0007	69.80	92.50	1.53	0.68	2.19	0.74	11.30	6.31
521	20.6	33.7	Dp–Ob	G/G–Mb	1.3 (1.1–1.5)	1.5 (1.5–1.6)	1.1	1.1	0.05	0.06	0.0007	0.0010	84.00	93.70	1.19	1.51	1.41	1.61	4.27	5.26
522	14.7	22.93	Dp–Ob	G/G–Mb	0.5 (0.1–0.9)	1.5 (1.3–1.8)	1.1	1.1	0.05	0.04	0.0003	0.0007	70.00	93.03	0.39	1.51	0.56	1.62	2.54	4.54
519	50.0	70.9	Dp–Ob	G/G–Mb	1.0 (0.7–1.2)	2.2 (1.7–2.6)	1.0	1.1	0.03	0.03	0.0003	0.0007	92.00	96.40	0.90	2.28	0.97	2.37	2.27	3.91
515A	50.2	—	M/G–Ob	—	⟨1.6 (1.6–1.7)⟩	—	—	—	0.33	—	0.0052	—	0.00	—	0.00	—	1.57	—	15.77	—
523	10.9	23.0	O–Dp	unclear	0.2 (0.1–0.3)	—	—	—	0.10	—	0.0002	—	90.00	—	0.16	—	0.18	—	2.31	—
518	21.6	33.6	Dp–Ob	G/G?–O	0.7 (0.7–0.8)	0.8 (0.7–0.9)	1.1	—	0.04	0.09	0.0003	0.0006	83.92	81.55	0.66	0.58	0.79	0.71	2.49	4.22
516A	10.4	17.7	—	G/G–Mb	—	0.8 (0.7–0.8)	—	1.1	0.12	0.08	—	—	—	94.42	—	—	—	—	—	1.58
517	30.5	42.0	—	G/G–Mb	—	1.9 (1.9–2.0)	—	—	0.12	0.12	—	—	82.50	83.80	—	—	—	—	—	6.01
514	41.4	—	M/G–Ob	unclear	⟨4.0 (4.0)⟩	—	—	—	0.30	—	0.0064	—	0.70	—	0.02	—	2.13	—	19.25	—
Caribbean																				
502C	57.0	84.2	M21–Dp	G/G–Mb	9.4 (8.4–10.5)	2.8 (2.5–3.1)	0.8	1.0	0.31	0.29	0.0218	0.0082	49.00	48.80	3.81	1.39	7.78	2.84	31.98	16.23
Pacific Ocean																				
V21-173	—	—	M/G–Ob	—	⟨4.4 (4.4)⟩	—	—	—	—	—	—	—	—	—	—	—	—	—	—	—
RC12-413	—	—	—	G/G–Mb	—	3.6 (3.6)	—	—	—	—	—	—	—	—	—	—	—	—	—	—
V21-148	—	—	M/G–Ob	G/G–Mb	⟨1.5 (1.5)⟩	5.2 (5.2)	—	—	—	—	—	—	—	—	—	—	—	—	—	—
RC10-203	—	—	M/G–Ob	—	⟨0.7 (0.7)⟩	—	—	—	—	—	—	—	—	—	—	—	—	—	—	—
DSDP580	—	—	M/G–Ob	—	⟨5.5 (5.5–5.4)⟩	—	—	—	—	—	—	—	—	—	—	—	—	—	—	—
RC12-415	—	—	M/G–Ob	—	⟨0.05 (0.05)⟩	—	—	—	—	—	—	—	—	—	—	—	—	—	—	—
V20-88	—	—	M/G–Ob	G/G–Mb	⟨0.2 (0.15)⟩	0.2 (0.2)	—	—	—	—	—	—	—	—	—	—	—	—	—	—
DSDP440B	388.6	—	Dp–Db	unclear	9.9 (15.3–4.4)	unclear	—	—	1.00	—	0.1200	—	0.00	—	—	—	12.20	—	114.36	—
DSDP579A	—	—	M/G–Ob	G/G–Mb	⟨3.8 (3.8–3.7)⟩	3.2 (3.9–2.6)	—	—	—	—	—	—	—	—	—	—	—	—	—	—
V20-89	—	—	M/G–Ob	G/G–Mb	⟨0.3 (0.25)⟩	0.1 (0.14)	—	—	—	—	—	—	—	—	—	—	—	—	—	—

DSDP310	29.8	46.6	O–Db	O–Gm	1.0 (1.2–0.9)	⟨2.1 (2.3–1.8)⟩	—	0.9	0.10	0.10	0.0009	0.0019	60.75	77.70	0.56	1.51	0.90	1.94	4.30	7.24
V32-127	—	—	Dp–Ob	—	0.3 (0.3–0.3)	—	—	—	—	—	—	—	—	—	—	—	—	—	—	—
DSDP578	—	—	M/G–Rb	G/G–Mb	⟨2.4 (2.35–2.4)⟩	1.1 (1.0–1.3)	—	—	—	—	—	—	—	—	—	—	—	—	—	—
DSDP577A	26.9	39.2	Dp–Ob	G/G–Mb	1.0 (0.8–1.2)	1.5 (1.4–1.6)	—	—	0.11	0.07	0.0010	0.0010	86.00	91.00	0.80	1.34	0.93	1.47	3.84	3.84
DSDP576	—	—	M/G–Ob	G/G–Mb	⟨0.4 (0.4–0.4)⟩	0.3 (0.3–0.35)	—	—	—	—	—	—	—	0.40	—	0.00	0.17	0.14	—	—
RC12-63	—	—	M/G–Ob	G/G–Mb	⟨0.2 (0.2)⟩	0.4 (0.4)	—	—	—	—	—	—	—	—	—	—	—	—	—	—
DSDP575	5.2	7.5	M/G–Ob	G/G–Mb	⟨0.2 (0.2–0.2)⟩	0.2 (0.1–0.2?)	0.5	0.4	0.15	0.13	0.0010	0.0001	58.80	37.20	0.05	0.02	0.09	0.06	1.81	1.49
V28-179	—	—	O–Ob	G/G–Mb	0.5 (0.5–0.5)	0.7 (0.7)	—	—	—	—	—	—	67.80	—	—	—	—	—	—	—
DSDP574	11.3	17.7	M/G–Ob	G/G–Mb	⟨0.5 (0.4–0.5)⟩	0.7 (0.6–0.8)	0.5	0.4	0.21	0.16	0.0006	0.0004	30.20	46.80	0.08	0.13	0.28	0.28	4.31	3.90
DSDP503B	47.8	70.0	M21–Op	G/G–Mb	2.7 (2.6–2.9)	2.3 (1.9–2.6)	0.4	0.4	0.60	0.32	0.0057	0.0026	13.60	44.10	0.13	0.36	0.95	0.81	3.97	2.41
V24-62	—	M21–Ob	G/G–Mb	0.5 (0.5)	0.3 (0.3)	—	—	—	—	—	—	—	—	—	—	—	—	—	—	—
V24-60	—	—	M/G–Ob	—	0.4 (0.4)	—	—	—	—	—	—	—	—	—	—	—	—	—	—	—
RC12-66	—	—	M21–Ob	G/G–Mb	0.4 (0.4–0.4)	0.2 (0.2)	—	—	—	—	—	—	—	—	—	—	—	—	—	—
V24-59	—	—	M21–Ob	G/G–Mb	0.3 (0.3–0.3)	0.2 (0.2)	—	—	—	—	—	—	40.00	44.00	—	—	—	—	—	—
KH68-4-18	—	—	M/G–Ob	—	⟨0.4 (0.4)⟩	—	—	—	—	—	—	—	—	—	—	—	—	—	—	—
DSDP504	87.6	152.5	O–Ob	G/G–O	0.6 (0.6–0.7)	imprecise	0.5	0.7	1.76	0.16	0.0047	—	33.70	66.30	0.09	—	0.27	—	14.73	8.43
DSDP573	34.6	46.9	M/G–Ob	G/G–Mb	⟨1.5 (1.4–1.6)⟩	1.3 (1.2–1.4)	0.7	0.6	0.18	—	0.0018	—	66.60	71.50	0.66	0.61	0.99	0.85	—	—
DSDP586B	56.4	74.9	Dp–Db	unclear	2.7 (2.4–3.0)	unclear	—	—	0.24	—	0.0064	—	88.00	—	2.34	—	2.65	—	7.94	—
DSDP157	129.3	211.0	O–Tc	unclear	20.2 (19.5–21.0)	unclear	0.6	—	0.70	—	0.0844	—	54.95	—	6.63	—	12.06	—	44.16	—
M70-39	—	—	—	G/G–Mb	—	0.6 (0.6)	—	—	—	—	—	—	—	—	—	—	—	—	—	—
KH68-4-20	—	—	—	G/G–Mb	—	1.1 (1.1)	—	—	—	—	—	—	—	—	—	—	—	—	—	—
M70-16	—	—	M/G–Ob	G/G–Mb	⟨0.2 (0.5)⟩	0.5 (0.5)	—	—	—	—	—	—	—	—	—	—	—	—	—	—
DSDP587	—	—	Dp–Db	G/G–Gs	0.4 (0.1–0.8)	1.9 (1.5–2.3)⟩	—	—	—	—	—	—	97.00	93.00	0.42	2.01	0.44	2.16	1.01	2.21
DP588	28.8	44.5	Dp–Rb	G/G–Mb	0.5 (0.1–0.8)	1.6 (1.2–2.1)	—	—	0.05	0.05	0.0003	0.0009	96.60	94.80	0.53	1.78	0.58	1.88	4.14	6.59
DSDP590A	46.1	75.4	O–Ob	G/G–Mb	2.1 (1.9–2.3)	5.1 (4.8–5.3)	—	—	0.13	0.11	0.0033	0.0069	91.00	92.60	2.28	5.78	2.50	6.24	7.41	11.30
DSDP591	64.2	95.8	Dp–Ob	G/G–Mb	2.8 (2.4–3.2)	6.0 (5.8–6.2)	1.1	1.1	0.13	0.12	0.0039	0.0076	85.80	93.50	2.55	5.91	2.97	6.32	2.10	—
DSDP592	—	—	Dp–Ob	G/G–Gs	0.6 (0.4–0.8)	0.9 (0.0–1.8)	—	—	—	—	—	—	—	—	—	—	—	—	2.64	2.04
DSDP593	44.3	—	Dp–Ob	unclear	2.0 (1.8–2.2)	unclear	—	—	0.06	—	0.0014	—	91.00	—	2.05	—	2.26	—	—	—
DSDP284	50.0	78.0	Dp–Db	O–Rp	1.6 (1.4–1.8)	⟨1.0 (0.0–2.0)⟩	—	—	0.10	0.10	—	—	87.00	90.00	—	—	—	—	—	—
DSDP594	—	—	unclear	G/G–Mb	—	3.2 (3.1–3.3)	—	—	—	—	—	—	—	—	—	—	—	—	—	—
E13-3	—	—	Hv–Ob	G/G–Mb	⟨0.4 (0.4)⟩	0.4 (0.4)	—	—	—	—	—	—	—	—	—	—	—	—	—	—
E14-8	—	—	Hv–Ob	G/G–Mb	⟨0.4 (0.4)⟩	0.3 (0.3)	—	—	—	—	—	—	—	—	—	—	—	—	—	—
Indian Ocean																				
V20-163	—	—	—	G/G–Mb	—	3.9 (3.9)	—	—	—	—	—	—	—	—	—	—	—	—	—	—
V29-40	—	—	—	G/G–Mb	—	9.4 (8.9–10.00)	—	—	—	—	—	—	—	—	—	—	—	—	—	—

Abbreviations used for datums: Db, *D. brouweri*; Dp, *D. pentaradiatus*; G/G, Gauss–Gilbert; Gm, *G. margaritae*; Gs, *G. seminulina*; Hv, *H. vema*; M/G, Matuyama–Gauss; Mb, Mammoth base; O, oxygen isotope shifts at *ca.* 2.4 and 3.15 Ma BP; Ob, Olduvai base; Rb, Reunion base; Rp, *R. pseudoumbilicata*; Tc, *T. convexa*.

2.43–2.33 Ma BP, which represent two climatic extremes separated by an average $\delta^{18}O$ rise of 1.0–1.4 ‰ over approximately 0.75 Ma. We expect the preliminary distribution patterns of shifting accumulation rates to indicate the history of CO_2 chemistry during this important phase of climatic deterioration.

Database and definitions

Figure 4 and table 1*a* show the location of 66 drill holes, in which we were able to define our two time intervals: time interval II from 3.4 to 3.18 Ma BP and time interval I from 2.43 to 2.33 Ma BP. Time interval (slice) II averages the time range from the $\delta^{18}O$ minimum (i.e. the warm phase) immediately after the Gauss–Gilbert magnetic reversal until the top of the $\delta^{18}O$ minimum near the Mammoth magnetic event. Time interval (slice) I starts at the base of the first extreme cold stage near the last occurrence of *Discoaster surculus* and reaches to the top of this stage, ending just after the last occurrence of *D. pentaradiatus*. In addition to oxygen-isotope curves and well-calibrated microfossil datums, detailed $CaCO_3^-$ content curves were studied to determine the position of time intervals I and II in the cores (figure 5). Those cores for which palaeomagnetic and micropalaeontological datums suggest a hiatus for time slices I and II were not included in this study (figures 6*b*–9*b*). Bulk sedimentation rates for the two time slices were determined from those first-order age datums lying closest to the respective intervals (table 2).

From both time intervals, a total of about 500 samples from 31 selected core profiles in the Atlantic and Pacific Ocean were investigated for dry bulk density (DBD), calcium carbonate content, and organic carbon (% C) content, by means of standard laboratory techniques (Coulomat 702, Leco CS 244). By incorporating the respective bulk sedimentation rates (S_b), for the two time slices, we calculated the mass accumulation rates of the bulk sediment, carbonate sediment, and organic carbon. Figures 6–10 present average values for each time interval.

Furthermore, the new palaeoproductivity (P_{new}) (or 'export' productivity, from the surface ocean to the deep sea) of the ocean was calculated, based on the following assumptions.

(i) $P_{new} = P^2/400$, where P = primary production. Because this is a saturation function, $P_{new} = \frac{1}{2}P$ at more than $P = 200$ g m^{-2} a^{-1} (Eppley & Peterson 1979).

(ii) The accumulation rates of organic carbon (C_A) are considered as a function of P_{new}, the water depth (z) (Suess 1980) and the 'sealing effect', S_{b-c}). The latter is included in the calculation by using the (organic) carbon-free bulk sedimentation rate (Mueller & Suess 1979). According to Sarnthein *et al.* (1987*a*, *b*) we use the following equation for P_{new} (where D is the DBD):

$$P_{new} = 0.24\, C^{0.64} S_b^{0.86} D^{0.54}\, z^{0.83} S_{b-c}^{-0.24}\ (\mathrm{g\ cm^{-2}\ a^{-3}}) \qquad (1)$$

Changes in the sediment budgets

From time interval II to time interval I the bulk sedimentation rates (figure 6*a*, *b*) and bulk accumulation rates (figure 7*a*, *b*) increased over large parts of the ocean. For example, in the north and equatorial Atlantic they increased by a factor of 1–3.5, in the equatorial eastern Pacific by a factor of 1–2, offshore from Japan by a factor of 1.2–2.5, and in the southern Pacific, south of 40° S, by a factor of 1–1.5. The rates generally decreased in the western

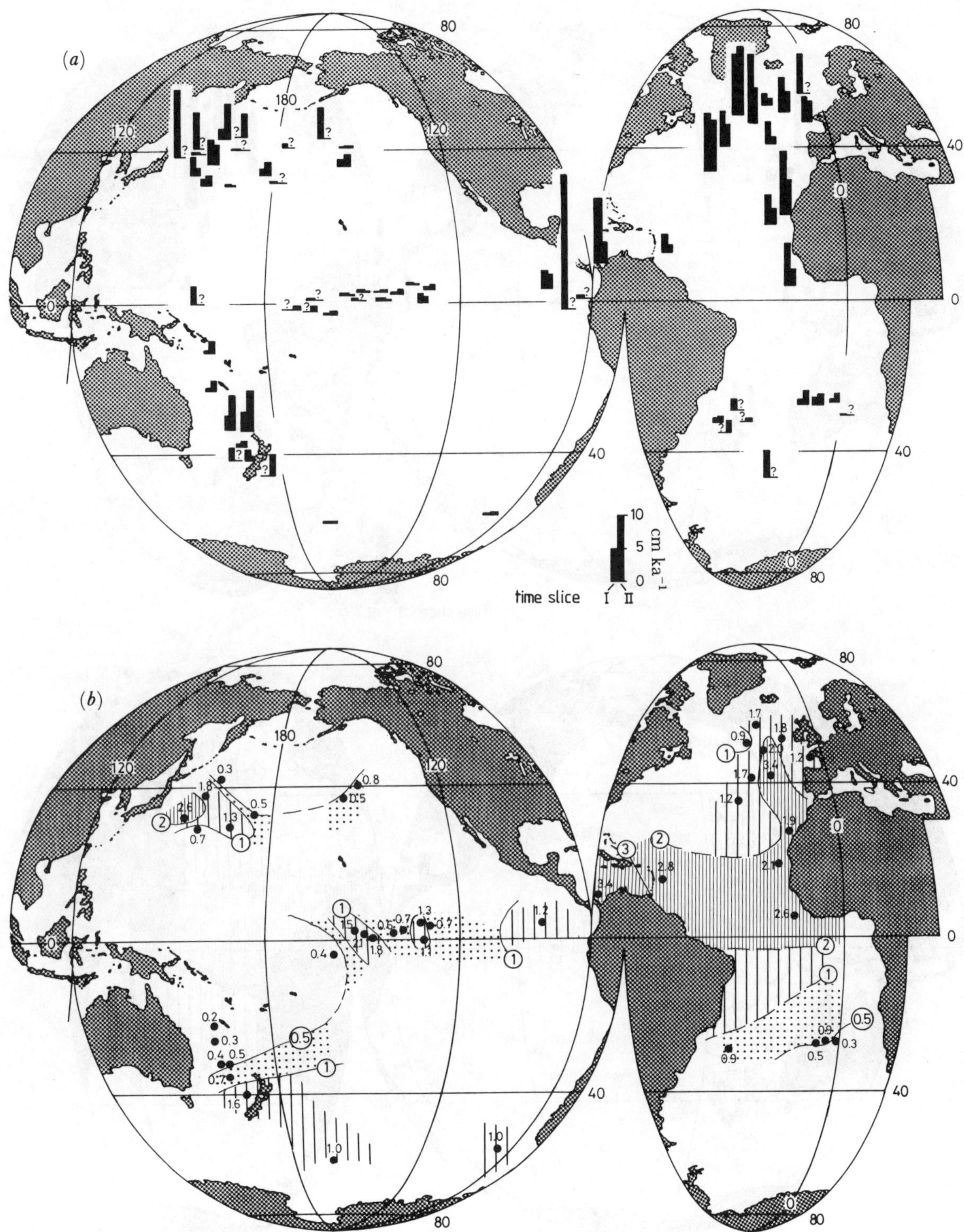

FIGURE 6. (*a*) Bulk sedimentation rates during time intervals I and II. ?, Hiatus and/or imprecise chronostratigraphy. (*b*) Change of bulk sedimentation rates from time interval II to time interval I (by factor x; > 1.0 = increase; < 1.0 = decrease).

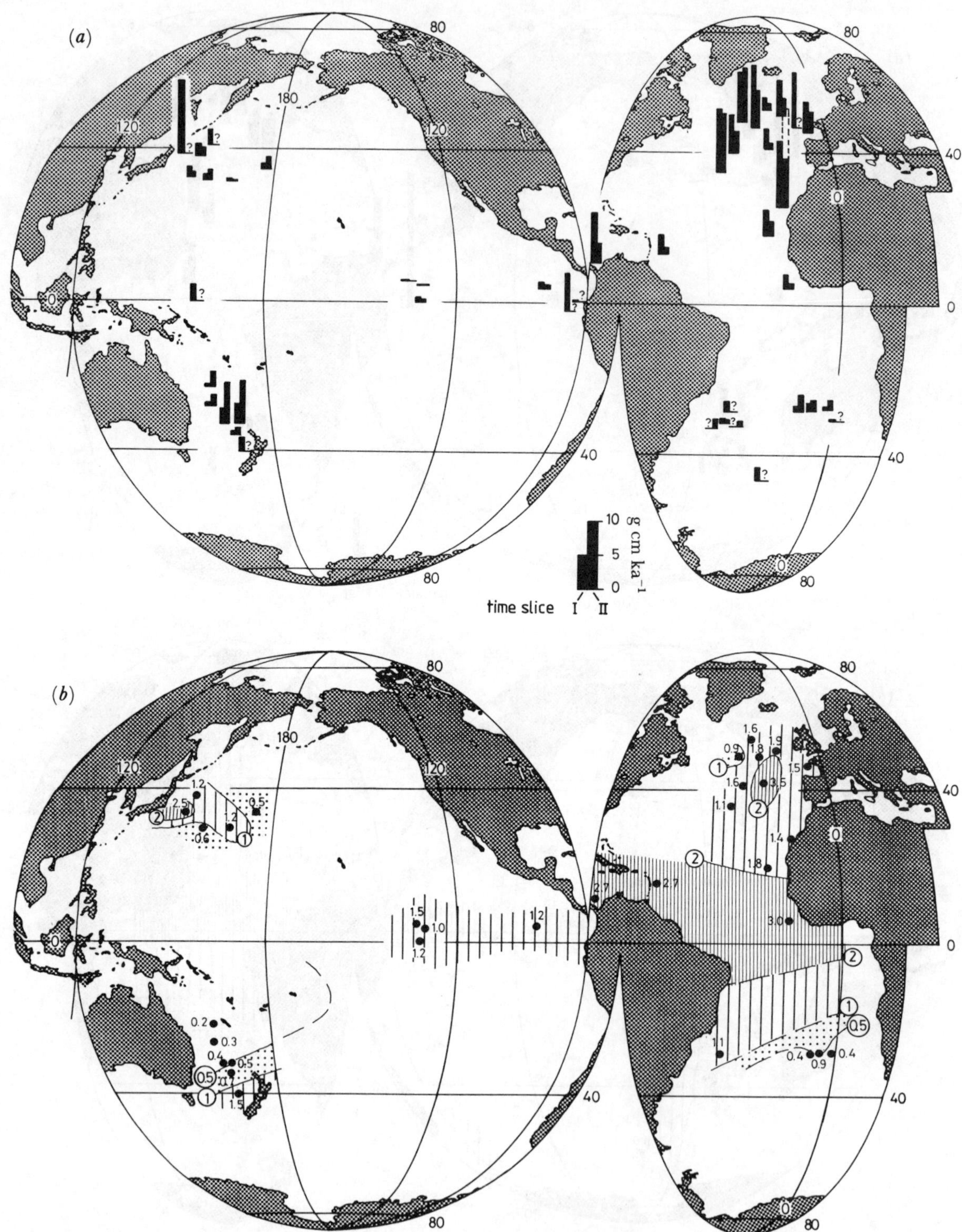

FIGURE 7. (*a*) Bulk accumulation rates during time intervals I and II. ?, Hiatus and/or unprecise chronostratigraphy. (*b*) Change of bulk accumulation rates from time interval II to time interval I (by factor *x*).

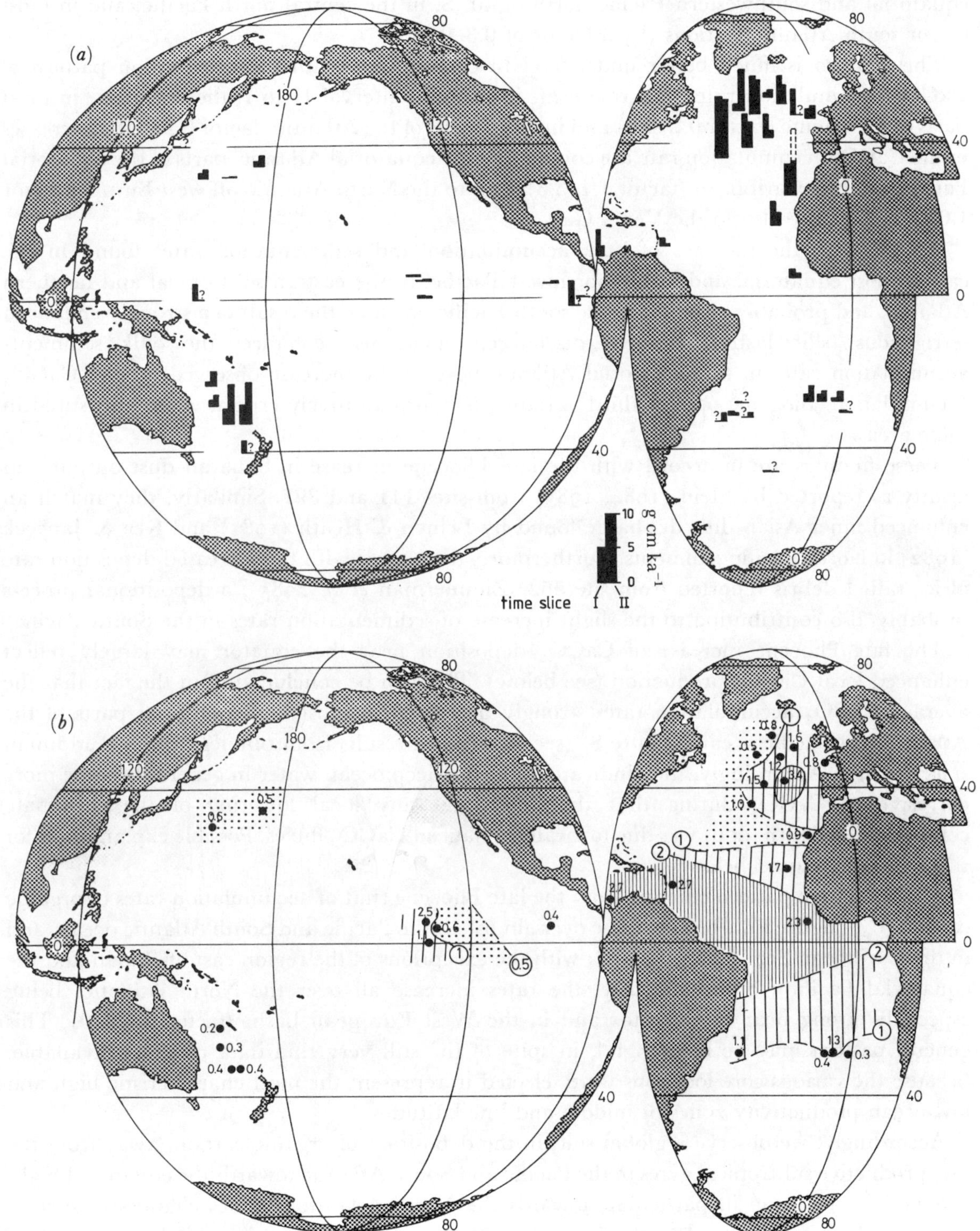

FIGURE 8. (*a*) $CaCO_3$ accumulation rates during time intervals I and II. ?, Hiatus and/or imprecise chronostratigraphy. (*b*) Change of $CaCO_3$ accumulation rates from time interval II to time interval I (by factor *x*).

equatorial and southwestern Pacific north of 40° S, in the central north Pacific, and in most of our south Atlantic sections (by a factor of 0.3–0.9).

This pattern is much better understood in combination with the distribution pattern of $CaCO_3$ accumulation rates (figure 8*a*, *b*). From time interval II to I, they decrease in most parts of the Pacific (factor 0.2–0.6) and in large parts of the Atlantic (factor 0.2–0.9). Increases in the $CaCO_3$ accumulation rate are confined to the equatorial Atlantic, parts of the equatorial Pacific, and the Caribbean (factor 1.1–3.3), and to the North Atlantic off west Europe (factor 1.0–3.4), possibly also to the Vema Gap.

Accordingly, the increase of bulk accumulation and sedimentation rates found in the easternmost equatorial and in the northwest Pacific, in the eastern subtropical and northern Atlantic, and probably also that in the south Pacific, must be the result of a strongly enhanced terrigenous (siliciclastic) sediment discharge, which also enhances the bulk sediment-accumulation rates in the equatorial Atlantic beyond the increase observed for the $CaCO_3$ accumulation. Biogenic opal, a third variable, is not quantitively crucial where measured in these areas.

These findings are in accord with the late Pliocene increase in Saharan dust output and aridity as reported by Stein (1984, 1985) from sites 141 and 397. Similarly, they match an enhanced inner Asian dust discharge found by Leinen & Heath (1981) and Rea & Janecek (1982) in North Pacific sediments. Furthermore, they agree with the increased deposition rate of ice-rafted debris reported from site 552 (Zimmerman *et al.* 1985), a depositional process probably also contributing to the slight increase of sedimentation rates in the South Pacific.

The late Pliocene increase of $CaCO_3$ deposition near the equator may largely reflect enhanced local $CaCO_3$ production (see below). This can be concluded from the fact that the average $CaCO_3$ accumulation rates strongly decreased in most non-equatorial parts of the Atlantic and Pacific oceans (figure 8) (see also recent results from ODP leg 108 in Ruddiman *et al.* (1987)). Accordingly, they indicated that the deep ocean water in general became more corrosive to $CaCO_3$ during that time and that any local lowering of the carbonate compensation depth (CCD) was due to locally enhanced $CaCO_3$ fluxes. Possible explanations for this will be discussed below.

A fairly simple pattern emerges from the late Pliocene shift of accumulation rates of organic carbon (C_A) (figure 9). They decrease over almost all the Pacific and South Atlantic ocean, and in the Caribbean Sea (factor 0.3–0.7; with the exceptions of the region east off Japan and the equatorial Pacific). Simultaneously, the rates increase all over the North Atlantic, being especially strong near the equator and in the West European basin (factor 2.5–6.0). This general pattern may be meaningful, in spite of the still very thin data coverage available, because the various core locations were selected to represent the most characteristic high and low ocean productivity zones in middle and low latitudes.

Accordingly, we observe a global shift in the deposition of organic carbon, away from the low-productive subtropical gyres in the Pacific and South Atlantic towards the equatorial high-productivity belt and in particular, towards the North Atlantic. Besides changes in surface ocean productivity (figure 10), this increase in C_A may be due to the increased sedimentation rates (figure 6), especially in the North Atlantic, which have led to an enhanced 'sealing effect' for the carbon flux arriving on the sea floor. On the other hand, reduced ventilation of the deep

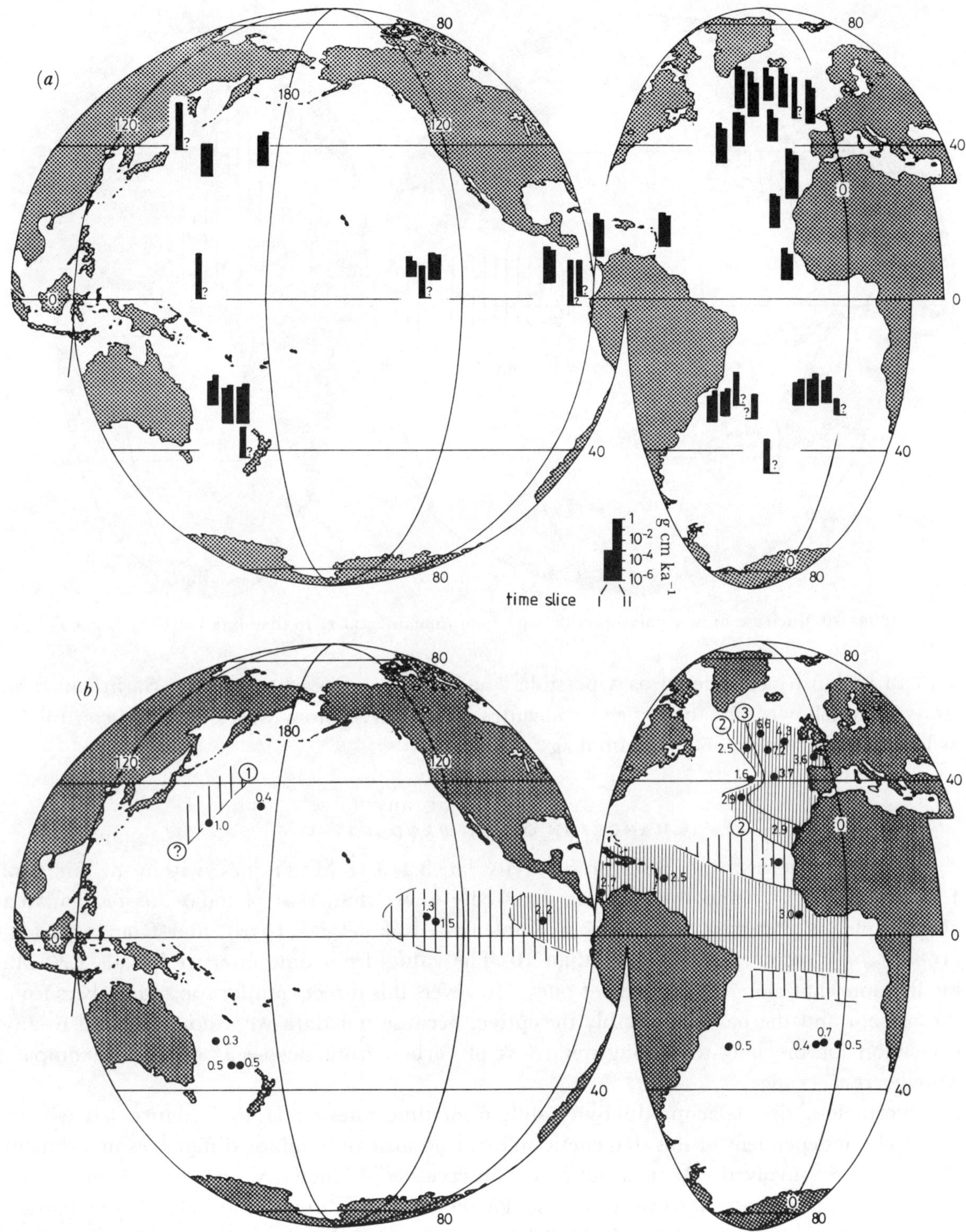

FIGURE 9. (*a*). Accumulation rates of organic carbon during time intervals I and II. ?, Hiatus and/or imprecise chronostratigraphy. (*b*) Change of C_{org} accumulation rates from time interval II to time interval I (by factor *x*).

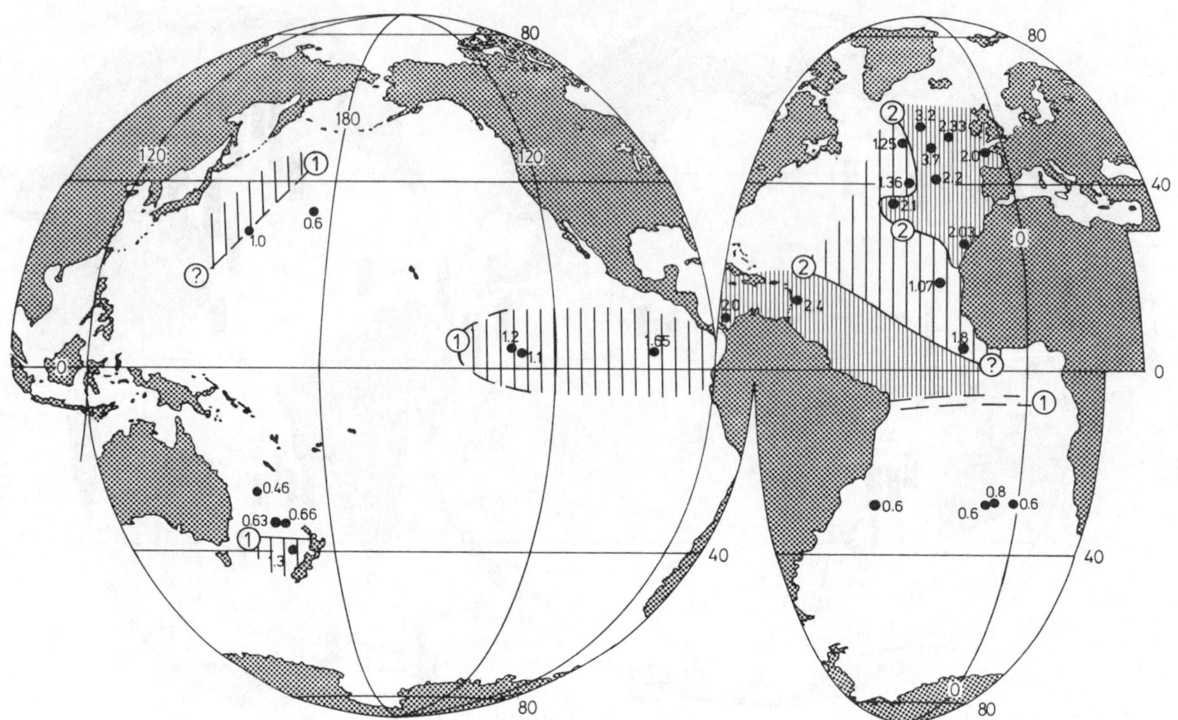

FIGURE 10. Increase in new palaeoproductivity from time interval II to time interval I (by factor x).

sea can be hardly considered as a possible cause, because Suess (1980) and Sarnthein *et al.* (1987) have shown that the oxygen concentration in the bottom water does not control C_A as long as it exceeds *ca.* 50–100 μmol kg^{-1}.

CHANGES IN OCEAN PRODUCTIVITY

In absolute terms, the new productivity at 3.4–3.18 Ma BP lies within a range of 1.5–71.0 g cm^{-2} ka^{-1} and thus appears markedly lower than that of today (as recalculated after Eppley & Peterson (1979), from Koblentz-Mishke *et al.* (1970), and Romankevitch, (1984)), by a factor of up to more than 10. The values from time interval I, 2.43–2.33 Ma BP, lie somewhat closer to the present ones. However, this direct comparison of numbers from the present and the past is probably deceptive, because our data were not subjected to any correction for the long-term diagenetic loss of carbon from deep-sea sediments (compare Mueller *et al.* (1983)).

Nevertheless, the palaeoproductivity shift from time interval II to I (figure 10) will be essentially independent of this diagenetic imprint because only minor differences in sediment thickness are involved. As a result, we observe a productivity increase covering the southwesternmost and northwesternmost Pacific and the upwelling belt of the eastern equatorial Pacific (factor of 1.0–1.65). Ocean productivity has increased much more (by a factor of 1.1–3.7) in the equatorial Atlantic, the Caribbean Sea, the west-European–Iberian Sea, and the coastal upwelling region off northwest Africa. The latter can be taken as a representative example for the four great, trade-wind driven near-shore upwelling belts along the eastern continental margins of the Atlantic and Pacific in the subtropics. On the other hand, regions with decreasing palaeoproductivity occur in the midlatitudinal south Atlantic, northern and southwestern Pacific (factor of 0.6–0.8). In summary this means that the low productivity of 'blue' ocean regions in the subtropical gyres further decreased and simultaneously, the high productivity in the 'green' ocean regions further increased, i.e. the whole productivity pattern was more polarized 3.2–2.4 Ma BP.

Discussion

From the distribution and pattern of palaeoproductivity increases, we may infer that the upwelling in the eastern equatorial Pacific and along the eastern margins of the Pacific and Atlantic Oceans was markedly intensified during the late Pliocene phase of global climatic deterioration. Based on the evidence from increasing grain sizes of the aeolian dust discharge in the low-latitudinal east Atlantic (Stein 1985) (figure 2), this intensification of upwelling was probably controlled by strongly enhanced meridional trade-wind speeds (by a factor of about 3), i.e. a wind régime already coming close to that of the Last Glacial Maximum. The wind-induced, intensified upwelling, in turn, led to reduced sea-surface temperatures in low latitudes, and hence to reduced evaporation. Accordingly, enhanced oceanic upwelling was probably crucial for the large-scale aridification of the Saharan belt, observed during that time (Sarnthein *et al.* 1982; Stein 1985) (figure 2).

The most important implication of the increased ocean productivity is to be expected in the field of ocean and atmospheric chemistry. We may conclude that the enhanced new production led to a globally increased rate of carbon extraction from the surface ocean (and hence the atmosphere) to the deep ocean on the order of magnitude of that observed about 20 ka ago, during the Last Glacial Maximum (Sarnthein *et al.* 1988). This late Quaternary shift resulted in a decrease of the atmospheric p_{CO_2} from an interglacial level of about 300 p.p.m (by volume) down to about 200 p.p.m. (by volume) (Delmas *et al.* 1980). A similar somewhat more modest shift may be now conceivable for the late Pliocene phase of climatic deterioration, implying a feedback loop of further climate cooling. This assumption is in accord with the independent evidence obtained from benthic $\delta^{13}C$ values, showing a lowering of about 0.25 ‰ and thus an increased storage of CO_2 in North Atlantic Deep Water during that time, as outlined in the Introduction (figure 3). Furthermore, this enhanced storage of CO_2 in the deep ocean is in general agreement with the record of stronger $CaCO_3$ dissolution in large parts of the ocean and also with the locally enhanced rates of $CaCO_3$ accumulation found immediately below the equatorial high productivity zones (figure 7). They indicate a local increase of $CaCO_3$ fluxes paralleled by a generally enhanced carbonate aggressivity of the bottom water.

We gratefully acknowledge the help of R. Tiedemann and Dr K. Winn in preparing the figures and tables. This work was generously supported by the funding of the Deutsche Forschungsgemeinschaft.

References

Delmas, R. J., Ascensio, J.-M. & Legrand, M. 1980 Polar ice evidence that atmospheric CO_2 20,000 yr BP was 50% of present. *Nature, Lond.* **284**, 155–157.

Duplessy, J. C. 1982 North Atlantic deep-water circulation during the last climatic cycle. *Bull. Inst. Geol. Bassin d'Aquitaine, Bordeaux* **31**, 379–391.

Eppley, R. & Peterson, B. J. 1979 Particulate organic matter flux and planktonic new production in the deep ocean. *Nature Lond* **282**, 677–680.

Koblentz-Mishke, O. J., Volkowinsky, V. V. & Kabanova, J. G. 1970 Plankton primary production of the World Ocean. In *Scientific exploration of the South Pacific* (ed. W. S. Wooster), pp. 183–193.

Leinen, M. & Heath, G. R. 1981 Sedimentary indicators of atmospheric activity in the northern hemisphere during the Cenozoic. *Palaeogeogr. Palaeoclim. Palaeoecol.* **36**, 1–21.

Mueller, P., Erlenkeuser, H. & Grafenstein, R. V. 1983 Glacial–Interglacial cycles in oceanic productivity inferred from organic carbon contents in eastern North Atlantic sediment cores. In *Coastal upwelling: its sediment record* (ed. J. Thiede & E. Suess), part B, pp. 365–398.

Mueller, P. & Suess, E. 1979 Productivity, sedimentation rate, and sedimentary organic matter in the oceans. I. Organic carbon preservation. *Deep Sea Res.* A**26**, 1347–1362.

Rea, D. K. & Janecek, T. R. 1982 Mass-accumulation rates of the non-authigenic inorganic crystalline (eolian) component of deep-sea sediments from the western Mid-Pacific Mountains, Deep Sea Drilling Project Site 463. *Init. Rep. DSDP*, vol. 62 (ed. J. Thiede, T. L. Vallier *et al.*), pp. 653–659. Washington, D.C.: U.S. Government Printing Office.

Romankevich, E. A. 1984 *Geochemistry of organic matter in the ocean.* Heidelberg: Springer-Verlag. (334 pages.)

Ruddiman, W., Sarnthein, M., Baldauf, J. *et al.* (eds.) 1987 *Init. Rep. DSDP*, vol. 108, part A. Washington D.C.: U.S. Government Printing Office.

Sarnthein, M., Thiede, J., Pflaumann, U., Erlenkeuser, H., Fuetterer, D., Koopmann, B., Lange, H. & Seibold, E. 1982 Atmospheric and oceanic circulation patterns off NW-Africa during the past 25 million years. In *Geology of the Northwest African Continental Margin* (ed. U. v. Rad, K. Hinz, M. Sarnthein & E. Seibold), pp. 545–604. Berlin: Springer-Verlag.

Sarnthein, M., Winn, K., Duplessy, J.-C. & Fontugne, M. R. 1988 Global variations of surface ocean productivity in low and mid latitudes: influence on CO_2 reservoirs of the deep ocean and the atmosphere during the last 21,000 years. *Palaeoceanography.* (Submitted.)

Sarnthein, M., Winn, K. & Zahn, R. 1987 Paleoproductivity of oceanic upwelling and the effect on atmospheric CO_2 and climatic change during glaciation times. In *Abrupt climatic change* (ed. W. H. Berger & L. D. Labeyrie), pp. 311–337. Dordrecht: D. Reidel Co.

Shackleton, N. J. & Hall, M. A. 1985 Oxygen and carbon isotope stratigraphy of Deep Sea Drilling Project Hole 552 A: Plio-pleistocene glacial history. *Init. Rep. DSDP*, vol. 81 (ed. D. G. Roberts, D. Schnitker *et al.*), pp. 599–610. Washington D.C.: U.S. Government Printing Office.

Shackleton, N. J. & Cita, M. B. 1979 Oxygen and carbon isotope stratigraphy of benthic foraminifers at site 397: detailed history of climatic change during the late Neogene. *Init. Rep. DSDP*, vol. 47, pp. 433–445. Washington, D.C.: U.S. Government Printing Office.

Stein, R. 1984 Zur neogenen Klimaentwicklung in Nordwest-Afrika und Palaeo–Ozeanographie im Nordost-Atlantik: Ergebnisse von DSDP Sites 141, 366, 397 und 544 B. *Berichte-Reports, geol. palaeont. Inst. Univ. Kiel*, vol. 4 (210 pages.)

Stein, R. 1985 Late Neogene changes of paleoclimate and paleoproductivity off Northwest Africa (DSDP Site 397). *Palaeogeogr. Palaeoclim. Palaeoecol.* **49**, 47–59.

Suess, E. 1980 Particulate organic carbon flux in the oceans surface productivity and oxygen utilization. *Nature, Lond.* **288**, 260–263.

Zahn, R., Winn, K. & Sarnthein, M. 1986 Benthic foraminiferal $\delta^{13}C$ and accumulation rates of organic carbon: *Uvigerina peregrina* group and *Cibicidoides wuellerstorfi*. *Paleoceanography* **1**(1), 27–42.

Zimmerman, H. B., Shackleton, N. J., Backman, J., Kent, D. V., Baldauf, J. G., Kaltenback, A. J. & Morton, A. C. 1985 History of Plio-Pleistocene climate in the northeastern Atlantic, Deep Sea Drilling Project Hole 552 A. *Init. Rep. DSDP*, vol. 81 (ed. G. D. Roberts, D. Schnitker *et al.*), pp. 861–876. Washington, D.C.: U.S. Government Printing Office.

Phil. Trans. R. Soc. Lond. B **318**, 505–522 (1988)
Printed in Great Britain

The record of the cold stages

By R. G. West, F.R.S.

Subdepartment of Quaternary Research, Botany School, University of Cambridge, Downing Street, Cambridge CB2 3EA, U.K.

Evidence for climatic conditions on the continents during the cold stages lies in a number of areas: glacial history, vegetational and faunal history, periglacial history, weathering horizons. Whereas glacial history provides evidence of periods of glacierization within cold stages, palaeontology provides evidence of a variety of climates within cold stages, as does the periglacial evidence, at times when glacierization was not extensive. The definition and history of cold stages is considered. The evidence for climatic change is reviewed, based on the occurrence of periglacial phenomena and fossil floras and faunas, and the problems of interpreting cold-stage climates are considered.

1. Introduction

With the advent and acceptance of the glacial theory in the last century the idea of a succession of glaciations formed the basis for the classification of Quaternary successions, implying the existence of interglacial events of climatic amelioration. Thus, the sequence of glacial deposits, the most obvious consequence of climatic change, was the measure of Quaternary stratigraphy, until it was later replaced by systems of alternating glacial and interglacial stages as knowledge of biostratigraphy accrued.

With the continued accession of terrestrial and marine evidence concerning Quaternary climates, this simple division is no longer sufficient for the classification and understanding of events. In northwest Europe, at least, a major division into temperate and cold stages is likely to be a useful basis for subdivision of the Quaternary sequence, with the temperate stages resulting from climatic amelioration to a state we see at the present day, to use one possible definition of the term interglacial. A summary of such a division in the later Pleistocene of northwest Europe is given in table 1.

It is also now clear that the cold stages are themselves complex, probably more so than the temperate stages, containing one or more periods of ice expansion and showing evidence for degrees of climatic change within them.

On the continents, phenomena related to physical conditions, such as permafrost, can be clearly used to interpret cold-stage climates and, as Dylik (1975) has pointed out, can be used more widely for climatic interpretation than the more locally developed glaciations. The biological evidence of climate is more difficult. Whereas we see in the development of the Flandrian (postglacial) a sequence of vegetation units or faunas which are recognizable to a large extent in living plant or animal communities, in the cold stages we are dealing with peculiar biota and environments largely outside our experience, even if they are now represented to some extent in cold regions. Interpretation of cold-stage biota and environments is a challenge, not least because with biota the present is by no means certainly the key to the

TABLE 1. STAGE NAMES IN NORTHWEST EUROPE IN THE LATER PLEISTOCENE

(Abbreviations: t, temperate stage; c, cold stage.)

Ireland[a]	Britain	continental northwest Europe	stage
Littletonian	Flandrian	(Holocene)	t
Midlandian	Devensian	Weichselian	last cold stage
Glenavian	Ipswichian	Eemian	t
Munsterian	Wolstonian	Saalian	c
Gortian	Hoxnian	Holsteinian	t
	Anglian	Elsterian	c

[a] From Mitchell (1986).

past. Present distributions, which might be used to interpret past environments, represent the organism's response to present climate, historical factors and competition. In cold stages, cold conditions reached to lower latitudes and the climates that occurred then are not necessarily similar to those of present cold regions. The response of identified organisms to such climates is not known.

The purpose of this discussion is to examine this complexity and the problems of interpretation in more detail in the area of northwest Europe.

2. DEFINITION AND OCCURRENCE

Cold stages are naturally not easy to define with any precision, because climatic variation is a continuous process. They are not synonymous with glacial periods, but include them. In northwest Europe, and in botanical terms, they are periods of time when climatic deterioration leads to the long-term prevalence of largely herbaceous vegetation with no thermophilous trees. As discussed below, there may be interruptions by short forested periods. Expansion of ice sheets and periods of permafrost occur from time to time, and there is evidence of lowered sea levels. Formally, it is desirable to define the lower limit of a stage; Zagwijn (1957) has suggested a definition based on the appearance of a subarctic vegetation type, when tree pollen is largely replaced by herbaceous pollen. The opposite change, with the subsequent development of a long period of temperate forest, is taken to indicate the beginning of a temperate stage and the end of a cold stage, by analogy with the changes seen at the beginning of the Flandrian (Holocene) temperate stage (Zagwijn 1957).

In the marine oxygen-isotope record cold stages are represented by periods of increased proportions of ^{18}O in foraminiferal tests compared with those of the present time and in the last temperate stage. The end of such periods may be marked by sharp terminations (Broecker & van Donk 1970), indicating rapid deglaciation. Such terminations appear to herald the beginning of a temperate stage, and there is evidence to link them to the beginning of the continental temperate stages. The beginnings of cold stages are more problematical, both in the oceans and on the continent, because they seem to be accompanied by more gradual and variable conditions, as will be discussed later. Shackleton (1986) discusses the problems of the subdivision of the isotope record.

Between temperate stage climates and the full expression of cold stage climates as seen in the so-called full-glacial or pleniglacial, there lies a variety of conditions, and it is periods with this variety that make for the difficulties of definition, particularly so when it seems possible that such periods may have occupied a larger fraction of Quaternary time than periods with more extreme climates.

The difficulties of definition make an assertion about the number of cold stages impossible. But in the Pacific cores V28-238 (Shackleton & Opdyke 1973) and V28-239 (Shackleton & Opdyke 1976) some ten cold stages were indicated in the past 800 ka and certainly there were earlier ones, as indicated by The Netherlands succession (Zagwijn 1985), with the Praetiglian cold stage at about 2.3 Ma BP. A similar age is suggested by Backman (1979) for evidence in the north Atlantic of the initial ice rafting and so extension of Northern Hemisphere ice sheets, and by the isotope record in the north Atlantic (Shackleton *et al.* 1984; Shackleton 1986), although there are signs of a slightly earlier refrigeration in the isotope record (Shackleton *et al.* 1984). There is little stratigraphical evidence for associating widespread ice advances in northwest Europe with these earliest cold stages, although indications are present from erratics in gravel trains arising in the Alps, Scandinavia and Britain. Elsterian and Anglian glacial deposits of the Middle Pleistocene thus appear to be the earliest extensive evidence of widespread glaciation, although there may be evidence for further earlier and less extensive development of ice sheets. The sequence of glacial deposits, starting with those of the Elsterian and Anglian, formed the original basis for the subdivision of the Quaternary in northwest Europe. Evidently in the Middle and Late Pleistocene extensive ice sheets were characteristic of cold stages in a way not so far discovered in the earlier Pleistocene.

In view of the difficulties of exact definition of cold stages (see the discussion by Shackleton (1986)), their length is consequently difficult to estimate. To judge by the course of the isotope curve, especially during at least the past 300 ka, cold stages in the Middle and Upper Pleistocene are considerably longer than the major recognized temperate stages. In the earlier Pleistocene this does not obviously seem to be so, except for the earliest very pronounced cold stage of the Pretiglian in The Netherlands. This implies rather different relations between cold and temperate stages in the Lower Pleistocene, also suggested by the evidence for glaciations discussed above.

Most information on length naturally comes from the most recent cold and temperate stages. Figure 1 outlines subdivisions and estimated ages for the past 130 ka in northwest Europe.

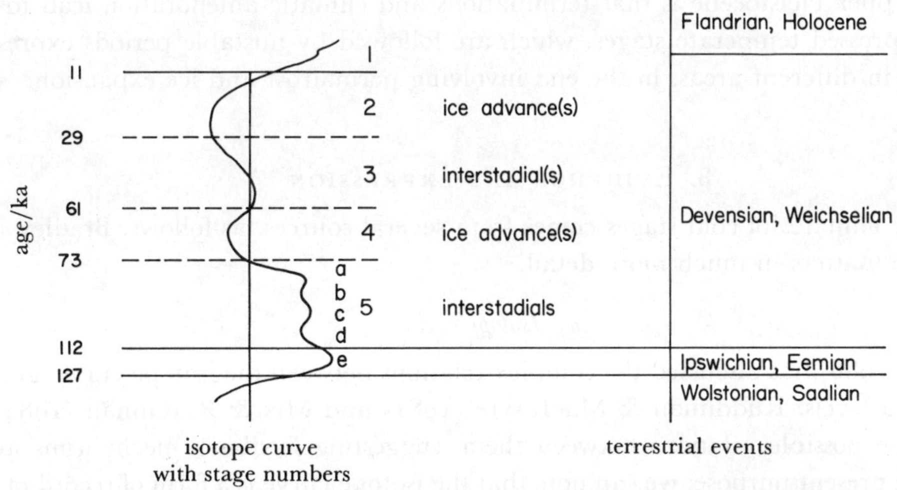

FIGURE 1. Generalized oxygen-isotope curve through the past 130 ka according to Shackleton (1969), using his and Emiliani's (1961) subdivisions, related to estimated ages (years B.P.) suggested by Woillard & Mook (1982), to major terrestrial events in northwest Europe, and to the latest Pleistocene stages.

Taking the length of the last temperate stage (Eemian, Ipswichian) as, say, 15 ka and the length so far of the present one as 10 ka, the intervening cold stage (Vistulian, Weichselian, Devensian) lasted some 102 ka. The complexity and instability of this period is indicated in figure 1, and is seen in both the marine and continental evidence. There must have been considerable climatic changes during the period, on a timescale not perhaps very different from those of a single temperate stage. The first part of the cold stage, as defined here, is some 40 ka in length. It is characterized by alternating so-called stadial and interstadial conditions, in The Netherlands (Staalduinen *et al.* 1979) by three periods of subarctic park or open landscape enclosing and following two periods with boreal forest. Further southeast, at Grande Pile (Woillard & Mook 1982), similar changes are expressed by short periods with herbaceous pollen increasing in frequency and by forest periods with temperate trees. The contrast is a result of degrees of climatic change and proximity of plant refuges, and exposes the difficulties of characterizing periods as interglacials or interstadials, and also of drawing the lower boundary of a cold stage. In this instance, should it be at 73 ka BP or 112 ka BP? I prefer the latter, and have used the latter, because the end of the preceding temperate stage can be clearly defined in palaeobotanical terms.

The remaining part of the cold stage, 63 ka in our estimate, is characterized by severer conditions, with clear evidence for permafrost, ice advances, polar desert in The Netherlands, and 'tundra' vegetation, with evidence for climatic amelioration from time to time. The details will be discussed in a later section.

The instability seen in the early part of the last cold stage also appears to have been present in the early part of the preceding cold stage, the Saalian, as expressed in The Netherlands sequence by the Hoogeveen and Bantega interstadials followed by the major Saalian ice advances (Zagwijn 1985). Even so there is no reason why the type of sequence seen in the last cold stage should be repeated in earlier cold stages; climatic changes and their causes are very complex. Indeed, it appears that certain pre-Weichselian cold stages have more than one extensive ice advance separated by an interval (e.g. the Saalian in northwest Germany (Grube *et al.* 1986)). Each major ice advance itself may result may result in a complex multiple till sequence, as in the Weichselian–Devensian cold stage.

Perhaps the only generalization that can currently be made about these matters in the Middle and Upper Pleistocene is that terminations and climatic amelioration lead to major and widely expressed temperate stages, which are followed by unstable periods expressed in different ways in different areas, in the end involving permafrost and ice expansion.

3. Evidence and expression

Evidence for climates of cold stages comes from several sources, as follows. Bradley (1985) considers these matters in much more detail.

(*a*) *Isotope curve*

Shackleton (1986) has discussed the complex relations between the isotope curve, global ice volume and sea levels. Ruddiman & MacIntyre (1981) and Mix & Ruddiman (1984) have investigated the possible relations between them, suggesting feedback mechanisms and lag effects. For the present purpose, we can note that the isotope curve is a form of record of global ice volumes at particular times in the cold stages. The details of more local expansion of ice sheets must come from the study of continental sequences. The contrast between the terminations and the instability of earlier parts of cold stages presumably reflects rapid widespread deglaciation on the one hand, and on the other hand the variable buildup of ice in the various Pleistocene ice sheets and the climatic changes associated with these ice-volume changes.

(*b*) *Geological evidence*

Geological evidence for cold-stage environments comes from the very variable lithology of the sediments. Historically, glacigenic sediments formed the basis for the designation of cold stages, but periglacial sediments are now seen to supply details of environmental changes not available from glacigenic deposits. The problem now is to place glacigenic sediments in their correct position in relation to the widespread deposition of fluviatile and aeolian sediments in periglacial sequences. Such sequences are divisible through periods of soil formation (see, for example, Kukla 1975) and organic deposition which occur within them. Each sediment type makes its own contribution to the interpretation of cold-stage environments: glacigenic sediments to the record of glaciation, fluviatile sands and gravels to the history of cold-stage rivers, cover sand and dunes to evidence for severe and often pleniglacial environments, loess

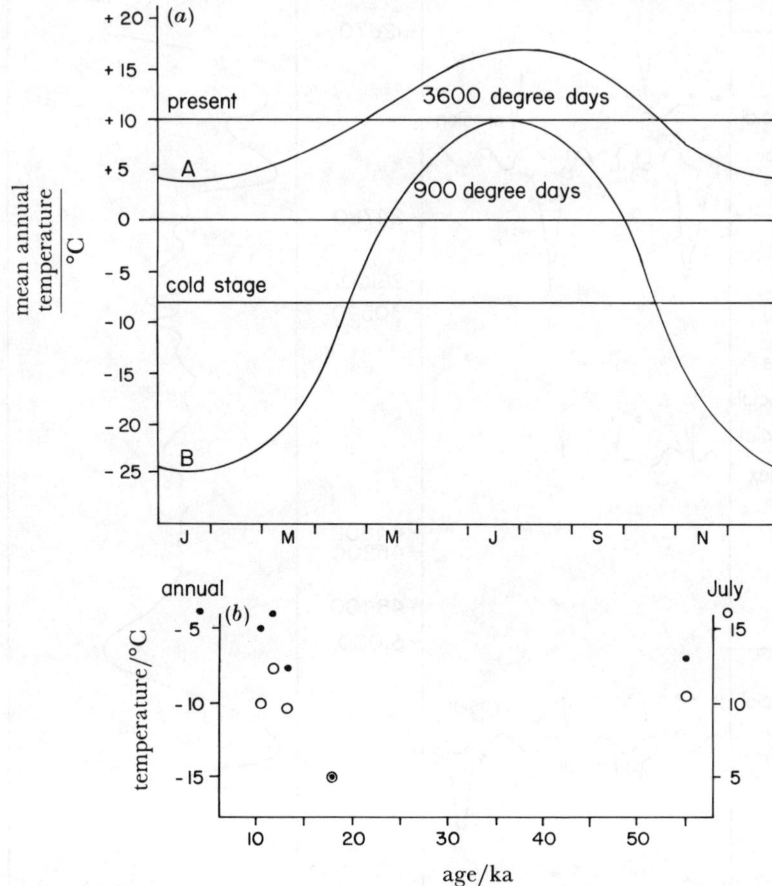

FIGURE 2 (*a*). Williams (1975) re-construction of monthly temperatures in central England at present (A) and in the coldest part of the last cold stage (B). (*b*). Watson's (1977) estimated mean air temperatures in the last cold stage.

to evidence for more continental environments, and solifluction sediments to periods of degradation of landscapes.

A very important component is the structural evidence for cold climates seen in periglacial phenomena, including involutions and features indicating permafrost, such as thermal contraction cracks. The observation that these latter occur at particular horizons within, for example, the last cold stage, underlines the climatic variability of cold stages, proving the presence of mean annual temperatures of at least *ca.* −5 °C at particular times. Thermal contraction cracks have been described from cold stages back to the Beestonian of East Anglia (West 1980*b*) and intra- or pre-Eburonian (?) in Normandy (Clet-Pellerin 1983).

Karte & Liedtke (1981) have discussed further the climatic interpretation of periglacial phenomena in some detail. They thus enlarge the point made by Dylik (1975), contrasting climatic evidence from the more local ice-sheet development in cool and oceanic areas with that from periglacial phenomena, expressed more widely and significant for drier and more continental climates. Williams (1975) has used periglacial phenomena of the last cold stage in

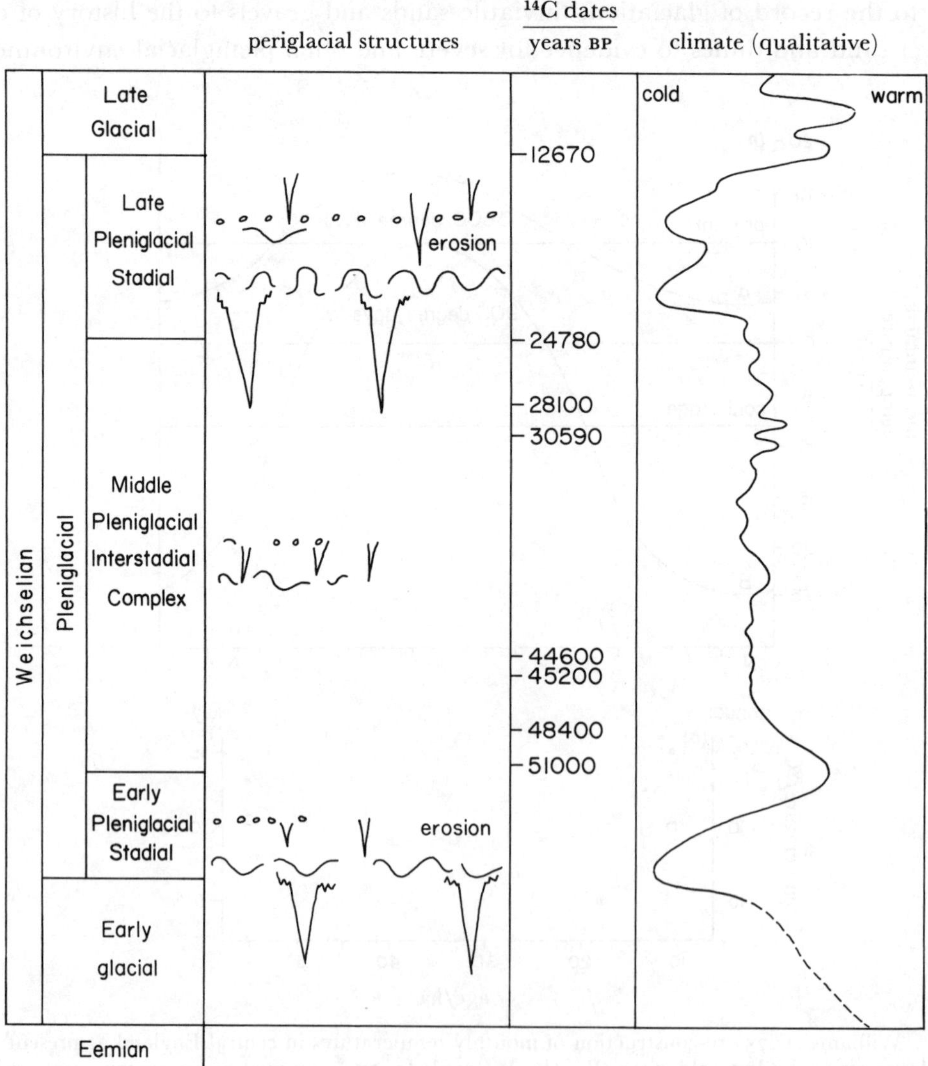

FIGURE 3. Outline of Vandenberghe's (1985) stratigraphy of the Weichselian in the southern Netherlands, showing periglacial structures and inferred climate. Sediments, mainly cover sands and loams, are omitted.

Britain to estimate mean annual temperature, monthly temperatures, rainfall and snowfall, and wind and pressure systems. Watson (1977) has considered the climatic significance of permafrost, continuous and discontinuous, in Britain, suggesting falls of mean annual air temperature of up to 25 °C and falls of July mean temperature of 5–10 °C. Figure 2 shows some of the suggested temperatures of Williams and of Watson. The opportunity offered by sediment and structure studies for reconstructing climates and environments of the last cold stage are excellent, as shown by studies of the Netherlands and Belgian periglacial sequences of the Weichselian by Vandenberghe (1985) (figure 3).

A further and different type of evidence for climatic change in cold stages derives from speleothem studies. Periods of low or no speleothem growth are thought to indicate cold and/or conditions of climate with warmer and wetter conditions resulting in greater growth (Ivanovich 1985). Cave sequences appear to yield information on climatic fluctuations in cold stages as well as between cold and temperates stages.

Glacioeustatic sea-level changes have already been mentioned above in relation to the isotope curve. From the stratigraphical point of view, they are important in our area as a factor (low sea level) in control of river development in cold stages and as a factor (high sea level) evident in the major widespread temperate stages.

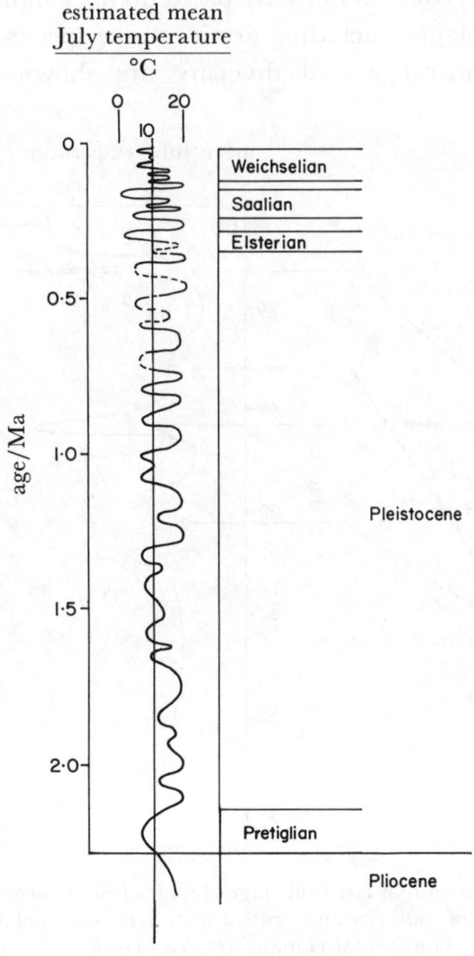

FIGURE 4. Zagwijn's (1985) climatic curve (estimated mean July temperature) for The Netherlands. Only the latest cold stages (which contain very extensive ice advances) and the earliest are named.

(*c*) *Biological evidence*

Fossil evidence of cold-stage biota is abundant; pollen, plant macroscopic remains, vertebrates, molluscs, ostracods and Coleoptera have all provided bases for climatic interpretation, on the basis of present-day distribution of either species or of communities. Here I consider floras, Coleoptera and vertebrates.

The most complete temperature curve for northwest Europe is that for The Netherlands Quaternary described by Zagwijn (1985), based on climatic deteriorations indicated mainly by a paucity of trees (treeline taken to be related to 10 °C isotherm in the warmest month) and by the presence of evidence for polar desert (temperature lower than 5 °C). The curve (figure 4) shows the marked cold stage at 2.3 Ma BP, the later and lesser fluctuations of the earlier Pleistocene and the more marked cold stages of the later Pleistocene. A similarly based curve for the last cold stage is discussed later. The use of particular plant species for climatic deductions, based on their present day ecology, is exemplified by Kolstrup (1979, 1980) in her estimations of last cold stage climates in The Netherlands.

The detailed work of Coope (1977) on Coleoptera led to his reconstructions of an average July temperature curve for the last cold stage in southern and central British Isles and of average monthly temperatures for interstadial climatic regimes in the last cold stage in the British Isles (figure 5). Such reconstructions are based on the combined distributional evidence of species from dated assemblages, including arctic–alpine species and continental east Asian species. Massive changes in range and diversity are shown by some species in these assemblages.

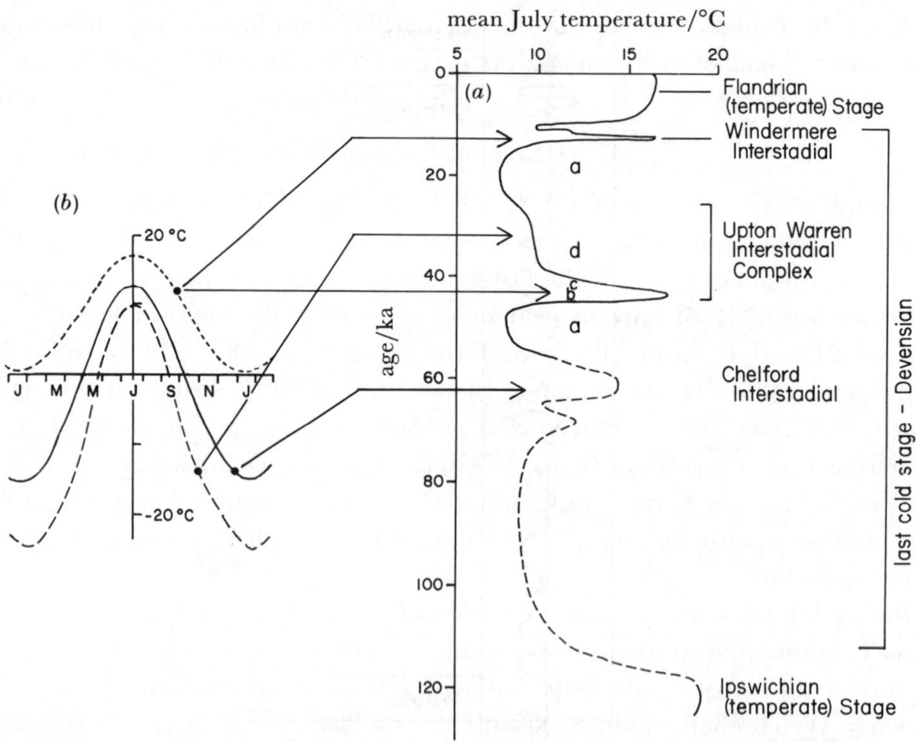

FIGURE 5. Coope's (1977) reconstruction of last cold stage climates from Coleopteran evidence. (*a*) Average July temperature in lowland areas of southern and central British Isles: a, polar climate; b, moderately oceanic climate; c, transitional; d, cold continental climate. (*b*) Average monthly temperatures for four interstadial climates in the British Isles.

On the basis of the plant assemblages, polar desert, tundra, 'steppe–tundra', and shrub tundra have all been described, again with mixtures of arctic–alpine and continental species (Bell 1969; Godwin 1975), with the occasional addition of halophytic species, a problematical group perhaps indicative of continental climate. Vertebrate faunas, although their taphonomy is more insecure, also show the presence of northern and continental species (Stuart 1977, 1982).

The question remains, however, how all this data can be used to determine palaeoclimates. The situation is far more complex than with ocean biota. The first problem is the relation today of the organism concerned to climate, with the additional effects of historical and competition factors now and then; the second is the unknown ecological tolerance of the organism at the present time, and the third is whether the organism at the time of the fossil occurrence was distributed according to an equilibrium with the climate (Prentice 1986; Webb 1986). Discrepancies in the evidence from different parts or the biota suggest that equilibrium was not always obtained by the plants, as discussed later with reference to the last cold stage.

4. The last cold stage

There is abounding evidence for last cold-stage environments in northwest Europe. In addition the chronology, relative in the earlier part and absolute in the later part, is also reasonably well established. The result is that we can survey the last cold stage as an example of the diversity of evidence for Pleistocene cold-stage climates. This is not to say that the course of each cold stage followed the same pattern, in terms of, say, glacial or permafrost events. Even though we have this evidence for an outline of environmental history and climatic change in the last cold stage, much is still conjectural or uncertain, as will be seen in the following discussion.

(*a*) *Geological evidence*

The till complex associated with the last cold stage in northwest Germany, Denmark, southern Sweden and southern Finland now appears to have resulted from ice advance and retreat (figure 6) in the later part of the last cold stage, although there is evidence for earlier advances further north. Radiocarbon evidence suggests the till complex was deposited in a period between 21 and 13 ka BP (Lagerlund 1983; Sjørring 1983; Ehlers 1983; Lundqvist 1986; Donner *et al.* 1986). The same appears to be true in eastern Britain (Penny *et al.* 1969), although, in Ulster, McCabe (in Bowen *et al.*, 1986) has suggested an early Midlandian ice advance. There is yet no certain evidence of extensive earlier ice advances into these areas in the last cold stage. Thus the period of greatest ice extension was short compared with the total length of the cold stage, although its effects on landscape through ice movement and glacigenic sedimentation were huge.

Outside the ice limit, the most complete record of the last cold stage has been found in The Netherlands, Belgium and northwest Germany, expressed as a sedimentary sequence of lacustrine, fluvial and aeolian deposits, interrupted by periglacial structures. A threefold division of early Weichselian, middle Weichselian or Pleniglacial, and late Weichselian has been established (figure 7). It should be noted that whereas the late Weichselian begins at *ca.* 13 ka BP, the late Devensian lower boundary has been placed at 26 ka BP(Mitchell *et al.* 1973).

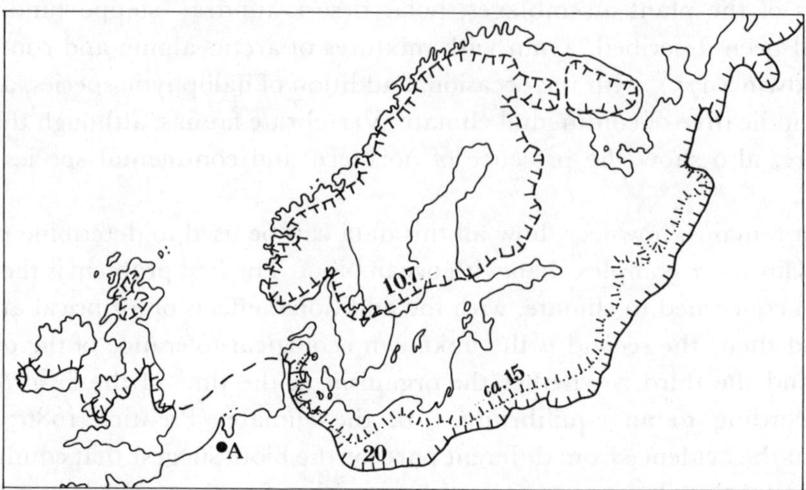

FIGURE 6. Major ice margins in northwest Europe in the last cold stage (Weichselian, Devensian) (after Starkel 1977). Point A marks the general area of the detailed Netherlands/Belgian Weichselian sequence (figures 3 and 7); figures refer to time (ka BP).

(*i*) *Early Weichselian*

A number of periods with climatic amelioration have been described from The Netherlands (Zagwijn 1961), Belgium (Woillard 1978), northwest Germany (Menke & Tynni 1984; Behre & Lade 1986) and Denmark (Andersen 1961) (figure 7). They are separated from the last temperate stage (Eemian) by a cold interlude. The Brørup and Odderade interstadials, with their correlatives at Grande Pile, St Germain I and II, are the firmest established of these, with the earlier Rodebaek and Amersfoort interstadials and later Keller interstadial of less certain status. Further south, at Les Echets, near Lyon, de Beaulieu & Reille (1984)have described a long pollen diagram which substantiates the evidence for these interstadials, as does the long sequence at Samerberg, Bavaria, described by Grüger (1979).

(ii) *Middle Weichselian or Pleniglacial*

Vandenberghe (1985) has described in detail the sediments and structures of this period. He proposes early and late Pleniglacial stades, characterized by permafrost features with thermal contraction cracks, and an intervening middle Pleniglacial interstadial complex with milder conditions. The age of the upper stade is *ca.* 18–25 ka BP., and the less certain age of the lower stade is *ca.* 60–70 ka BP. The upper stade thus bears a relation to the time of ice expansion. Climatic variation within the middle Pleniglacial interstadial has been much discussed. On the one hand there is the view that amelioration is associated with deposition of organic sediments within the sequence of cover-sands and fluvial sediments (see, for example, van der Hammen *et al.* 1967; Zagwijn & Paepe 1968; Zagwijn 1974), resulting in the formulation of the Moershoofd, Hengelo and Denekamp interstadials. On the other hand, Vandenberghe (1985) has pointed out that peat formation is not necessarily related to amelioration, but is controlled by geomorphology and hydrology. He has considered in detail, as has Maarleveld (1976), the climatic implications of periglacial structures and sediments in The Netherlands (see figure 3).

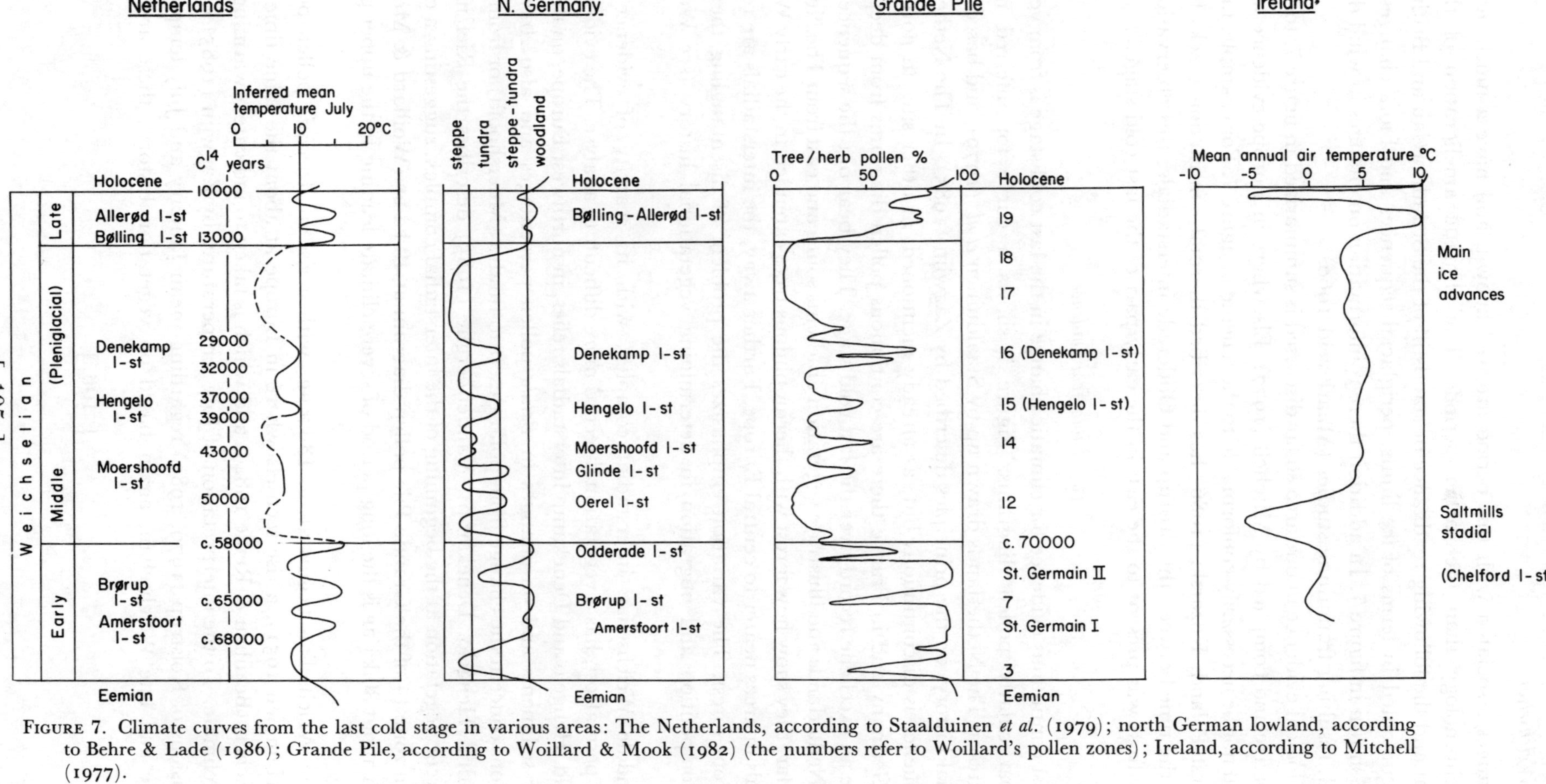

FIGURE 7. Climate curves from the last cold stage in various areas: The Netherlands, according to Staalduinen *et al.* (1979); north German lowland, according to Behre & Lade (1986); Grande Pile, according to Woillard & Mook (1982) (the numbers refer to Woillard's pollen zones); Ireland, according to Mitchell (1977).

(iii) *Late Weichselian*

The changes associated with this period are well known, and have a much more certain absolute chronology than the earlier periods. The climatic amelioration of the Bølling interstadial and the following readvance or standstill of the Scandinavian and British ice-caps is well documented in terms of ice limits, periglacial sequences and biota changes. Climatic curves are shown in figure 7. In addition there is information on changes in wind direction in this period, resulting from dune studies (Maarleveld 1960).

The geological evidence in our area so far discussed is summarized in figure 7, together with a curve for Ireland complied by Mitchell (1977). Elsewhere in Europe evidence for climatic change will not be necessarily conformable to the scheme in figure 7. For example, further east, in northwest Poland, Kozarski (1986) has described thermal contraction cracks believed to date from the time between the Brørup and Odderade interstadials; this observation suggests that permafrost was present to the east in the early part of the last cold stage.

(*b*) *Biological evidence*

The most continuous evidence for climatic change in the last cold stage is from vegetational history, particularly pollen diagrams. Figure 7 shows a curve for inferred mean July temperatures in The Netherlands drawn up by Staalduinen *et al.* (1979), and based mainly on vegetational history, on the principles described by Zagwijn (1985). In The Netherlands the pollen sequence is discontinuous but, as already mentioned, at other sites in northwest and southern Germany and in France there are continuous pollen diagrams from deep lake sites which give an excellene record over the last cold stage. They bear out the sequence described from The Netherlands and illustrate vegetation gradients south and east from The Netherlands. In particular, they show how forest with thermophilous trees extended in the early Weichselian interstadials in areas nearer to central Europe. Further away, the interstadials are represented by coniferous forest. The variation emphasizes the problem of disentangling the factors of climate, competition and migration in determining vegetational history (see West 1980*a*, p. 615).

The Middle Weichselian interstadial complex, with its scarcity of evidence for well-developed permafrost, has a vegetational record more difficult to analyse. The evidence for the Moershoofd, Hengelo and Denekamp interstadials relies, in northwest Europe, on the presence of organic sediments and on changes in *Betula* pollen percentages and also *Artemisia*. The ameliorations concern the change from polar desert to tundra (Moershoofd) or from tundra to shrub–tundra (Hengelo, Denekamp). Vandenberghe (1985) describes the Riel interstadial with subarctic vegetation at the beginning of the interstadial complex, suggesting a correlation with pollen zone 14 of the Grande Pile pollen diagram at 49 ka BP (Woillard & Mook 1982). Later than about 30 ka BP is the long period of severe climate leading to the upper permafrost episode.

The Late Weichselian begins at *ca.* 13 ka BP with a rise in *Artemisia* pollen percentages (van der Hammen 1951), a rise seen elsewhere in Europe at about the same time or earlier (e.g. Les Echets (Beaulieu & Reille 1984)) and which is taken to indicate a warming and drying of the climate. To the interpretation of July temperatures by Zagwijn (1985) we can add the conclusions of Kolstrup (1979, 1980) regarding mean January and July temperatures in the middle and late Weichselian, again based on vegetation history; these are shown in figure 8.

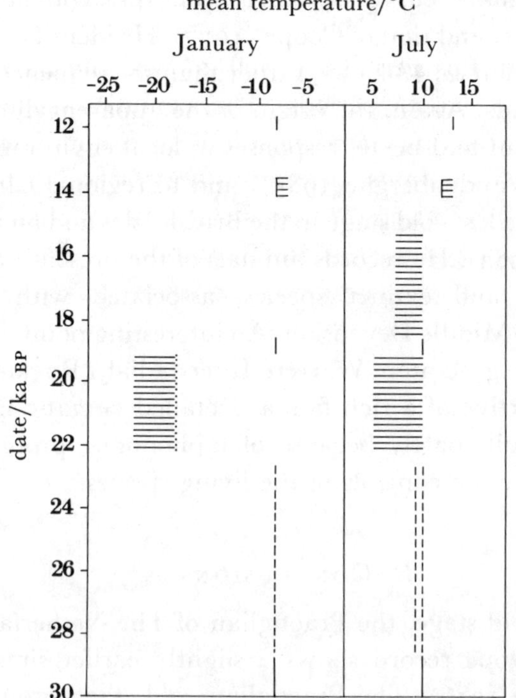

FIGURE 8. Kolstrup's (1980) reconstruction of January and July temperatures in the middle and late Weichselian in The Netherlands, based on vegetational history and periglacial phenomena. Solid line, temperature approximately as indicated; solid line with hatching, temperature possibly higher than indicated by line; horizontal lines, temperature estimated to fall between the limits; dotted line with hatching, temperature possibly lower than dotted line; single or double broken line, temperature approximately as shown, but age uncertain.

Turning now to further evidence of last cold-stage climates we must consider the extremely valuable evidence obtained by fossil Coleoptera by Coope (1977). He has made a detailed study of last cold stage faunas and has outlined curves for average July temperatures in lowland areas of the southern and central British Isles and for average monthly temperatures during the Chelford Interstadial, the thermal maxima of the Upton Warren and Windermere Interstadials, and the latter part of the Upton Warren interstadial complex. These are shown in figure 5, and portray striking changes in July temperatures and the coldness of winters. More recently Coope and his colleagues have extended the detailed interpretation of seasonal changes in the past 22 ka (Atkinson *et al.* 1987). The evidence for seasonal changes is particularly valuable. The warming at 13–14 ka BP compares with evidence for the same effect from pollen studies mentioned above. Similarly the evidence for warming in the Chelford Interstadial (Brørup Interstadial?) of the early Devensian compares well with the palaeobotanical evidence (Simpson & West 1958).

In the intervening middle Devensian, however, there is a strong contrast in the Upton Warren Interstadial between herbaceous pollen spectra of 'full-glacial' type (see, for example, Kerney *et al.* 1982), normally thought to indicate colder conditions, and beetle faunas indicating temperate conditions possibly warmer than those of the present day (Coope & Angus 1975). The origin of the contrast is fully discussed by Coope (1977). It may lie in rapid climatic change and the response times of different parts of the biota. The contrast poses fundamental questions about the interpretation of palaeoclimates from fossil biota. A second contrast is between the climate curve obtained from palaeobotanical evidence in The Netherlands, considered above, and that of Coope (1977). He identifies on convincing evidence a single temperate interstadial at *ca.* 43 ka BP, rather than the sequence of Middle Weichselian interstadial of The Netherlands. Again, the origin of the apparent discrepancy may lie in the variable behaviour of the plant and beetle responses to local environmental changes (related, for example, to hydrology (Vandenberghe 1985)) and to regional climatic changes.

The vertebrate faunas of the last cold stage in the British Isles and on the continent have been discussed by Stuart (1977, 1982). He records animals of the present day tundra, boreal forest and steppe, with southern and extinct species, associated with the herbaceous plant communities of the Early and Middle Devensian. An interesting point is the abundance of *Bison* bones in at certain times (e.g. Upton Warren Interstadial (Rackham 1978)), indicating productive vegetation supportive of a rich fauna. Detailed climatic interpretations from the vertebrate fauna seem difficult, partly because of taphonomic problems, partly because of difficulties of interpreting climatic controls of the living species.

5. Conclusions

The first clearly defined cold stage, the Praetiglian of The Netherlands, occurred at about 2.3 Ma BP, although the isotope record shows a slightly earlier similar event of the much shorter duration (Shackleton 1986). The Praetiglian cold stage brought to an end the long period in northwest Europe of Tertiary forests with their rich floristic diversity.

The cold stages show a great variety of geological events within glacial and periglacial régimes, accompanied by faunal and vegetational changes, all controlled by climatic change. The position of northwest Europe on the fringe of the continent is likely to result in a sensitivity to complex climatic changes, involving degrees of continentality–oceanicity (e.g. seasonal changes, snowfall) and mean annual temperatures and precipitation. The distance of refuges for temperate-stage species also complicates the biological evidence for climatic change.

Taking the last cold stage as an example, climatic change in terms of temperatures has been investigated through the geological and biological evidence. Variable precipitation and snowfall are less amenable to investigation, except generally through evidence for continentality and for ice-sheet growth. The identification of climatic change through the postulation of interstadials is problematical, because definitions vary, and the evidence from different parts of the biota is at first sight conflicting, indicating a complexity in the processes of climatic control and migration. Evidence from periglacial phenomena, especially permafrost, is more secure, with indications of two important times of permafrost development. The main ice advance appears to have taken place over a few thousand years in the later parts of the last cold stage. Nevertheless, the glacigenic sediments left behind show a complex development of the Scandinavian, British and Irish ice sheets, with sharp changes of direction of flow, a succession of ice margins and multiple tills.

The complexity seen in the last cold stage is likely to have occurred in the earlier cold stages, but not necessarily with the same organization in time and space. The extent of ice-sheet expansion in earlier cold stages in the Middle Pleistocene is approximately known, but much less is known of the timing within cold stages, or of periglacial sequences, both basic to the reconstruction of climate change within cold stages and necessary to understand the causes of climatic change.

I thank Ms S. M. Peglar for assistance with the illustrations for this discussion.

References

Andersen, S. T. 1961 Vegetation and its environment in Denmark in the Early Weichselian Glacial (Last Glacial). *Danm. geol. Unders.* ser. 2, no. 75.

Atkinson, T. C., Briffa, K. R. & Coope, G. R. 1987 Seasonal temperatures in Britain during the past 22000 years, reconstructed using beetle remains. *Nature, Lond.* **325**, 587–592.

Backman, J. 1979 Pliocene biostratigraphy of DSDP sites 111 and 116 from the North Atlantic ocean and the age of Northern Hemisphere glaciations. *Stockh. Contr. Geol.* **32**, 115–137.

de Beaulieu, J.-L. & Reille, M. 1984 A long Upper Pleistocene pollen record from Les Echets, near Lyon, France. *Boreas* **13**, 112–132.

Behre, K.-E. & Lade, U. 1986 Eine Folge von Eem und 4 Weichsel-Interstadialen in Oerel/Niedersachsen und ihr Vegetationsablauf. *Eiszeitalter Gegenw.* **36**, 11–36.

Bell, F. G. 1969 The occurrence of southern, steppe and halophyte elements in Weichselian (last glacial) floras from southern Britain. *New Phytol.* **68**, 913–922.

Bowen, D. Q., Rose, J., McCabe, A. M. & Sutherland, D. G. 1986 Correlation of Quaternary glaciations in England, Ireland, Scotland and Wales. *Quat. Sci. Rev.* **5**, 299–340.

Bradley, R. S. 1985 *Quaternary paleoclimatology*. Boston: Allen & Unwin.

Broecker, W. S. & van Donk, J. 1970 Insolation changes, ice volumes, and the ^{18}O record in deep see cores. *Rev. Geophys. Space Phys.* **8**, 169–198.

Clet-Pellerin, M. 1983 Le Plio-Pleistocene en Normandie: apports de la palynologie. Thesis, C.N.R.S., Caen.

Coope, G. R. 1977 Fossil Coleopteran assemblages as sensitive indicators of climatic changes during the Devensian (Last) cold stage. *Phil. Trans. R. Soc. Lond.* B **280**, 313–340.

Coope, G. R. & Angus, R. B. 1975 An ecological study of a temperate interlude in the middle of the last glaciation, based on fossil Coleoptera from Isleworth, Middlesex. *J. Anim. Ecol.* **44**, 365–391.

Donner, J., Korpela, K. & Tynni, R. 1986 Veiksel-Jääkauden alajaotus Suomessa. *Terra* **98**, 240–247.

Dylik, J. 1975 The glacial complex in the notion of the Late Cenozoic cold ages. *Biul. peryglac.* **24**, 219–231.

Ehlers, J. 1983 The glacial history of north-west Germany. In *Glacial deposits in north-west Europe* (ed. J. Ehlers), pp. 229–247. Rotterdam: Balkema.

Emiliani, C. 1961 Cenozoic climatic changes as indicated by the stratigraphy and chronology of deep-sea cores of Globigerina-ooze facies. *Ann. N. Y. Acad. Sci.* **95**, 521–536.

Godwin, H. 1975 *History of the British flora.* 2nd edn. Cambridge University Press.

Grube, F., Christensen, S. & Vollmer, T. 1986 Glaciations in north west Germany. *Quat. Sci. Rev.* **5**, 347–358.

Grüger, E. 1979 Spätriss, Riss/Würm und Frühwürm am Samerberg in Oberbayern – ein vegetationsgeschichter Beitrag zur Gliederung des Jungpleistozäns. *Geol. Bavarica* **80**, 5–64.

van der Hammen, T. 1951 Late Glacial Flora and periglacial phenomena in the Netherlands. *Leidse geol. Meded.* **17**, 71–183.

van der Hammen, T., Maarleveld, G. C., Vogel, J. C. & Zagwijn, W. H. 1967 Stratigraphy, climatic succession and radiocarbon dating of the last glacial in the Netherlands. *Geologie Mijnb.* **46**, 79–95.

Ivanovich, M. 1985 Application of Uranium series dating to palaeoclimatic studies. *Mod. Geol.* **9**, 249–260.

Karte, J. & Liedtke, H. 1981 The theoretical and practical definition of the term 'periglacial' in its geographical and geological meaning. *Biul. peryglac.* **28**, 123–135.

Kerney, M. P., Gibbard, P. L., Hall, A. R. & Robinson, J. E. 1982 Middle Devensian river deposits beneath the 'Upper Floodplain' terrace of the River Thames at Isleworth, West London. *Proc. Geol. Ass.* **93**, 385–393.

Kolstrup, E. 1979 Herbs as July temperature indicators for parts of the Pleniglacial and Late-glacial in the Netherlands. *Geol. Mijnb.* **58**, 377–380.

Kolstrup, E. 1980 Climate and stratigraphy in northwestern Europe between 30000 B.P. and 13000 BP., with special reference to the Netherlands. *Meded. Rijks geol. Dienst* **32** (15), 181–253.

Kozarski, S. 1986 Early Vistulian permafrost occurrence in north-west Poland. *Biul. peryglac.* **31**, 163–170.

Kukla, G. J. 1975 Loess stratigraphy of central Europe. In *After the Australopithecines* (ed. K. W. Butzer & G. L. Isaac), pp. 99–188. The Hague: Mouton.

Lagerlund, E. 1983 The Pleistocene stratigraphy of Skane, southern Sweden. In *Glacial deposits in north-west Europe* (ed. J. Ehlers), pp. 155–159. Rotterdam: Balkema.

Lundqvist, J. 1986 Stratigraphy of the central area of the Scandinavian glaciation. *Quat. Sci. Rev.* **5**, 251–268.

Maarleveld, G. C. 1960 Wind directions and cover sands in the Netherlands. *Biul. peryglac.* **8**, 50–58.

Maarleveld, G. C. 1976 Periglacial phenomena and the mean annual temperature during the last glacial time in the Netherlands. *Biul. peryglac.* **26**, 37–78.

Mitchell, G. F. 1977 Periglacial Ireland. *Phil. Trans. R. Soc. Lond.* B **280**, 199–209.

Mitchell, G. F. 1986 *The Shell Guide to reading the Irish landscape.* Dublin: Country House.

Mitchell, G. F., Penny, L. F., Shotton, F. W. & West, R. G. 1973 A correlation of Quaternary deposits in the British Isles. *Geol. Soc. Lond. Spec. Rep.* no. 4, 99 pp.

Mix, A. C. & Ruddiman, W. F. 1984 Oxygen isotope analyses and Pleistocene ice volumes. *Quat. Res.* **21**, 1–20.

Penny, L. F., Coope, G. R. & Catt, J. A. 1969 Age and insect fauna of the Dimlington Silts, East Yorkshire. *Nature, Lond.* **224**, 65–67.

Prentice, I. C. 1986 Vegetation responses to past climatic variation. *Vegetatio* **67**, 131–141.
Rackham, D. J. 1978 Evidence for changing vertebrate communities in the Middle Devensian. *Quat. Newsl.* **25**, 1–3.
Ruddiman, W. F. & MacIntyre, A. 1981 The mode and mechanism of the last deglaciation: oceanic evidence. *Quat. Res.* **16**, 125–134.
Shackleton, N. J. 1969 The last interglacial in the marine and terrestrial records. *Proc. R. Soc. Lond.* B **174**, 135–154.
Shackleton, N. J. 1986 the Plio-Pleistocene ocean: stable isotope history. *Mesozoic and Cenozoic Oceans, Geodynamics Series*, **15**, 141–153.
Shackleton, N. J., Backman, J., Zimmerman, H. *et al.* 1984 Oxygen isotope calibration of the onset of ice rafting and history of glaciation in the North Atlantic region. *Nature, Lond.* **307**, 620–623.
Shackleton, N. J. & Opdyke, N. D. 1973 Oxygen isotope and palaeomagnetic stratigraphy of equatorial Pacific core V28-238: oxygen isotope temperatures and ice volumes on a 10^5 and 10^6 year scale. *Quat. Res.* **3**, 39–55.
Shackleton, N. J. & Opdyke, N. D. 1976 Oxygen isotope and palaeomagnetic stratigraphy of equatorial Pacific core V28-239, late Pliocene to latest Pleistocene. In *Investigation of late Quaternary paleoceanography and paleoclimatology* (ed. R. M. Cline & J. D. Hays), pp. 449–464. Geological Society of America Memoir no. 145.
Simpson, I. M. & West, R. G. 1958 On the stratigraphy and palaeobotany of a late Pleistocene organic deposit at Chelford, Cheshire. *New Phytol.* **57**, 239–250.
Sjørring, S. 1983 The glacial history of Denmark. In *Glacial deposits in north-west Europe* (ed. J. Ehlers), pp. 163–179. Rotterdam: Balkema.
van Staalduinen, C. J. *et al.* 1979 The geology of the Netherlands. *Meded. Rijks geol. Dienst* **31**, 9–49.
Starkel, L. 1977 The palaeogeography of mid- and east Europe during the last cold stage, with west European comparisons. *Phil. Trans. R. Soc. Lond.* B **280**, 351–372.
Stuart, A. J. 1977 The vertebrates of the Last Cold Stage in Britain and Ireland. *Phil. Trans. R. Soc. Lond.* B **280**, 295–312.
Stuart, A. J. 1982 *Pleistocene vertebrates of the British Isles.* London: Longman.
Vandenberghe, J. 1985 Paleoenvironment and stratigraphy during the Last Glacial in the Belgian–Dutch border region. *Quat. Res.* **24**, 23–38.
Watson, E. 1977 The periglacial environment of Great Britain during the Devensian. *Phil. Trans. R. Soc. Lond.* B **280**, 183–198.
Webb, T. 1986 Is vegetation in equilibrium with climate? How to interpret late-Quaternary pollen data. *Vegetatio* **67**, 75–91.
West, R. G. 1980*a* Pleistocene forest history in East Anglia. *New Phytol.* **85**, 571–622.
West, R. G. 1980*b* *The Pre-glacial pleistocene of the Norfolk and Suffolk coasts.* Cambridge University Press.
Williams, R. B. G. 1975 The British climate during the Last Glaciation: an interpretation based on periglacial phenomena. In *Ice ages: ancient and modern* (ed. A. E. Wright & F. Moseley), pp. 95–120. Liverpool: Seel House Press.
Woillard, G. M. 1978 Grande Pile peat bog: a continuous pollen record for the last 140000 years. *Quat. Res.* **9**, 1–21.
Woillard, G. M. & Mook, W. G. 1982 Carbon-14 dates at Grande Pile: correlation of land and sea chronologies. *Science, Wash.* **215**, 159–161.
Zagwijn, W. H. 1957 Vegetation, climate and time-correlations in the Early Pleistocene of Europe. *Geologie Mijnb.* **19**, 233–244.
Zagwijn, W. H. 1961 Vegetation, climate and radiocarbon datings in the Late Pleistocene of the Netherlands. I. Eemian and Early Weichselian. *Meded. geol. Sticht.* N.S. **14**, 15–45.
Zagwijn, W. H. 1974 Vegetation, climate and radiocarbon datings in the Late Pleistocene of the Netherlands. II. Middle Weichselian. *Meded. Rijks geol. Dienst* N.S. **25**, 101–110.
Zagwijn, W. H. 1985 An outline of the Quaternary stratigraphy of the Netherlands. *Geologie Mijnb.* **64**, 17–24.
Zagwijn, W. H. & Paepe, R. 1968 Die stratigraphie der weichselzeitlichen Ablagerungen der Niederlande und Belgiens. *Eiszeitalter Gegenw.* **19**, 129–146.

Discussion

J. Rose (*Department of Geography, Birkbeck College, University of London, U.K.*). One of the problems with the identification of cold stages is the differing levels of sensitivity of the types of evidence used to identify 'cold' conditions and in particular to identify stage boundaries. This is seen specifically in the Cromerian Interglacial Stage, to which Professor West made reference, where plant evidence in zone CrIVc suggests interglacial status, whereas soil evidence for the same episode, in the form of ice-wedge casts, suggested that permafrost was in existence and the environment was more typical of the conditions that are generally attributed to the Anglian Glacial Stage (West 1980*b*). Similar problems exist when sensitive and mobile insect faunas are used as a basis of climatostratigraphy in parallel with less sensitive arboreal plant assemblages, as is well understood for the middle and late Devensian. Should priority or a ranking be given to different types of evidence in order that a degree of consistency can be maintained in subdividing the Quaternary on a climatic basis and deriving generally acceptable stage and substage boundaries?

R. G. West. Of course, at present, permafrost underlies forest over large areas. There will always be a problem in devising boundaries where none exist in nature; the question of definitions will have to await agreement on the most useful and informative method of division. It will have to be realized that any division is artificial and should not obscure the variety of climatic change or overshadow the variety of evidence for such change. The problems revealed by the wealth of evidence for climatic change is one of the attractions of Quaternary research. Such problems must also exist further back in geological timescale, but cannot be revealed by the time resolution available.

P. Coxon (*Department of Geography, Trinity College, Dublin, Ireland*). The terrestrial record of the penultimate interglacial (Gortian–Hoxnian–Holsteinian) from Ireland is known from numerous sites around the country (Watts 1985); one common feature of these sites is that the interglacial cycle is abruptly truncated during its latter part. Recent detailed sampling at one Gortian site at Derrynadivva, County Mayo (P. Coxon, G. Hannon & P. J. Foss, unpublished results) has shown that the interglacial is terminated within a phase of heath (especially *Empetrum*), *Pinus*, *Abies* and *Picea* dominating the vegetation. The upper sedimentological units of the interglacial deposit at Derrynadivva record increased inwashing of inorganic sediment in discrete horizons and re-worked pollen (especially of *Alnus*) becomes increasingly frequent.

The Gortian sites appear to contain a record that is the western equivalent of the Hoxnian Interglacial except that the later stages of the Gortian Interglacial appear to be truncated or missing. Is it possible that the sudden end of the Gortian Interglacial was due to the more rapid onset of climatic deterioration in these western sites that was caused by their proximity to the polar front in the North Atlantic?

Reference

Watts, W. A. 1985 Quaternary vegetation cycles. In *The Quaternary history of Ireland* (ed. K. Edwards & W. P. Warren), pp. 155–185. London: Academic Press.

R. G. West. This idea could be a valuable starting point for realizing the variation of climates at the end of a temperate stage on a transect across Europe.

H. Osmaston (*Department of Geography, University of Bristol, U.K.*). Professor West has contrasted the apparently greater complexity of glacials as compared with interglacials, and Mr Rose has commented on the differing sensitivity of the various indicators we use. I suggest that many of the indicators, from glaciers to beetles, are generally more sensitive to 'equal' changes in the macroclimate (e.g. a change of 1 °C in mean temperature) of cold periods than to those of warm periods. This affects the apparent complexity of such periods and makes it necessary to distinguish between complexity of indicator behaviour and complexity of the underlying factors.

H. H. Lamb (*Climate Research Unit, University of East Anglia, Norwich, U.K.*). If there is anything relevant to this discussion to be learnt from studies of the last thousand years, it surely is that the year-to-year, decade-to-decade, and century-to-century variability increased greatly during the development of the colder phases. This also applies to the spatial variability. The colder phases and their onset are associated with blocking anticyclones, whose occurrence is bound up with the dynamics of the upper westerlies (which they distort and interrupt), and they characteristically turn up in different positions, especially different longitudes, from one year to the next.

This variation suggests that all through the long development of the Pleistocene cold stages, as long as there was no extensive ice-sheet present, there may have been considerable variability in where the coldest climates in Eurasia were found to be developing. But once an extensive ice-sheet had formed, it presumably introduced a greater stability of position. It is probably implied that, because the cold trough in the upper westerlies in the North American–western Atlantic sector would be to a great extent anchored in the lee of the Rocky Mountains–Laurentide ice-sheet massif, the spatial and temporal variability there should have been less than over Eurasia.

Phil. Trans. R. Soc. Lond. B **318**, 523–537 (1988)
Printed in Great Britain

The sedimentary record of climatic variation in the southern North Sea

By D. Long[1], C. Laban[2], H. Streif[3], T. D. J. Cameron[1]
and R. T. E. Schüttenhelm[2]

[1]*British Geological Survey, Murchison House, West Mains Road, Edinburgh EH*9 3*LA, U.K.*
[2]*Rijks geologische Dienst, Spaarne* 17, 2000 *AD Haarlem, The Netherlands*
[3]*Niedersächsisches Landesamt für Bodenforschung, Alfred-Bentz-Haus, Postfach* 51 01 53,
3000 *Hannover* 51, *F.R.G.*

The sedimentary sequence on the shelf of the southern North Sea records Quaternary climatic changes in two ways. They are indicated directly by moraine and glaciofluvial deposits from the Elsterian, Saalian and Weichselian glacial periods when the British and the Scandinavian ice sheets covered parts of the area. An indirect response to the climate is indicated by sea-level changes. Phases of cooling are characterized by regressions and low sea-level stands; phases of warming are indicated by marine transgressions and high sea levels during the Holsteinian, Eemian and Holocene periods. The seismic characteristics of the different lithological units, the sedimentary sequences and their fossil content are described for the offshore area and the adjacent coastal zone. This provides a record of the interaction of sedimentary processes and the palaeogeographic development as a response to climatic changes.

1. Introduction

In the study of Quaternary deposits, the marine environment has a major advantage over terrestrial localities, where the Quaternary record is often fragmentary and is strongly influenced by local geographical factors. Although the deep ocean record is usually complete, it is normally very condensed and gives little opportunity to identify small-scale fluctuations in the environmental history of an area. Such sites also tend to lie a great distance from the major regional factor in sedimentation and control on climatic history, namely the continental glaciations. The Quaternary deposits on the continental shelves are much thicker, provide a more continuous record than those found on land, and are influenced to a much greater extent by nearby glaciations than deep-ocean sediments. In the North Sea they also provide essential evidence for the cross-correlation of the Quaternary stratigraphies already identified on land, between the U.K. and continental northwestern Europe.

The present state of knowledge on the geology of the southern North Sea is highly variable. In the British and Dutch sectors, systematic geological surveys are in progress, and the initial results have been published in several papers and in a set of geological maps at a scale of 1:250000, depicting the solid geology, the Quaternary geology and the sea bed sediments. In the Belgian, German, and Danish sectors there is a less dense grid of observations and systematic investigation of the Quaternary deposits is at an early stage. Nevertheless, the present information can be used to interpret the sedimentary sequences in the southern North Sea with regard to their evidence for Quaternary climatic changes. These changes are indicated in two ways. During the Elsterian (Anglian), Saalian (Wolstonian) and Weichselian (Devensian) glacial periods, the British and the Scandinavian ice sheets covered parts of the

North Sea area and deposited moraine, glaciofluvial, and glaciolacustrine deposits. An indirect response to the climate is indicated by sea-level changes. Phases of cooling are characterized by regressions and low sea-level stands; phases of warming are indicated by marine transgressions and high sea levels during the Holsteinian (Hoxnian) and Eemian (Ipswichian) interglacial periods and the Holocene. Similar marine regressions and transgressions have been interpreted from the early Pleistocene deposits. By the interaction of these processes, complicated patterns and sequences of sedimentary units were formed. The stratigraphy and the suggested correlation of the sedimentary units in the southern North Sea is illustrated in figure 1; the stratigraphic range of the Quaternary facies units in the southern, central and northern North Sea is shown in figure 2.

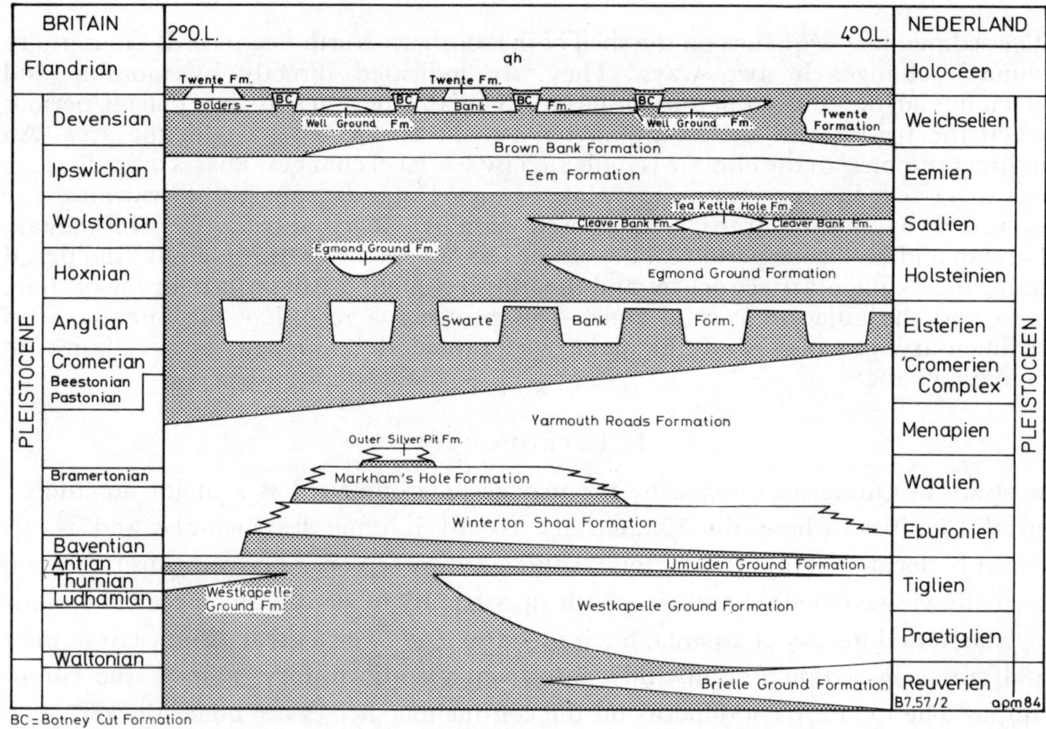

FIGURE 1. Suggested correlation of the Pleistocene formations in the southern North Sea with the Quaternary Stages of Britain and The Netherlands.

The climatic history of the North Sea has been deduced from the interpretation of seismic profiles of the offshore Quaternary deposits and the analyses of many boreholes and sea bed samples. The borehole samples have been subjected to micropalaeontological analyses, relative and absolute age determinations (amino-acid and carbon-14), and microfabric examination. Their geotechnical properties have been tested where possible and argillaceous horizons, where appropriate, have been tested for their primary palaeomagnetic polarity. These analyses were aimed at correlating the seismostratigraphy with borehole evidence from the offshore and onshore areas and at determining the age and depositional environments of the sediments, their subsequent geological history, and to identify periods of non-deposition and possible erosion.

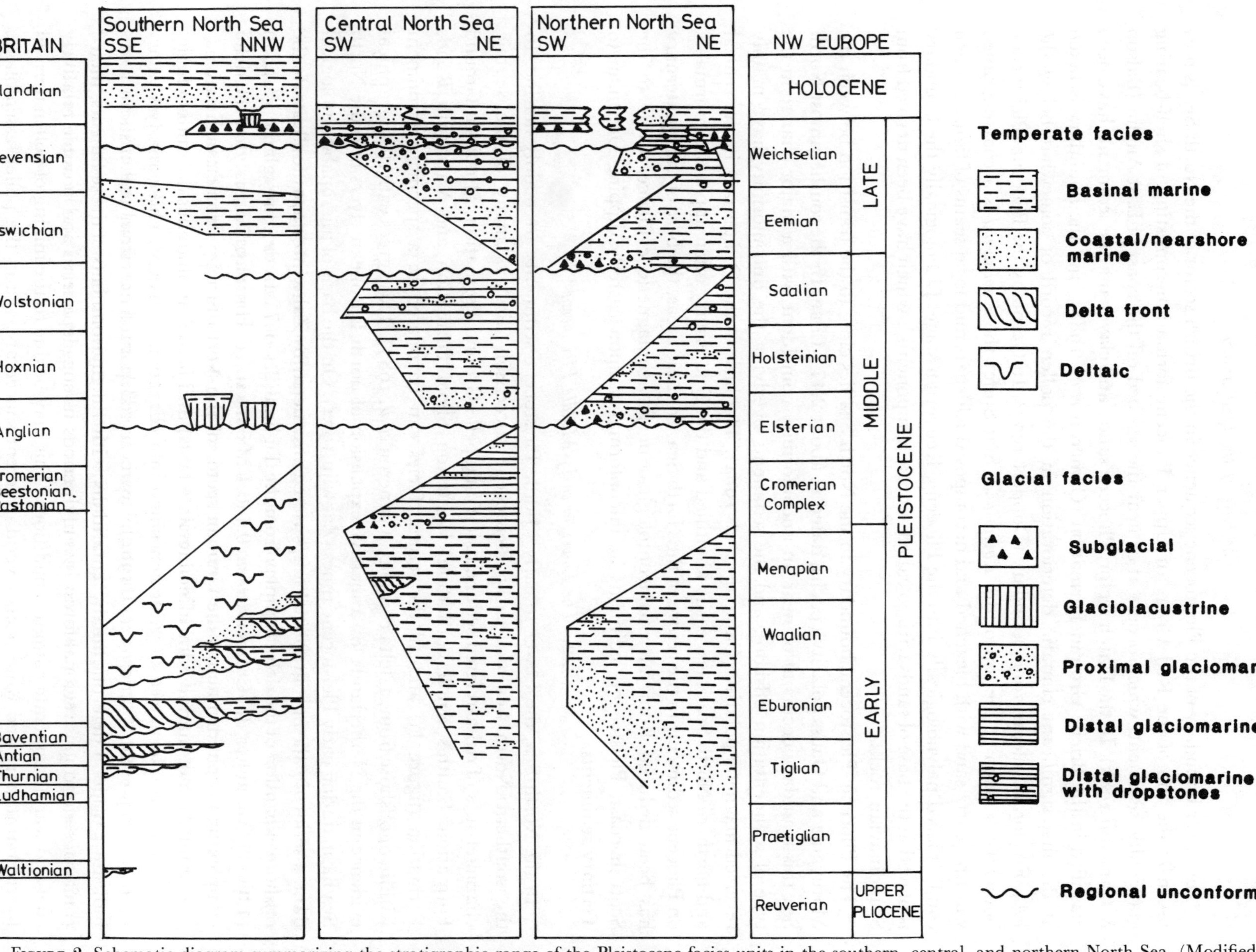

FIGURE 2. Schematic diagram summarizing the stratigraphic range of the Pleistocene facies units in the southern, central, and northern North Sea. (Modified from Cameron *et al.* 1987.)

2. The Quaternary stratigraphy

2.1. *The base of the Quaternary*

The lower boundary of the Pleistocene sequence in the British sector of the North Sea is now taken at the top of the Red Crag, an Upper Pliocene formation consisting of shell-bearing glauconitic sediment, which occurs at or near the sea bed off the coast of East Anglia (Balson & Cameron 1985). In the Dutch sector, Pliocene sandy and clayey marine sediments have been classified in the Brielle Ground Formation (Cameron *et al.* 1988), and the boundary is taken at the first significant climatic deterioration in the pollen record or somewhat below the FA_1–FA_2 foraminiferan zone boundary (Doppert 1975). In Germany the Pliocene–Pleistocene boundary is taken at the top of the *Kaolinsand*, Sylt-Stufe; the *Kaolinsand* is a coarse-grained whitish quartz sand with bleached and decomposed feldspars and local seams of lignite, which can be placed palynologically into the Pliocene, Reuverian stage. Lithologically the boundary is placed at the base of sands with so called 'nordic components', which have been derived from Scandinavian rocks.

The Pliocene–Pleistocene boundary in the North Sea lies close to the transition between the Matuyama and Gauss polarity epochs, dated at about 2.47 Ma BP. In the southernmost North Sea this boundary occurs at or near an unconformity coincident with a major change in the type of sedimentation offshore, and the sediments overlying the unconformity have mainly reversed magnetic polarity (Cameron *et al.* 1984). The base of the Quaternary in the central and northern North Sea is less easily identified and the Quaternary often appears conformable on Pliocene sediments in the centre of the North Sea. In these areas, the base of the Quaternary has been deduced from micropalaeontological or palaeomagnetic evidence. Close to the Scottish coast, Pleistocene sediments lie unconformably upon easterly dipping Devonian to Tertiary sediments.

2.2. *The Lower to early Middle Pleistocene*

In the North Sea, the Lower to early Middle Pleistocene sediments are mainly marine. In the southern North Sea the early Pleistocene succession displays many of the seismic characteristics of delta-related sediments (Westkapelle Ground Formation, IJmuiden Ground Formation, Smith's Knoll Formation, Winterton Shoal Formation and Yarmouth Roads Formation (figure 1)) and all the seismic facies can be related to a specific environment within, or offshore from, a delta complex (Cameron *et al.* 1987). This agrees with stratigraphical evidence in the Netherlands for a massive expansion of north European rivers into the North Sea Basin during early Pleistocene times (Zagwijn 1974). On the basis of micropalaeontological data, a series of palaeogeographical maps were constructed by Zagwijn (1979) to illustrate the offshore expansion of the delta complex from the Tiglian (2.1–1.7 Ma BP) through the Waalian (1.3–0.95 Ma BP) to the Cromerian (*ca.* 0.7–0.4 Ma BP) stage. His maps suggest that the delta complex had expanded across the German sector of the North Sea by Cromerian times.

Examination of marine flora (dinoflagellate cysts) and fauna (foraminifera) from boreholes in the southern North Sea suggests deposition in a climate as warm as, and possibly warmer than, that at present (Cameron *et al.* 1984). However, redeposited terrestrial pollen assemblages indicate a wider range of climatic conditions with fluctuations between boreal and mixed coniferous–deciduous regional forest cover; these results suggest alternation between relatively cool and warm temperate climatic conditions. This may be due to the mixing of pollen derived by river transport and wind from several different source areas, and the pollen assemblages may have been further complicated by resedimentation offshore. The Lower Pleistocene succession appears to represent a series of marine transgressions and regressions, perhaps recording the effects of eustatic sea-level changes on sedimentation on the periphery of the North Sea Basin, with periods of marine regression corresponding to periods of maximum climatic deterioration. However, differential isostatic movements between the North Sea Basin and peripheral areas may also have caused geographic and quantitative changes in sediment supply by various west-flowing rivers. Periods of maximum climatic deterioration may be expected to correspond with regional lowering of sea level and a transfer of marine sedimentation to the outer margins of the continental shelf. This may account for the absence of cool-temperature influences on the marine fossil assemblages in the southern North Sea.

The central North Sea deposits also contain evidence for deltaic sedimentation in the early Pleistocene (Stoker & Bent 1988), but include a greater proportion of prodelta and fully marine sediments and are believed to represent a more complete stratigraphic sequence than that preserved in the southern North Sea. Stoker & Bent (1988) suggest that major deltas of the North European Plain eventually coalesced with the much smaller local deltas of eastern Britain, most notably of an ancestral Tay–Forth estuarine system. The fossil assemblages contain evidence for a more varied climate here than in the southern North Sea, although, in general, warm temperate climatic conditions prevailed during deposition of the sedimentary sequence (Stoker *et al.* 1985); warm temperate deposits of Tiglian age have also been recorded from the Danish sector (Bertelsen 1972).

2.3. *The late Middle and Upper Pleistocene*

The Lower and early Middle Pleistocene deposits of the North Sea Basin are truncated by a regional unconformity of Elsterian age. Above the Middle Pleistocene Brunhes–Matuyama magnetic reversal, there is sedimentological and marine floral and faunal evidence in the central and northern North Sea for a dramatic deterioration in climatic conditions before the major seismic unconformity evident throughout the U.K. sector (the tops of the Yarmouth Roads, Aberdeen Ground and Shackleton Formations).

Sedimentation in the North Sea Basin has been dominated by glacial erosional and depositional processes since mid-Quaternary times. The base of the Upper Pleistocene succession in the Southern Bight is a gently undulating erosional unconformity less than 40 m below sea bed (less than 80 m below present sea level). Three regional glaciations affected northwest Europe to a variable extent during the Elsterian (Anglian), Saalian (Wolstonian) and Weichselian (Devensian) glacial stages, when ice sheets covered much of Britain, Scandinavia, Denmark, northern Germany and the Netherlands. Likewise, three major glacial episodes have been identified within the Middle and Upper Pleistocene sediments of the North Sea. The landforms and deposits associated with each episode are broadly similar but are separated by interglacial marine sediments.

2.3.1. *Elsterian glacial period*

In the southern North Sea, the earliest glacial conditions are represented by local ice-push deformation of the Lower to early Middle Pleistocene deposits in the Brown Bank area, and at the eastern margin of the Flemish Bight sheet in the Netherlands sector of the North Sea (Laban *et al.* 1984; Cameron *et al.* 1988). These features demonstrate that the margin of the Elsterian ice sheet extended southwards to at least 52° 20′ N. The Anglian tills cover much of

East Anglia and extend as far south as Ipswich (52° N). However, the Elsterian (Anglian) tills of East Anglia do not extend east of the present British coastline or have been eroded since Elsterian times.

In north Germany and in the Netherlands, a complex system of anastomosing subglacial valleys was eroded and infilled during the Elsterian glaciation (Kuster & Meyer 1979; Ehlers *et al.* 1984); similar valleys are eroded into early Middle and Lower Pleistocene and Tertiary sediments in the British and Dutch sectors of the North Sea, mainly to the north of 53° N. These valleys are typically aligned north–south or, in the UK sector, NNW–SSE. They are a few kilometres wide, and the largest has a maximum depth of 450 m (510 m below sea level) at 53° 16′ N and 3° E. All the valleys have a conspicuously undulating *thalweg*. The infill of these valleys, the Swarte Bank Formation, comprises a complex sequence of sediments. The basal unit, which is structureless or has a chaotic reflector configuration, has zones of reduced acoustic penetration and may consist of gravelly coarse sand, slump beds or perhaps, locally, till deposited penecontemporaneously with valley incision. The second well-bedded unit contains rare subparallel, high-amplitude reflectors at similar intervals and depth in adjacent valleys; this observation suggests that some of the valleys formed an interconnecting network of depressions during the stage of filling. This layered unit consists of glaciolacustrine clay with beds of silty clay and fine grained sand. The third unit of infill is mostly less than 30 m thick and is characterized by gently inclined, low- or moderate-amplitude reflections on the seismic profiles. This unit may represent a delta-related influx of fine-grained sands and clays into a glaciolacustrine environment comparable to the *potklei* in the Netherlands and the Lauenburg Clay of northwestern Germany.

It is not easy to interpret the morphology of these Elsterian channels in the terms of fluvial erosion or of scouring by the base of an ice sheet. It has been assumed by Cameron *et al.* (1987) that the features were most likely to have been eroded subglacially, under very high hydrostatic pressure by outbursts of meltwater beneath a continuous cover of land ice. The filling processes started under the cover of the ice, with slumping of moraine and the deposition of glaciofluvial sediments. After the ice had retreated from the area, thick accumulations of glaciolacustrine sediments, locally overlain by fluvial deposits, infilled most of the Elsterian valleys. Similar valleys and sedimentary sequences are well known from the onshore area of the Netherlands and of Lower Saxony, Germany, where they have been systematically mapped in the course of the search for groundwater resources (Kuster & Meyer 1979; Ehlers *et al.* 1984). The valley systems and the infill record the earliest extension of the inland ice into both the southern North Sea and its hinterland. Together with the ice-pushed deposits, they demonstrate that there was a continuous cover of inland ice from eastern England across the Southern Bight to the Netherlands and to Germany during Elsterian times (figure 3).

Elsewhere, beyond the offshore extent of land ice, smaller-scale valleys in the central and northern North Sea and the English Channel may have been formed by catastrophic proglacial fluvial events (Smith 1985; Long & Stoker 1986*a*) and are filled by postglacial marine deposits and by glaciomarine sediments from the succeeding glaciation.

2.3.2. *Holsteinian interglacial period*

By the onset of the Holsteinian (Hoxnian) interglacial period, most of the Elsterian valleys of the southern North Sea had been totally filled by non-marine sediments. The few remaining depressions contain an additional component of marine sediments from the Holsteinian transgression, which extended across most of the North Sea area and locally smoothed the pre-existing relief.

Holsteinian marine deposits are rarely preserved in the southern U.K. sector of the North Sea (Cameron *et al.* 1987), but borehole BH81/52A recovered 14 m of shallow-water marine clay of interglacial aspect, believed to be Holsteinian and comparable with Hoxnian dated deposits (Egmond Ground Formation) cropping out in Silver Pit (Balson & Cameron 1985). Further east, in the Dutch sector, Holsteinian interglacial deposits (Egmond Ground Formation) are represented by sparsely shelly very fine-, fine- and medium-grained sands with thin beds of silt and clay (Laban *et al.* 1984). Outside the former valleys these deposits are less than 25 m thick.

Examination of the molluscs and microfauna of the Holsteinian sediments in the Danish North Sea sector and in the German onshore area (Grahle 1939; Woszidlow 1962; Lange 1962; Knudsen 1980) as well as palynological studies (Menke 1970; Müller 1974) indicate an increasing amelioration of the water temperature during this interglacial period. The fauna and flora indicate that the arctic climate of the early Holsteinian gave way to boreal conditions and eventually to a temperate climate with water temperatures in the southern North Sea similar to those of the present day.

Electron spin resonance (ESR) spectroscopy of 27 samples of aragonitic mollusc shells from the Hamburg area, from the lower Elbe river valley, and from Cuxhaven on the German North Sea coast (Linke *et al.* 1985) have yielded three groups of ages for the different localities with average ages of 195 ± 25 ka BP, 223 ± 25 ka BP and 218 ± 25 ka BP; these ages indicate correlation with stage 7 of the oxygen-isotope deep-sea record (Shackleton & Opdyke 1973). On the basis of three uranium–thorium dates, selected from a total of seven dates, Sarnthein *et al.* (1986) concluded that the Holsteinian deposits at Wacken, Schleswig–Holstein, are older than 350–370 ka. Therefore, these authors correlate the Holsteinian interglacial period with $\delta^{18}O$ stage 11 on the CARTUNE and SPECMAP scale (Herterich & Sarnthein 1984). Also outside Germany there is no agreement on a correlation between dated Holsteinian (Hoxnian) deposits and the deep-sea record.

On the margins of the southern North Sea, Holsteinian marine deposits occur in marine terraces 25 m above sea level in East Anglia (Mitchell 1977), at an elevation of 10–12 m near Sangatte on the French coast (Sommé 1979) and in Belgium (Paepe & Baeteman 1979). In The Netherlands, the surface of marine Holsteinian deposits occurs at -25 to -40 m (van Staalduinen 1977) and in northern Germany they fill a '*Förde*'-like system to an elevation of *ca.* -20 to -30 m (Ehlers *et al.* 1984; Linke 1986). The difference in elevation can perhaps be attributed to post-Holsteinian isostatic processes.

2.3.3. *Saalian glacial period*

At the maximum extent of the Saalian glaciation, an extensive ice-sheet covered the entire lowland area of Lower Saxony, Germany (Meyer 1983) and parts of the adjacent mountainous regions to the south. The ice-thrusted ridges of Itterbeck-Uelsen, Emsland, and of Krefeld–Kleve at the Lower Rhine River, mark the southern limit of the ice sheet in northwestern Germany (figure 3). In The Netherlands, this ice advance formed the ice-pushed ridges of Nijmegen, Arnhem and Rhenen and formed deep glacial basins in the Amsterdam–Haarlem area. In the Dutch North Sea sector, till has been found up to 40 km off the coast on the Terschelling Bank sheet, and an ice-marginal valley and ice-pushed structures have been

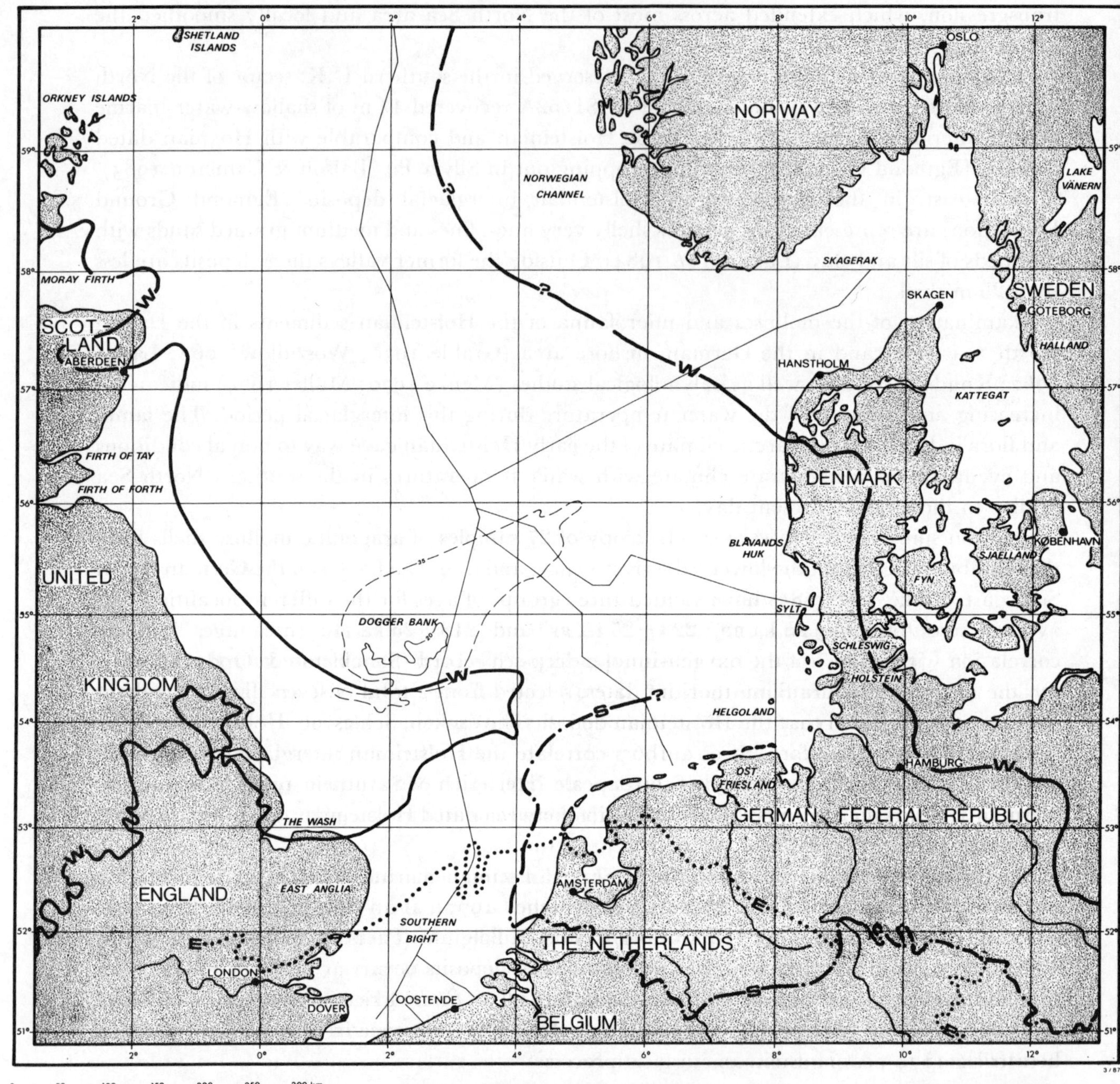

FIGURE 3. Palaeogeographic reconstruction of the ice margins of the Elsterian (Anglian), Saalian (Wolstonian), and Weichselian (Devensian) glacial periods in the North Sea and in the adjacent onshore areas.

detected in the northwestern Flemish Bight sheet (Laban *et al.* 1984). The available evidence suggests that the Saalian ice has not extended west of 4° E into the Indefatigable or Silver Well sheets (figure 3). Saalian proglacial deposits (Cleaver Bank Formation) and periglacial deposits (Tea Kettle Hole Formation) occur on the eastern halves of both sheets (Cameron *et al.* 1986; Jeffery *et al.* 1988).

The Tea Kettle Hole Formation consists of well-sorted, very fine- or fine-grained, and only sparsely micaceous wind-blown sands up to 6 m thick. The Cleaver Bank Formation consists of proglacial silty clays with silt and sand laminae and locally (especially on Indefatigable) fluvioglacial, very fine- to fine-grained, micaceous outwash sands with beds of silt and clay, which were deposited beyond the ice front.

Opposing views have been expressed by Straw (1983), Sumbler (1983), and Rose (1987) as to whether a contemporary ice sheet covered eastern England. Till-like sediments of Saalian age have neither been sampled in the U.K. sector of the North Sea nor in the western half of the Dutch sector. The offshore evidence in the Dutch sector indicates that periglacial conditions prevailed over most of the southern North Sea during this period. Therefore, it seems likely that there was no connection between the Scandinavian and possible Scottish ice sheets across the Central or northern North Sea during the Saalian glaciation (figure 3).

2.3.4. *Eemian interglacial period*

Marine sediments of the Eem Formation (115–125 ka BP) have been identified at many sites in the southern North Sea. In the British sector, the Eem Formation comprises up to 20 m of intertidal and shallow marine sands and clays. The sediments become more fully marine towards the east; in the Dutch sector, the formation is composed of fine- or medium-grained, locally gravelly sands with a markedly higher shell content than the deposits of the Holsteinian and Holocene transgressions (Cameron *et al.* 1988). Fully marine Eemian deposits occur as the basal fill of some of the Saalian valleys in the central and northern North Sea.

No marine beds of the E1 and E2 pollen zones have so far been recorded from the southern North Sea (Jelgersma 1979), but marine deposition has been demonstrated for the E3 to E6 pollen zones by Oele & Schüttenhelm (1979), and offshore from Sylt and Borkum on the German North Sea coast (Ludwig *et al.* 1981). The investigations of the molluscan and microfaunal assemblages (Lafrenz 1963; Hinsch 1985; Knudsen 1985*a*, *b*) and of the pollen assemblages (Menke 1985) of the deposits have indicated climatic amelioration during Eemian transgression from subarctic or high boreal conditions towards temperatures which were at least as warm as they are at present in the southern North Sea.

In eastern Britain, marine deposits of the Eemian interglacial *ca.* 7 m above present sea level occur in the inner part of the Thames estuary and around the Wash, and at +2 m in the seaward part of the Humber estuary and in East Anglia (Mitchell 1977; Jardine 1979). In Belgium they occur at an average depth of −5 m below the western coastal plain (Paepe & Baeteman 1979). The highest elevation of the surface of marine Eemian deposits is at −8 m in the western part of the Netherlands and at −13 m in the Groningen area (Jelgersma 1979; Roeleveld 1974). Along the German North Sea coast the surface lies at −9 to −7 m in Lower Saxony and at −5 m in western Schleswig–Holstein. The Eemian coastline intersects with the present North Sea coast at Blåvands Huk, Denmark. The range of elevations of marine Eemian deposits shows a similar pattern to the Holsteinian units; this observation suggests that both have been influenced by post-Eemian isostatic processes.

2.3.5. *Weichselian glacial period*

Climatic deterioration towards fully glacial conditions caused regional sea level in the North Sea to fall to at least 110 m below present by late Weichselian times (Jansen *et al.* 1979). Local effects of glacial depression and rebound have not been studied so far. The uppermost Eemian sediments in the Southern Bight of the North Sea are widely overlain by grey-brown, brackish-marine to freshwater silty clays with silt and very fine sand laminae (Brown Bank Formation). The major part of these sediments has been deposited in a lagoonal environment during the early stages of regression. The deposits are extensively bioturbated and locally cryoturbated (Cameron *et al.* 1988). According to pollen analyses, sedimentation occurred in the Dutch pollen zone EW 1A, which ranges from Late Eemian to Early Weichselian times.

In the Dutch sector, the brackish-marine beds are overlain by lacustrine clays, deposited as the sea fell to below −40 m in the area. Glaciomarine clays, deposited in front of the ice in the Dogger Bank area, are now at −55 m (Jeffery *et al.* 1988). Probably there existed at one time a restricted and elongated marine embayment over a glacially depressed part of the North Sea just in front of the British Weichselian land ice. This embayment extended from the northern North Sea as far south as SE of the Dogger Bank.

A blanket deposit of Weichselian till (Bolders Bank Formation) extends northeastwards from the coast of East Anglia into the northwestern part of Indefatigable sheet (Cameron *et al.* 1988) and contains boulders derived mainly from the Upper Palaeozoic and Mesozoic rocks of eastern England (figure 3). Glaciolacustrine clays and diamictons occur below and east of the southern Dogger Bank (Cameron *et al.* 1986; Jeffrey *et al.* 1988). Along the eastern coast of the North Sea, the Weichselian end-moraines run through eastern Schleswig-Holstein, Germany, and southern Jutland, in a north–south direction. At the Limfjord, Denmark, the moraines turn to the west and trace into the North Sea; their offshore continuation is uncertain (figure 3). Offshore evidence, however, suggests that there was no connection between the British and the Scandinavian ice sheets across the southern North Sea during the Weichselian glacial period (Cameron *et al.* 1988).

South of the maximum extent of the ice sheet, terrestrial and fluvial sedimentation took place in vast tracts of the southern North Sea (Cameron *et al.* 1988). Aeolian fine-grained sands (Twente Formation) are preserved locally on the Flemish Bight and Indefatigable sheets, and in the German sector of the North Sea. Fluvioglacial micaceous, very fine- or fine-grained Weichselian sands with intercalations of silt and clay occur adjacent to the till margin on Indefatigable sheet. Late Weichselian subglacial valleys in the north of Indefatigable have been partly or completely infilled by poorly sorted, gravelly coarse sands overlain by very soft sandy lacustrine clay.

During the Weichselian the rivers discharged into the central North Sea. The most important drainage system of the north German hinterland was the so-called 'Elbe–Urstromtal', which acted as an ice-marginal valley during the Weichselian glacial maximum. Figge (1980) has mapped the course of this 30–40 km wide valley system on the basis of Boomer profiling between Helgoland and the White Bank area. During the Weichselian late-glacial period, the central North Sea, comprised a shallow brackish marine sea, with a cover of sea ice, surrounded by flat periglacial tundra dissected by braided rivers downstream of the Middle and North European rivers (Stoker & Long 1984). As the temperature and sea level rose, sea-ice keels became smaller and the sediments at the bottom of the shallow seas ceased to be reworked; this resulted in the deposition of acoustically well-layered formations. Dinoflagellate cyst assemblages indicate that sea ice was absent from this area during the Bølling interstadial (Long *et al.* 1986), during which there was an influx of warmer Atlantic waters. Low salinity values (based on foraminiferal and molluscan assemblages) and evidence of rapid sedimentation together with possible water stratification suggest significant input continued from meltwater sources. During the Younger Dryas, the connection with the North Atlantic remained open and there was only limited pack-ice. These processes are summarized in figure 4. The onset

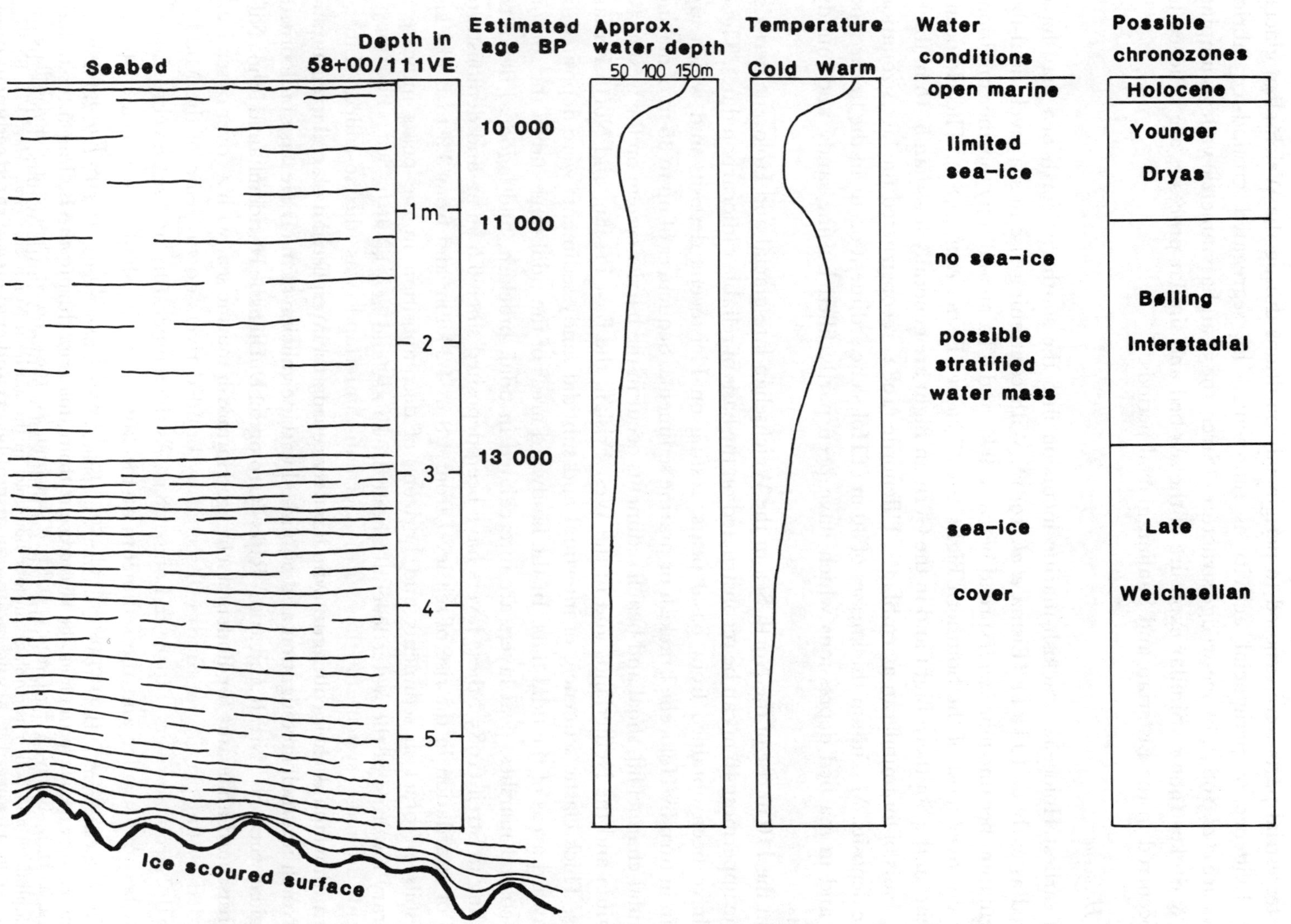

FIGURE 4. Estimated environmental conditions in the central North Sea for the Late Weichselian to Holocene periods, based on vibrocore 58+00/111VE from the Witch Ground Basin, with schematic seismic section. (Based on Long *et al.* 1986.)

of the Holocene is indicated by the sudden rise in water depths and the near cessation of sedimentation in the central North Sea.

The terrestrial parts were exposed to periglacial conditions during the Weichselian glacial period. Evidence of periglacial activity is represented by segregated ground-ice fabrics (Derbyshire *et al.* 1985), by ice-wedge structures (Streif 1985) and asymmetric valley slumping (Long & Stoker 1986*b*). Similar exposure of the sea bed and similar processes are believed to have occurred in the Elsterian and Saalian glacial periods.

2.3.6. *Holocene*

The earliest Holocene brackish-marine incursion into the southern North Sea may have occurred as early as 10 ka BP (Eisma *et al.* 1981). With continuing rise in sea level, tidal-flat sedimentation became more widespread between 9 ka and 8 ka BP and fully marine conditions spread out over most of the Southern Bight after 7 ka BP (Eisma *et al.* 1981). The Holocene sediments in the Southern Bight and in the German Bight are generally less than 5–15 m thick. On the basis of foraminiferan assemblages, Uffenorde (1982) reconstructed the facies zonation of these deposits. Maximum thicknesses of 30 m of Holocene sediment occur in the linear sand ridges and in sea bed depressions which have been partly filled by fine sandy and muddy deposits.

From the 110 m rise of the North Sea in the Weichselian late-glacial and Holocene periods, only the uppermost 46 m can be reconstructed on the basis of reliable radiocarbon dates. These dates have been obtained from basal peats, resting on Pleistocene deposits and which are overlain in turn by Holocene brackish or marine sediments. Sequences of up to 35 m of marine sand, tidal-channel fill, shoal and beach sediments occur in the barrier system on the west coast of Belgium and the Netherlands and in the West Frisian, the East Frisian, and North Frisian Islands. Thick clastic sequences of intertidal and subtidal sandy sediments were deposited in the seaward areas of the tidal flats. In the landward areas of the tidal flats and in the subsoil of the coastal marshes, peat layers are intercalated in tidal, brackish, and lagoonal deposits. According to Streif (1985), these layers have been deposited since 6.5 ka BP and demonstrate phases of retardation in the rise of sea level from 4.8 to 4.2 ka BP and from 3.3 to 2.3 ka BP. Fossil soils on brackish sediments and horizons of decomposition in the peats indicate a temporary lowering of the water level at about 2.7 ka BP, and at 2 ka BP.

The authors thank their colleagues who have retrieved and interpreted data collected as part of the North Sea shelf investigation and published with permission of the Directors of the British Geological Survey (N.E.R.C.), the Rijks Geologische Dienst, Haarlem, and the Niedersächsisches Landesamt für Bodenforschung, Hannover.

References

Balson, P. S. & Cameron, T. D. J. 1985 Quaternary mapping offshore East Anglia. *Mod. Geol.* **9**, 221–239.

Bertelsen, F. 1972 *Azolla* species from the Pleistocene of the central North Sea. *Grana* **12**, 131–145.

Cameron, T. D. J., Bonny, A. P., Gregory, D. M. & Harland, R. 1984 Lower Pleistocene dinoflagellate cyst, foraminiferal and pollen assemblages in four boreholes in the southern North Sea. *Geol. Mag.* **121**, 85–97.

Cameron, T. D. J., Laban, C. & Schüttenhelm, R. T. E. 1984 Flemish Bight sheet. 52° N–02° E. Quaternary Geology (Geologie van het Kwartair) BGS/RGD 1:250000 map series.

Cameron, T. D. J., Laban, C., Mesdag, C. S. & Schüttenhelm, R. T. E. 1986 Indefatigable sheet. 53° N–02° E. Quaternary Geology (Geologie van het Kwartair) BGS/RGD 1:250000 map series.

Cameron, T. D. J., Schüttenhelm, R. T. E. & Laban, C. 1988 Middle and Upper Pleistocene and Holocene stratigraphy in the Southern Bight of the North Sea. In *The Quaternary and Tertiary geology of the Southern Bight, North Sea* (*Int. Colloquy, Ghent, 1984*) (ed. J. P. Henriet & G. de Moor). Belgische Geologische Dienst. (In the press.)

Cameron, T. D. J., Stoker, M. S. & Long, D. 1987 The history of Quaternary sedimentation in the UK sector of the North Sea Basin. *J. geol. Soc. Lond.* **144**, 43–58.

Derbyshire, E., Love, M. A. & Edge, M. J. 1985 Fabrics of probable segregated ground-ice origin in some sediment cores from the North Sea Basin. In *Soils and Quaternary landscape evolution* (ed. J. Broadman), pp. 261–280. Chichester: J. Wiley & Sons.

Doppert, J. W. C. 1975 Foraminiferenzonering van het nederlanse Onderkwartair en Tertiair. In *Toelichting bij geologische overzichtskaarten van Nederland*, pp. 114–118. Haarlem: Rijks Geologische Dienst.

Ehlers, J., Meyer, K.-D. & Stephan, H. J. 1984 Pre-Weichselian glaciations of North-West Europe. *Quat. Sci. Rev.* **3** (1), 1–40.

Eisma, D., Mook, W. & Laban, C. 1981 An early Holocene tidal flat in the Southern Bight. In *Holocene marine sedimentation in the North Sea Basin.*, Special Publication of the International Association of Sedimentologists (ed. X, Nio, R. T. E. Schüttenhelm & van Weering), vol. 5, pp. 229–237.

Figge, K. 1980 Das Elbe-Urstromtal im Bereich der Deutschen Bucht (Nordsee). *Eiszeitalter Gegenw.* **30**, 203–211.

Grahle, O. 1936 Die Ablagerungen der Holstein-See (Mar. Interglaz. I), ihre Verbreitung, Fossilführung und Schichtenfolge in Schleswig-Holstein. *Abh. preuss. geol. Landesanst.* **172**, 1–110.

Herterich, K. & Sarnthein, M. 1984 Brunhes time scale: Tuning by rates of calcium carbonate dissolution and cross spectral analyses with solar insolation. In *Milankowitch and climate* (ed. A. Berger & J. Imbrie), part 1, pp. 446–466. Dordrecht: Reidel.

Hinsch, W. 1985 Die Molluskenfauna des Eem-Interglazials von Oldenbüttel-Schnittlohe (Nord-Ostsee-Kanal, Westholstein). *Geol. Jb.* A **86**, 49–62.

Jansen, J. H. F., van Weering, T. C. E. & Eisma, D. 1979 Late Quaternary sedimentation in the North Sea. In *The Quaternary history of the North Sea* (ed. E. Oele, R. T. E. Schüttenhelm & A. J. Wiggers) *Acta univ. ups. symp. univ. ups. a. quingent. celebr.*, vol. 2, pp. 175–187.

Jardine, G. W. 1979 The western (United Kingdom) shore of the North Sea in Late Pleistocene and Holocene times. In *The Quaternary history of the North Sea* (ed. E. Oele, R. T. E. Schüttenhelm & A. J. Wiggers) *Acta univ. ups. symp. univ. ups. a. quingent. celebr.*, vol. 2, pp. 156–174.

Jeffery, D. H., Laban, C., Mesdag, C. S. & Schüttenhelm, R. T. E. 1988 Silver Well Sheet. 54° N–02° E. Quaternary Geology (Geologie van het Kwartair) BGS/RGD 1:250000 map series.

Jelgersma, S. 1979 Sea-level changes in the North Sea basin. In *The Quaternary history of the Noth Sea* (ed. E. Oele, R. T. E. Schüttenhelm & A. J. Wiggers) *Acta univ. ups. symp. univ. ups. a. quingent. celebr.* **2**, 233–248.

Jelgersma, S., Oele, E. & Wiggers, A. J. 1979 Depositional history and coastal development in the Netherlands and the adjacent North Sea since the Eemian. In *The Quaternary history of the North Sea* (ed. E. Oele, R. T. E. Schüttenhelm & A. J. Wiggers) *Acta univ. ups. symp. univ. ups. a. quingent. celebr.*, vol. 2, pp. 115–142.

Knudsen, K. L. 1980 Foraminiferal faunas in Holsteinian Interglacial deposits of Hamburg-Hummelsbüttel. *Mitt. geol. paläont. Inst. Univ. Hamb.* **49**, 193–214.

Knudsen, K. L. 1985*a* Foraminiferal stratigraphy of Quaternary deposits in the Roar, Skjold and Dan fields, central North Sea. *Boreas* **14**, 311–324.

Knudsen, K. L. 1985*b* Foraminiferal faunas in Eemian deposits of the Oldenbüttel area near Kiel Canal, Germany. *Geol. Jb.* A **86**, 27–47.

Kuster, H. & Meyer, K.-D. 1979 Glaziäre Rinnen in mittleren und nordöstlichen Niedersachsen. *Eiszeitalter Gegenw.* **29**, 135–156.

Laban, C., Cameron, T. D. J. & Schüttenhelm, R. T. E. 1984 Geologie van het Kwartair in de Zuidelijke Bocht van de Noordzee. In *Mededelingen Werkgroep Tertiaire en Kwartaire Geologie*, **21**, 139–154.

Lafrenz, H. R. 1963 Foraminiferen aus dem marinen Riss–Würm Interglazial (Eem) in Schleswig-Holstein. *Meyniana* **13**, 10–45.

Lange, W. 1962 Die Mikrofauna einiger Störmeer-Absätze (I-Interglazial) Schleswig-Holsteins. *Neues Jb. Geol. Paläont. Abh.* **115**, 222–242.

Linke, G. (ed.) 1986 *Guidebook to the excursions of September 22, 23 and 26, 1986 Holstein-Symposium.* (89 pages.) Hamburg.

Linke, G., Katzenberger, O. & Grün, R. 1985 Description and ESR dating of the Holsteinian Interglaciation. *Quat. Sci. Rev.* **4**, 319–331.

Long, D., Bent, A., Harland, R., Gregory, D. M., Graham, D. K. & Morton, A. C. 1986 Late Quaternary palaeontology, sedimentology, and geochemistry of a vibrocore from the Witch Ground Basin, central North Sea. *Mar. Geol.* **73**, 109–123.

Long, D. & Stoker, M. S. 1986*a* Channels in the North Sea: the nature of a hazard. In *Advances in underwater technology, Ocean Science and Offshore Engineering*, vol. 6 (*Oceanology*), pp. 339–351. (*Proc. Oceanology International 1986*) Brighton, U.K.

Long, D. & Stoker, M. S. 1986*b* Valley asymmetry: evidence for periglacial activity in the central North Sea. *Earth Surf. Process. Landforms* **11**, 525–532.
Ludwig, G., Müller, H. & Streif, H. 1981 New dates on Holocene sea-level changes in the German Bight. *Holocene Marine Sedimentation in the North Sea Basin* (ed. S. D. Nio, R. T. E. Schüttenhelm & T. C. E. van Weering). *Int. Ass. Sedimentol. Spec. Publ.* **5**, 211–219.
Menke, B. 1970 Ergebnisse der Pollenanalyse zur Pleistozänstratigraphie und zur Plio-Pleistozän-Grenze in Schleswig-Holstein. *Eiszeitalter Gegenw.* **21**, 5–21.
Menke, B. 1985 Palynologische Untersuchungen zur Transgression des Eem-Meeres in Raum Offenbüttel/Nord-Ostsee-Kanal. *Geol. Jb.* A**86**, 19–26.
Meyer, K.-D. 1983 Saalian end moraines in Lower Saxony. In: *Glacial deposits in north-west Europe* (ed. J. Ehlers), pp. 335–342. Rotterdam: Balkema.
Mitchell, G. F. 1977 Raised beaches and sea-levels. In *British Quaternary studies: recent advances* (ed. F. W. Shotton), pp. 169–186. Oxford: Clarendon Press.
Müller, H. 1974 Pollenanalytische Untersuchungen und Jahresschichtenzählungen an der eem-zeitlichen Kieselgur von Hetendorf. *Geol. Jb.* A**21**, 87–105.
Oele, E. & Schüttenhelm, R. T. E. 1979 Development of the North Sea after the Saalian glaciation. In *The Quaternary history of the North Sea* (ed. E. Oele, R. T. E. Schüttenhelm & A. J. Wiggers). *Acta univ. ups. symp. univ. ups. a. quingent. celebr.*, vol. 2, pp. 191–215.
Paepe, R. & Baeteman, C. 1979 The Belgian coastal plain during the Quaternary. In *The Quaternary history of the North Sea* (ed. E. Oele, R. T. E. Schüttenhelm & A. J. Wiggers). *Acta univ. ups. symp. univ. ups. a. quingent. celebr.*, vol. 2, pp. 143–146.
Roeleveld, W. 1974 The Groningen coastal area. Thesis, Free University of Amsterdam. (252 pages.)
Rose, J. 1987 Status of the Wolstonian glaciation in the British Quaternary. *Quaternary Newsletter* no. 53, 1–9.
Smith, A. J. 1985 A catastrophic origin for palaeovalley systems of the eastern English Channel. *Mar. Geol.* **64**, 65–75.
Sommé, J. 1979 Quaternary coastlines in northern France. In *The Quaternary history of the North Sea* (ed. E. Oele, R. T. E. Schüttenhelm & A. J. Wiggers). *Acta univ. ups. symp. univ. ups. a. quingent. celebr.*, vol. 2, pp. 147–158.
Staalduinen, C. J. (ed.) 1977 *Geologisch onderzoek van het Nederlandse Waddengebied.* (77 pages.) Haarlem: Rijks Geologische Dienst.
Stoker, M. S. & Bent, A. 1988 Lower Pleistocene deltaic and marine sedimentation in the UK sector of the central North Sea. *J. quat. Sci.* (In the press.)
Stoker, M. S. & Long, D. 1984 A relict ice-scoured erosion surface in the central North Sea. *Mar. Geol.* **61**, 85–93.
Stoker, M. S., Long, D. & Fyfe, J. A. 1985 A revised Quaternary stratigraphy for the central North Sea. *Rep. Br. geol. Surv.* 17/2, pp. 1–35.
Straw, A. 1983 Pre-Devensian glaciation of Lincolnshire (Eastern England) and adjacent areas. *Quat. Sci. Rev.* **2**, 239–260.
Streif, H. 1985 Southern North Sea during the Ice Ages – inundations and ice-cap movements. In *German research: reports of the DFG 1985*, pp. 29–31. Weinheim: VCH-Verlagsgesellschaft.
Sumbler, M. G. 1983 A new look at the type Wolstonian glacial deposits of Central England. *Proc. Geol. Ass.* **94**, 33–44.
Uffenorde, 1982 Zur Gliederung des klastischen Holozäns im mittleren und nordwestlichen Teil der Deutschen Bucht (Nordsee) unter besonderer Berücksichtigung der Foraminiferen. *Eiszeitalter Gegenw.* **32**, 177–202.
Woszidlow, H. 1962 Foraminiferen und Ostrakoden aus dem marinen Elster-Saale-Interglazial in Schleswig-Holstein. *Meyniana* **12**, 65–96.
Zagwijn, W. H. 1974 The palaeogeographic evolution of the Netherlands during the Quaternary. *Geologie Mijnb.* **53**, 369–385.
Zagwijn, W. H. 1979 Early and Middle Pleistocene coastlines in the southern North Sea basin. In *The Quaternary history of the North Sea* (ed. E. Oele, R. T. E. Schüttenhelm & A. J. Wiggers). *Acta univ. ups. symp. univ. ups. a. quingent. celebr.*, vol. 2, pp. 31–42.

Discussion

R. Paepe (*Belgian Geological Survey, Brussels, Belgium*). How does Dr Streif explain that gullies on the Ostend Sheet, south of the Flemish Bight sheet, are less deep (30 m maximum) than in the area he studied (300 m depth)?

H. Streif. The channels seen in the Indefatigable map area are the largest observed in the North Sea and are up to 400 m deep; elsewhere channels of 100 m or more are normally the maximum. So an exceptional reason may need to be found for the Indefatigable area rather than elsewhere. The dramatic difference in channel depth over such a short distance may also be related to differing methods of formation.

If the most widely held view of channel formation is taken (subglacial erosion by meltwater) then it could be assumed that the large channels in the Indefatigable area represent the area with the maximum volumes of meltwater (i.e. near the glacier front) and the much smaller channels in the Ostend area may represent proglacial fluvial erosion by the release of the meltwater. However, if they too were formed by subglacial erosion, then it might be assumed that ice covered the Ostend area for only a short period of time, before channels deeper than 30 m could develop, but this seems particularly unlikely as there is no other evidence for glacial cover.

Such methods of channel formation assume that the channels observed in the Ostend and Indefatigable map areas are related. However, the usual explanation for the channels in the Ostend area is fluvial erosion during periods of low sea level. A proto-Thames channel of about 20 m depth has been identified in the U.K. sector of the Ostend map area (P. Balson, personal communication).

Phil. Trans. R. Soc. Lond. B **318**, 539–557 (1988)
Printed in Great Britain

Soils of the Plio-Pleistocene: do they distinguish types of interglacial?

By J. A. Catt
Soils Division, Rothamsted Experimental Station, Harpenden, Herts AL5 2JQ, U.K.

Soils form on land surfaces by the actions of physical, chemical and biological processes on the lithosphere, and are influenced by climate, parent material, relief, organisms and duration of formation. Remnants of Plio-Pleistocene soils may be buried beneath younger deposits or persist on present land surfaces. Their potential for rigorously differentiating interglacials by climatic characteristics is limited by problems of:
(i) precise dating of the beginning and end of soil-forming periods;
(ii) distinguishing characteristics attributable to climatic factors from those related to parent material, relief, etc;
(iii) calculating mathematical relations between measurable soil features and climatic variables;
(iv) diagenetic changes in buried soils;
(v) recognition and dating of relict features in unburied soils;
(vi) loss of many soils by erosion.
Some of these problems may be overcome if sequences of buried soils in periglacial loess deposits are used to compare the climates of successive interglacials in Europe and Asia. With the use of the length of interglacials derived from the oceanic record, the interglacials of the past million years are ranked according to approximate rate of soil development in loess. Two provisional equations relating soil development to time and climate are used; a linear relation probably overestimates the effect of time, and a logarithmic one seems to underestimate it. I tentatively suggest that oceanic oxygen-isotope stage 5e was warmer and wetter than the Holocene, stages 7 and 9 were cooler and drier than 5e, and 13–23 were generally warmer and wetter than 1–11.

1. Introduction

Soils are formed on land surfaces by processes dependent on the proximity of the uppermost layers of the earth's crust to the atmosphere and biosphere. Soil-forming processes, such as incorporation of humus, oxidative weathering of rock-forming minerals, leaching of soluble weathering products, and the downwashing (illuviation) of fine soil particles, affect various thicknesses of the crust, usually from a few centimetres to a few metres, although locally tens or even hundreds of metres. Generally, upper layers are modified more strongly and in different ways from deeper ones, thus producing a sequence of 'horizons' roughly parallel to the land surface but often disconformable with rock structures such as inclined bedding.

Simple climatic factors, such as mean annual air temperature, mean annual precipitation, and seasonal variations in temperature and rainfall, influence both inorganic soil processes (e.g. weathering and leaching) and the type and amounts of plant and animal life associated with the land surface, and also the rates and ways in which these organic materials decay after death. As the glacial–interglacial cycles of the past three million years are thought to have been reflected essentially in the same simple climatic factors, we might reasonably expect the sequence of soil horizons (or 'soil profile') at any place to preserve a history of climatic fluctuations similar to that recorded, for example, in deep oceanic sediments. However, this is rarely true, because land surfaces are hardly ever stable for very long; they are subject to episodes of erosion by streams, glaciers, the sea, the wind, or downslope mass movement, and are also buried periodically beneath new deposits of various types. As a result, the history of soil development over periods longer than the past 10^4 years is rarely complete; the most recent episodes are quite widely recorded, but earlier evidence is either lost by erosion or preserved partly and often in diagenetically modified form as buried soils.

In addition to the direct effects of climate, soils are also influenced by four other factors or groups of factors (Jenny 1941). These are the time over which they have been forming, the physical and chemical properties of the geological materials in which they have formed, the effects of relief on water flow and profile drainage, and the activities of organisms, including plants growing in the soil, animals living in and on it, and man. The many aspects of these five soil-forming factors, and also the numerous ways in which they interact, has led to the intense lateral variation observed in the present soil cover of the earth's surface. In past periods this variation would have been equally intense, but the patterns of variation in any region were probably different from the present pattern, because it is likely that one or more of the soil-forming factors was different. On the worldwide scale climate is probably the most important soil-forming factor, but the other four are just as important in determining regional, national and local patterns of soil variation (Harris 1968).

The oceanic isotopic record provides little evidence for climatic differences between successive warm stages over the past two million years or so (Shackleton & Opdyke 1976), but this could be because of the temperature-buffering effect of the oceans and the limited precision of the isotopic determinations themselves. If we accept that the main climatic fluctuations of the Quaternary were controlled by the Milankovitch orbital cycles of 100, 41 and 23 ka duration (Hays *et al.* 1976), we might expect interglacials of different lengths and climatic intensities, because superimposition of the three harmonic cycles would have produced non-harmonic cycles of variable wavelength and amplitude. Such climatic variation is indeed suggested by the foraminiferal assemblages in north Atlantic cores, which indicate that subtropical water penetrated further north in some warm stages than in others (Ruddiman & McIntyre 1976), and by differences between local interglacial assemblages of the more climatically sensitive fossil groups, such as the Coleoptera (Coope 1977), although it is less well expressed in the pollen records from long sequences such as the Macedonian peat bogs (Van der Hammen *et al.* 1971). The extent to which it can be discerned in the palaeopedological record has yet to be examined in full.

To obtain reliable palaeoclimatic evidence from soils (whether buried or remaining at the surface), it is necessary first to disentangle climatic effects from those of time, parent material, relief and organisms, and second to date the features attributable to climate. Before discussing the pedological evidence for palaeoclimatic differences between interglacials, it is necessary to examine critically the ways in which these two distinct problems are approached.

2. Climofunctions

If we accept the general hypothesis (Jenny 1941, 1961) that soil (S) or a specific soil property (s) is a product of the five factors: climate (cl), organisms (o), relief (r), parent material (p) and time (t), it follows that we can measure the effects of any one factor only if the remainder are constant or if any variations in them can be shown to have negligible effects. A climofunction is a specific mathematical solution of the general relation

$$S \text{ or } s = f(cl)_{o,r,p,t}$$

and may be expressed either graphically or as a quantitative function. However, there are many difficulties in solving such functions, including (i) many of the factors are not completely independent variables (e.g. climatic factors are influenced by elevation, slope aspect and other relief features), (ii) some factors cannot readily be quantified, and (iii) the constancy of some factors is difficult to establish.

As an example, consider the climofunction of the effect of mean annual rainfall on the depth to which calcium carbonate has been removed by acid dissolution. This would be determined by measuring the depth to carbonate in several profiles, each situated in areas where the rainfall has been consistently different but other climatic factors have been the same, and each developed for the same period of time, in parent materials of similar composition, and occurring in similar geomorphological situations, where the vegetation history and effects of animals (including man) also have all been the same. Some of these prerequisites may be met by simple field observation (e.g. similarity of geomorphological situation) or specialist studies (e.g. palynological studies of nearby lake deposits or the profiles themselves to establish similarity of vegetation history), and others (e.g. similarity of non-human animal influences) can perhaps be assumed. But other preconditions are very difficult to establish. In particular, many soil-forming processes, including decalcification, are so slow that their effects are measurable only over long periods (10^3–10^5 years), and during this time the climatic factor of interest (annual rainfall) is likely to have varied considerably at each site, certainly more than is indicated by recent meteorological records. This may not matter if all the profiles are from the same region, as they would probably have experienced similar climatic variations, and the present rainfall differences between sites could well have persisted while the various profiles developed. But if the selected profile sites are from different regions, it is less likely that their present rainfall differences have persisted for long; consequently they are unlikely to yield a reliable climofunction.

Another problem in evaluating climofunctions is establishing similarity of time over which the soil profiles have formed. The age of a soil cannot be measured directly by any known method, and must be estimated from two main lines of evidence.

(*a*) *Stratigraphic evidence*

A soil must obviously have begun to form some time after deposition of its parent material, but this may not help very much in dating if (as is very common) the soil has formed on an erosion surface truncating the parent material. It is then more useful to know the age of the erosion surface, but this can be determined only by tracing the surface laterally as far as possible and applying the rule that it is younger than the youngest deposit it truncates. With a soil on the present land surface the time of soil development extends from its initiation to the present day, although there could have been an intervening period when pedogenesis temporarily ceased because the soil was buried and subsequently exhumed. In a buried soil the period of pedogenesis was terminated by burial, and this is indicated by the age of the oldest deposit overlying the soil.

(*b*) *Internal evidence*

Features developed within the profile can indicate how long it took to form, providing we know the rates at which the processes responsible for those features occurred. Such rates, known as chronofunctions, are derived in a similar fashion to climofunctions, that is they are mathematical solutions of the general relation

$$S \text{ or } s = f(t)_{cl,o,r,p}.$$

They are determined from soil chronosequences (Vreeken 1975), which are sets of profiles whose individual members differ in age but have similar climatic histories, similar parent materials, similar geomorphological situations, and similar histories of vegetation, animal and human influences. Establishing soil age from chronosequences is obviously subject to many of the same constraints as our original objective, that of determining an accurate climofunction. Many chronofunctions are not straight-line relations; instead, the soil property changes rapidly in the early stages of soil development, then later slows down and eventually reaches a 'steady state' with little or no further change over long periods.

Because of the difficulties involved in calculating how long individual profiles have taken to form and in establishing similarities of other soil-forming factors, we have as yet few reliable and generally valid climofunctions (Yaalon 1975). Many relate to limited regions, often in the tropics, and are based on assumptions or imperfect assessments of other (non-climatic) soil-forming factors. Another problem is that most are for soil properties which are easily modified and therefore unlikely to survive unaltered during burial or exposure to climatic change (% organic C, %N, % base saturation, pH, cation exchange capacity). However, Bockheim (1980) showed that useful qualitative information concerning climatic effects on soil properties can be obtained by comparing chronofunctions for the same property in two or more chronosequences from areas of markedly different climate. Also, we may be able to infer climatic effects by comparing topofunctions derived from toposequences (soils differentiated only by geomorphic situation) in regions of different macroclimate (Yaalon 1975).

3. Dating of soil features

The dating of soil profiles and of individual features within them (e.g. accumulations of illuvial clay) involves two distinct problems: (i) calculating how long the profile or feature took to form (as discussed above in § 2), and (ii) deciding when that period of formation was. The stratigraphic evidence outlined above provides some information for both (i) and (ii) although it is often rather imprecise. In contrast, internal evidence on the extent of soil development provides information only on the length of the soil-forming period, and this can be related to years BP for the beginning of soil development only if the soil is still at the surface, or if the date of burial of a buried soil is known.

Several methods are available for approximate dating of buried interglacial soils, but they usually provide some date during the period of soil development rather than the dates when soil development started and burial occurred. These methods include thermoluminescence studies (Wintle *et al.* 1984), palaeomagnetic measurements (Heller & Liu Tungsheng 1984), amino acid analysis (Limmer & Wilson 1980), U–Th disequilibrium relations of carbonate concretions (Ku *et al.* 1979) and palaeontological studies.

Where sequences of buried soils are preserved in thick Plio-Pleistocene loess accumulations, as in eastern Europe (Kukla 1977), Soviet Central Asia (Dodonov 1979) and China (Heller & Liu Tungsheng 1982), fairly precise dates for older Pleistocene and even Pliocene soils have been inferred by matching the sequences to the dated oceanic isotope curves, assuming that loess was deposited in cold stages and that the soils formed in warm stages. As Pye (1984) pointed out, the correlation between these long loess–soil sequences in Europe and Asia is far from perfect, and this does cast doubt upon some of the inferred dates. The assumptions that the Plio-Pleistocene history of these areas was simply alternating loess deposition and pedogenesis, and that the climatic cycles recognized in deep sea sediments had similar effects throughout the world, are probably too simple to provide a completely reliable means of dating these older buried soils. But when combined with palaeontological and palaeomagnetic evidence, they have given a much better indication of the age of older Plio-Pleistocene soils than any other method.

Some buried soils may represent single major climatic episodes, such as interglacials, but many could have formed over longer periods because it is possible for parts of some land surfaces to remain stable through major climatic changes without suffering erosion or burial beneath fresh sediment. In soils on the present land surface, features inherited from earlier periods, when soil-forming conditions were different, are termed relict features, and the soils, relict soils. Buried soils showing a similarly complex history are usually termed polycyclic or composite soils. The most useful technique for identifying polycyclic soils and dating pedological features relative to one another is micromorphology. Thin sections show, for example, the relations between successive episodes of clay illuviation, iron and manganese mobilization, soil faunal activities and disruption by frost action, so that a history of climatic changes during soil development can be reconstructed (Kemp 1985*a*). Where two or more soils, perhaps partly truncated, are separated by thin sediment layers showing pedogenetic alteration, the resulting profile is termed a compound soil or pedocomplex.

4. Effects of climate on soil properties

Although there are few rigorously evaluated soil–climate relations (climofunctions), much is known in qualitative and sometimes semi-quantitative terms about the influence of rainfall and temperature on some soil characteristics. However, problems arise in the application of these generalized relations to the reconstruction of past climates because of the modifications that many soil features undergo on burial (diagenesis) or when soil-forming conditions change. Some features are more likely to persist than others, but much depends on the conditions to which they are subjected. For example, the humus in a buried soil is likely to be oxidized if it is above the groundwater table and the overlying deposits are permeable, but it may be preserved indefinitely in anaerobic conditions below the groundwater table.

The following climate-related soil features are especially relevant to the various types of interglacial soil found in loess sequences, which will be discussed in §§ 5 and 6.

Organic matter content

The amount of total organic matter in soils is determined by the balance between input and decomposition rates: thus soils of hot and cold deserts have little or no organic matter, because there are few plants and input is very small; tropical forest soils usually have fairly small amounts because, although input is rapid, so also is decomposition; the most organic soils are those of humid temperate regions, where high rainfall and weak evapotranspiration encourage lush vegetation but the low temperature prevents rapid decomposition. The distribution of organic matter with depth can suggest general climatic conditions because it is often affected by vegetation type: under forest the organic matter decreases rapidly with depth below a thin humic horizon, but under grassland organic matter is fairly evenly distributed over a greater thickness of soil, as in chernozems.

Clay content and type

In areas where there is little or no mineral weathering because of low rainfall or low temperatures, the clay (equivalent particle diameter less than 2 μm) content of soils derived from most parent materials is not increased, even over long periods of soil development. In contrast, the clay content of soils in warmer, wetter regions is often greatly increased by weathering, but it is very difficult to infer climatic conditions from a simple measure of clay content. This is because many soils are derived from clay-rich parent materials, and in subsurface horizons the clay content may be increased by physical illuviation from overlying horizons, although changes of clay content with depth may also be inherited from an inhomogeneous, stratified parent material. Clay produced by weathering must therefore be distinguished in each horizon from clay inherited from the parent material, a problem that usually involves careful quantitative granulometric, mineralogical, geochemical and micromorphological comparisons between horizons.

Clay in subsurface horizons which has been illuviated from above can usually be recognized in thin section because it forms birefringent coats (argillans) on sand grains or the walls of channels. The process of clay illuviation, typical of parabraunerde, is favoured by a soil pH of 4.5–6.5 (or higher if associated with much exchangeable sodium), small amounts of cementing and flocculating agents (carbonates, humus, sesquioxides, exchangeable Al, Mg, Ca), a system of fissures and other voids such as those formed by dissolution of limestone clasts, and a seasonal rainfall distribution (McKeague 1983). It is associated with woodland rather than open vegetation, but is usually a slow process, so large amounts of redeposited clay imply a long period (several thousand years or more) of soil development in cool or temperate, humid conditions. In mid- and high- latitude regions this implies that argillans formed in interglacials rather than in other Quaternary periods.

In clay-rich soils, birefringent aggregates of oriented clay within the soil matrix are also formed by seasonal shrink–swell processes. Lehm soil microfabrics are characterized by clayey matrices strongly reorganized in this way and by physical re-incorporation of clay from illuvial accumulations. The distinction between these two types of oriented clay in the matrix is often difficult. However, in soils formed in loess or other sediments originally containing little clay, a lehm-type fabric implies considerable clay enrichment by prolonged interglacial weathering and illuviation.

The main climatic factor influencing clay-mineral transformations is the excess of rainfall over evapotranspiration, which determines the extent of leaching. Increased leaching results mainly in the loss of bases and silica, with consequent accumulation of alumina and iron oxides. The main effect of increasing temperature is to increase the rate at which a particular mineral type is formed, and thus its abundance in profiles of a particular age. In the subpercolative soils of semi-arid regions, where evapotranspiration exceeds rainfall for much of the year so that there is little or no leaching of silica or bases, the clay fractions are progressively enriched in 2:1 minerals (Brown 1984), such as smectite. In the percolative soils of more humid regions without a marked dry season, 1:1 minerals, such as kaolinite and halloysite, are formed in preference to 2:1 minerals. Further leaching of silica, especially in the humid tropics, results in an excess of residual alumina, which forms gibbsite. Impeded drainage, even in hot humid regions, causes retention of silica and bases in the profile, and thus favours formation of smectite.

Soil colour

Soil colours are either inherited from the parent material, or result from soil-forming processes, such as incorporation of humus and oxidation or reduction of inherited or newly formed iron compounds. Any red or brown colours in subsoil horizons not derived from similarly coloured parent materials usually indicate oxidation of iron compounds in a fairly warm climate. Brown colours (Munsell hue 10YR) often result from crystallization of goethite from the amorphous iron released by biodegradation of iron–organic complexes (Schwertmann *et al.* 1974) and are typical of brown earths formed in cool humid regions with little seasonal variation of climate. Redder colours (hues of 7.5YR to 10R) often correlate with increasing abundance of haematite (Kemp 1985*b*; Barron & Torrent 1986), formation of which requires neutral conditions, little organic matter, and a strongly seasonal climate with hot, dry summers (Guillet & Souchier 1982). They occur in some parabraunerde, the haematite often being illuviated with silicate clays to form red argillans in subsoil horizons, and also in soils with lehm-type fabrics (rotlehm).

In temperate regions, such as the northern U.S.A. and northwestern Europe, reddened soils occur mainly on pre-Eemian deposits or land surfaces, whereas younger soils, formed since the Eemian, are usually brown. This observation suggests that in these regions reddening occurred in the Eemian and some earlier interglacials, but rarely in the Holocene. Either the Holocene was too short or its climate in the northern U.S.A. and northwestern Europe did not fully favour haematite formation. Reddening in undoubted Holocene soils (those formed on late Weichselian or younger sediments) is known from Israel (Dan *et al.* 1968), Morocco (Sabelberg 1977), southern France (Bresson 1974), southern Spain (Torrent *et al.* 1980), the alpine foreland in southern Germany (Schwertmann *et al.* 1982), and a few isolated sites in western Britain (Clayden 1977), but many of these soils are sandy or gravelly, and probably have a warmer pedoclimate than finer, more water-retentive soils. In south Germany, soils on silty or clayey Würm moraines adjacent to the reddened soils on glaciofluvial gravels have yellowish-brown subsurface horizons, and the redness of the coarser soils increases westwards with a decrease in mean annual precipitation from 1200 to 500 mm and an increase in mean annual temperature from 7 to 11 °C. This suggests (i) that there are at least two constraints on reddening, climate and parent material, both of which affect the pedoclimate; and (ii) that during the Holocene the summers in most of northwest Europe were just too cool and wet for reddening to occur in most water-retentive soils.

In the American system of soil classification (Soil Survey Staff 1975), unburied interglacial soils with reddening as a presumed relict feature are separated as various 'Pale-' great soil groups, and in the system used in England and Wales (Avery 1980) they are in 'paleo-argillic' subgroups. Many paleo-argillic soils also have a lehm microfabric. Pale-great groups and paleo-argillic subgroups both include soils with a wide range of particle-size distribution, so in Eemian and some other interglacials in these countries there was less parent material constraint on reddening than during the Holocene. This observation suggests that the summer climate during those interglacials was warmer and drier than it was in the Holocene. The pedological contrast may have been accentuated by soil development over longer periods in the interglacials than in the Holocene, but it is unlikely that time alone accounts for the much more widespread occurrence of reddened soils in interglacials.

Carbonate content

The amounts of calcium carbonate in soils depend on the original composition of the soil parent material, any continuing additions of carbonate from aeolian deposition or rainfall, and the balance between leaching losses and reprecipitation from the soil solution. Reprecipitation of carbonate occurs in the still calcareous subsoil beneath decalcified horizons, or nearer to the surface in soils subject to a persistent soil-moisture deficit in a prolonged dry season. However, secondary carbonate deposited close to the surface in a dry season may be redissolved and leached downwards in the wet season if the rainfall exceeds the field capacity of the soil for any significant period. In seasonally dry climates the depth below the soil surface at which carbonate first appears therefore reaches an equilibrium determined by the ratio between the mean number of days per year when soil-moisture deficit exceeds a certain value and the mean days when the soil received rainfall in excess of that required to maintain field capacity. As these periods depend on the type and density of the vegetation cover and on soil particle-size distribution, structure and porosity, as well as on the seasonal distribution of rainfall and air temperature, calculation of palaeoclimatic variables from the distribution of calcium carbonate in a soil profile is fairly complex. However, it is clear that the presence of secondary carbonate close to the soil surface results from low annual rainfall; the carbonate concretions (loess 'dolls') often distributed throughout beds of loess probably originated in this way, showing that even as it was being deposited loess was often subject to arid-region pedogenesis.

5. Important sequences of Quaternary soils in loess

Although precise dating and palaeoclimatic interpretation of soils formed in past Quaternary periods presents considerable problems, we can avoid the complications of different parent materials and geographical variations of climate within each interglacial by comparing buried interglacial soils formed within limited areas in which only loess accumulated for much of the Quaternary. In some periglacial regions, loess was deposited throughout most of each cold stage, because extensive glacial grinding or frost shattering of rock was necessary to produce large quantities of silt. However, silt production decreased in the intervening warm stages, and soils developed in areas where the loess surface remained stable. Although much loess must have been eroded from these regions, sequences of numerous loess-soil cycles were preserved in suitable sediment traps, such as subsiding basins or the sides of valleys above the level of later fluvial activity. Terrestrial gastropods and sporadic pollen preserved in these sequences show that the loess accumulated in cold, dry conditions, and that the soils usually represent periods of warm, humid climate with interglacial forest development (Kukla 1977; Lazarenko *et al.* 1981; Liu Tungsheng *et al.* 1982). Another advantage of using soils buried within loess sequences for palaeoclimatic distinction between interglacials is that deposition of loess usually results in a level land surface, so slope, the main component of the relief factor (*r*) in pedogenesis, is virtually uniform in space and time.

The main sites at which Quaternary loess successions with multiple buried soils have been studied are in central and eastern Europe, Germany, southern Ukraine, northern China and Soviet Central Asia. Palaeomagnetic and other datings of the soils and intervening loess layers in these areas often allow fairly clear correlations to be drawn with the worldwide sequence of oceanic oxygen-isotope stages, but only for about the past million years. Earlier parts of the terrestrial successions are often less complete, and earlier oceanic stages are less clearly defined.

In other regions, such as northwestern Europe and North America, loess deposition was apparently much more sporadic and often confined to later Quaternary cold stages, so that only a few of the later warm stages are represented by buried soils developed in loess. Buried soils developed in other parent materials (tills, aeolian sands, river-terrace gravels, volcanic ashes, colluvial and gelifluction deposits) are also common in these regions, and some of them even form fairly continuous sequences (Rohdenburg & Sabelberg 1973). However, most are less closely related to interglacial stages than the soils in more continuous successions of periglacial loess, because deposition in the intervening episodes was not as closely controlled by cold conditions. Even glacial deposits such as tills were deposited by relatively short-lived ice advances within long cold stages; consequently a soil between the tills of two different cold stages could have formed over a much longer time interval than just the warm interglacial between the cold stages. The same problem exists with soils developed in the 'hot' loess deposits associated with hot deserts. This type of loess is not dependent on frost or glaciations; the silt it contains was probably formed mainly by salt-weathering (Goudie *et al.* 1979) and was then concentrated by short-lived fluvial transportation followed by wind action. Consequently any buried soils within it represent wetter periods, which may or may not coincide with the warm stages of the oceanic oxygen isotope record; a good example is provided by the Netivot section in Israel, which contains six similar semi-arid calcareous soils dating from various times over the past 130 ka approximately (Bruins & Yaalon 1979).

I shall examine the palaeopedological evidence for interglacial climates from five European and Asian areas of semi-continuous periglacial loess deposition and three areas in northern Europe with shorter sequences. At present it is not worth considering the evidence for interglacial climates from soils in other parent materials; Quaternary sediments apart from periglacial loess are too variable for the parent-material factor to be eliminated, relief is less closely controlled, and the soil-forming periods are often not closely related to warm stages. Interglacial soil types at each site are given numerical values 1–7 indicating increasing degree of profile development (table 1). This sequence is based on evidence from the area considered first (central and eastern Europe). Apart from the climatic variables mean annual temperature and mean annual rainfall, the main factors affecting soil type were probably time (lengths of interglacials) and organisms. During pre-Holocene warm stages, organisms were limited to vegetation and animals (i.e. man had negligible effects) and were probably strongly related to climate, so that they can be combined with climate as a single major factor in genesis of loess soils. To minimize the effects of geographic variations in this combined climate–organisms factor within each interglacial, and thus compare the combined effects of climate–organisms plus time between interglacials, average numerical soil values for each interglacial were calculated for three broad regions of loess deposition (table 3).

The effects of man during the Holocene were probably quite variable, both geographically and in terms of soil type. Deforestation and man-induced erosion could both result in some Holocene soils' having characteristics indicating cooler, drier conditions than the actual Holocene climate.

TABLE 1. TYPES OF INTERGLACIAL LOESS SOIL AT SELECTED SITES

(Numbers indicate: 1, chernozem; 2, slightly decalcified brown earth; 3, strongly decalcified brown earth; 4, parabraunerde with weak argillic horizon; 5, parabraunerde with strong argillic horizon; 6, braunlehm; 7, rotlehm.)

			central-eastern Europe									northern Europe								Asia			
Glacial cycles	European loess marklines	Oceanic warm stages	Cerveny Kopec, Czechoslovakia	Kutna Hora, Czechoslovakia	Krems, Austria	Tutrakan, Bulgaria	Costinesti, Rumania	Paks, Hungary	Dunaföldvár, Hungary	Stari Slankamen, Yugoslavia	Kährlich, Nordrhein-Westfalen	Terrace sites of the lower and middle Rhein	Bad Soden, Hessen	Buggingen, Baden-Württemberg	Heitersheim, Baden-Württemberg	Normandy, NW France	Achenheim, NE France	Eastern Poland	Lochuan, N. China	Charvak, Tashkent	Chasmanigar, Tajikistan	Ukraine
A	I	1	5	5	1	1	1	1	1	1	4	4	4	4	4	4	4	3	1	3	3	1
B	II	5e	5	5	—	3	4	4	3	3	—	4	5	4	4	5	4	4	3	4	3	4
C	III	7	5	—	—	3	—	3	3	6	—	4	—	4	4	5	4	4	3	4	3	1
D	IV	9	5	—	—	5	1	—	—	3	4	4	—	4	—	4	—	—	2	3	3	2
E	V	11	5	—	—	5	5	2	3	3	—	6	—	—	4	5	—	—	5	3	3	2
F	VI	13	7	—	—	7	7	3	—	7	4	—	—	6	—	—	—	—	6	1	—	3
G	VII	15	6	—	—	5	5	—	—	7	4	—	5	3	—	—	—	—	6	4	3	4
H	VIII	17	6	—	—	7	7	3	—	—	—	—	5	4	—	7	—	—	3	2	—	3
I	IX	19	6	—	5	7	7	3	4	—	—	—	4	—	—	7	—	—	5	3	3	1
J	X	21	3	—	5	—	—	—	—	—	—	—	5	—	—	5	—	—	3	3	3	3
K	XI	23	7	—	6	—	7	3	—	7	—	7	5	—	—	—	—	—	3	2	3	2

Central and eastern Europe

For much of the Quaternary the Carpathian basin (parts of Czechoslovakia, Austria, Hungary, Yugoslavia and Rumania drained mainly by the Danube and its tributaries) experienced strong climatic fluctuations, from cold dry periglacial continental conditions with loess deposition, to warm wet Atlantic (interglacial) conditions (Kukla 1977). Different types of soil were formed under all conditions between these extremes. Many have been dated by radiocarbon, magnetostratigraphy, thermoluminescence, and the assumption that the 'marklines' of Kukla (1969), which are boundaries between (i) thick layers of loess, with gastropods of cold, dry conditions, and (ii) overlying interglacial soils or hillwash, indicating abrupt ameliorations of climate, are equivalent to the dated 'terminations' of the oceanic oxygen-isotope curve (horizons of rapidly decreasing ^{18}O content, indicating rapid warming from a glacial to an interglacial stage). The marklines delimit glacial cycles (Kukla 1970), designated A, B, C, D, etc. backwards in time; the present incomplete (Holocene) cycle (A) began with markline I *ca.* 10 ka BP. Within each glacial cycle, submarklines form boundaries between deposits and overlying soils which are less well developed than the interglacial soils immediately above the marklines.

The various soil types form the following development sequence.

cold, dry: loess steppe soils
frost gley
grassland soil with little humus
grassland soil rich in humus (chernozem) (1)
slightly decalcified brown earth (2)
strongly decalcified brown earth (3)
parabraunerde with weak argillic horizon (4)
parabraunerde with strong argillic horizon (5)
braunlehm (6)
warm, wet: rotlehm (7)

(loess steppe soils to slightly decalcified brown earth: interstadial soils; grassland soil rich in humus to rotlehm: interglacial soils)

This has been confirmed as a weathering sequence by micromorphological and mineralogical analyses of selected Holocene and buried profiles (Bronger 1976, 1979; Bronger *et al.* 1976), and its interpretation as a climatic sequence is based on the present mean annual rainfall and temperature for nearby areas where analogous loess-derived soils have developed during the Holocene. Although this ignores other soil-forming factors, it is similar to the climatic sequence expected from the properties discussed in §4. The parabraunerde, braunlehm and rotlehm soils often contain molluscs and plant remains (Frenzel 1964; Lozek 1969; Urban 1984) indicating formation under typical interglacial forest, but the brown earths and chernozems were probably formed in areas that were too dry for forest development even in interglacials. Soil types known to have been formed during interglacials or the Holocene in what can be inferred as increasingly warm and wet conditions are designated 1–7.

Germany

Sequences of buried interglacial soils in loess are also common in central Germany, at sites such as Bad Soden (Semmel & Fromm 1976) in Hesse, Heitersheim (Bronger 1966) and

Buggingen (Bronger 1969) in Baden-Württemberg, and Kärlich (Brunnacker 1978) and other sites on terraces of the Lower Rhine in Nordrhein-Westfalen (Paas 1982, Figure 17). Palaeomagnetic dating is available for some of the sequences, such as Kärlich and Bad Soden, but the exact dating of many pre-Eemian soils is still uncertain. Also there are long gaps in some successions, and some interglacials in valley successions are represented by poorly drained soils (*Nassböden*) or alluvial soils (*Auenböden*), which cannot easily be correlated with the approximate climatic sequence of well-drained soils established in central and eastern Europe. Despite these problems, it is possible to assign many buried interglacial loess soils to oceanic warm stages and to members of the soil development sequence established in central and eastern Europe (table 1).

Most of the buried loess soils in Germany are parabraunerde, which are at least as strongly developed as nearby unburied Holocene loess soils. Some of the older interglacial soils along the Rhine valley are weak braunlehms (Bronger 1969), and in Rheinhessen rotlehms are developed in older loess with reversed magnetization (Plass *et al.* 1977).

Southern Ukraine

Veklich (1979) recognized eight loesses and eight soils (including that of the Holocene) formed in southern Ukraine within the past million years. The dates he proposed for each loess and soil do not match those of the cold and warm oceanic stages, so correlation with the dated oceanic sequence initially seems impossible, especially as fewer soil horizons have been recognized than there are known warm stages over the past million years. However, at least four of the Ukrainian soil horizons (Priluky, Kaydak, Zavadovka and Lubny) are pedocomplexes, and each probably represents two warm interglacial stages with minor deposition of loess between. This makes it easier to correlate individual soils (rather than named soil horizons) with the oceanic succession (table 2) assuming that Veklich's date for the base of the Priazovye loess (1 Ma BP) is approximately correct and that the youngest buried soil at least as strongly developed as the Holocene soil was formed in stage 5e. In at least four of the Ukrainian profiles, the youngest soil meeting this criterion (the Vitachev horizon) is a weak parabraunerde. The soils of other warm stages equate approximately with the chernozems and brown earths of central and eastern Europe. In table 1 these are assigned to oceanic warm stages according to the correlation and revised dating proposed in table 2.

Northern China

The loess plateau of northern China includes large basins with long sequences of interbedded loess and soils, forming loess *Yuans* (flat uplands), such as the *Yuan* at Lochuan. The loess layers in these basins contain herbaceous pollen, remains of dry steppe mammals, and molluscs indicating cold, dry steppe conditions, but the intervening soils contain pollen of mainly broadleaved trees and molluscs of warm, humid habitats (Liu Tungsheng *et al.* 1982). Palaeomagnetic dating of the Lochuan succession (Heller & Liu Tungsheng 1982, 1984) showed that the number of buried soils above the Brunhes–Matuyama boundary is the same as the number of warm stages in the oceanic oxygen-isotope record.

An Zhisheng *et al.* (1982) classified the buried soils in the Lochuan section into the following climatic sequence.

Table 2. Proposed correlation of Ukrainian loess and soil horizons with oceanic oxygen-isotope stages

Ukrainian horizons (Veklich 1979)	dating (ka BP) (Veklich 1979)	deposits and soil types (Veklich 1979)	oceanic oxygen–isotope stages	revised dating (ka BP) based on oceanic stages
Holocene	0–10	chernozems, grey forest soils	1	0–12
Prichernomorye	10–22	loess with a brown semi-desert soil locally	2,3,4	12–71
Dofinovka	22–30	weak chernozems, brown semi-desert soils	5a	71–88
Bug	30–50	loess	5b, c, d	88–122
Vitachev	50–60	weak parabraunerde, brown earths (often reddish)	5e	122–128
Uday	60–70	loess	6	128–186
Priluky	70–100	chernozems	7	
			8	186–339
		chernozems, grey–brown forest soils, chestnut brown soils	9	
Tyasmin	100–115	loess	10	339–362
Kaydak	115–175	chernozems, chestnut brown soils	11	
		—	12	362–524
		grey forest soils, podzolized brown forest and grassland soils, chernozems	13	
Dnieper	175–250	loess	14	524–565
Zavadovka	250–370	chernozems, brown forest soils, weak parabraunerde	15	
		loess	16	565–689
		chernozems, brown forest and grassland soils	17	
Tiligul	370–470	loess	18	689–726
Lubny	470–650	chernozems	19	
			20	726–763
		brown forest and meadow soils, chernozems	21	
Sula	650–700	loess	22	763–795
Martonosha	700–920	brown grassland soils	23	795
Priazovye	920–1000	loess	24	

cold, dry:	
weak pedogenic loess (*Cathaica* assemblage)	interstadial soils
medium pedogenic loess	
pedogenic loess (*Metodontia* assemblage)	
black loessial soil (1)	interglacial soils
carbonate drab (cinnamon) soil (2)	
drab (cinnamon) soil (3)	
luvic drab (cinnamon) soil (5)	
warm, wet:	
drab (cinnamon) brown earth (6)	

The black loessial soil probably corresponds approximately with the chernozem of central Europe, the drab (or cinnamon) soil with decalcified brown earth, the luvic drab soil with parabraunerde, and the drab brown earth with braunlehm. Table 1 shows the types of interglacial soil formed during oceanic warm stages 1–23 at Lochuan.

Comparisons with modern analogue soils elsewhere in China enabled Liu Tungsheng *et al.*

(1986) to propose annual average temperature and rainfall values for the Lochuan area during the various cold and warm stages over the past 900 ka. However, these palaeoclimatic inferences ignored the effect of time on development of the soils. In cold stages, when there was little or no weathering and clay illuviation, this is probably unimportant. But in the warmer and more humid interglacial climates, the length of each soil-forming episode was perhaps almost as important as temperature and rainfall in determining the type of soil produced.

Soviet Central Asia

Several long Plio-Pleistocene loess soil successions have been reported from parts of Soviet Central Asia, including southern Tajikistan (Dodonov 1979) and the Tashkent region (Lazarenko 1980; Lazarenko *et al.* 1981). In both these areas there is palaeomagnetic evidence for the age of some soil horizons; for example, the Saylyk Horizon of the Charvak section near Tashkent and Pedocomplex V at Chasmanigar in Tajikistan are both firmly related to oceanic stage 5e, and the Azadbash horizon at Charvak and Pedocomplex X at Chasmanigar are both immediately beneath the Brunhes–Matuyama boundary, so they were probably formed in oceanic stage 21. But the ages of intervening soils in these two regions are less certain, because there appear to be fewer typical interglacial soils than warm oceanic stages. This could be because of gaps in the successions, but it is more likely that some of the cooler and drier interglacials resulted in soils more like the interstadial soils of Europe, because the Holocene and most interglacial soils in Soviet Central Asia are less strongly developed than those in Europe. The most obvious point in the Charvak succession lacking an interglacial soil is between the Khandaylyk horizon (probably oceanic stage 15) and the Barrazh b horizon (probably 11); the loess between these two soils is much thicker than others at Charvak, and 1–2 'rudimentary soils' occur within it. I therefore conclude that the weakest interglacial soil at Charvak is that of oceanic stage 13.

Unfortunately the Chasmanigar section must have at least two weak interglacial soils, as there are only five typical interglacial brown earths between the Brunhes–Matuyama boundary and Pedocomplex V. It is again unlikely that there is a break in the Chasmanigar sequence, because the same number of brown earths is found at three other sites spanning this time period (Kayrubak, Lakhuti and Khonako 2). In table 1, I have provisionally allocated the soils between the Brunhes–Matuyama boundary and Pedocomplex V (all brown earths) to the interglacials which produced the strongest soils at Charvak.

Northern Europe

In Normandy the loess contains seven interglacial soils and has been divided into three formations (Lautridou *et al.* 1986). The youngest (Saint Pierre les Elbeuf) formation contains four parabraunerden (Elbeuf I–IV), the most recent of which is dated by thermoluminescence to oceanic stage 5e (Wintle *et al.* 1984). The next older (Mesnil-Esnard) formation contains two red clayey (rotlehm) soils (Mesnil-Esnard V and VI), and the oldest (Saint Prest) formation a parabraunerde (Bosc Hue VII). The sandy loess of the Saint Prest Formation has a reversed geomagnetic polarity (Biquand & Lautridou 1979), so the Bosc Hue VII soil was probably formed in oceanic stage 21 at the latest. The Mesnil-Esnard V and VI soils are tentatively correlated with the Netherlands Cromerian Interglacials and IV and II respectively (Lautridou 1977) and were therefore probably formed in oceanic stages 17 and 19. The Saint Pierre les Elbeuf formation is correlated with the Tourville formation (part of the Low Terrace of the River Seine), which contains Saalian fossils. The soil Elbeuf IV is thus Holsteinian, and soils II, III and IV were probably formed in oceanic stages 7, 9 and 11 respectively (table 1).

At Achenheim in Alsace, loess equivalent to the Saint Pierre les Elbeuf formation is divided into four layers by three parabraunerde soils (Heim *et al.* 1982). From the distribution of mammal remains and Palaeolithic artefacts in this succession, it is likely that the three soils are equivalent to Elbeuf I to III of Normandy (table 1).

On the Lublin Uplands of eastern Poland, the loess overlying weathered till of the Cracovian (south Polish) Glaciation contains two buried soils of parabraunerde type (Jersak 1977). The younger (Nietulisko Soil) is correlated with the Eemian (oceanic stage 5e), and the older (Tomaszów Soil) with the Lublin interglacial. The Tomaszów Soil is developed in 'lower older loess', which is correlated with the Odra glaciation. Odranian deposits further west have been dated to oceanic stage 8 by thermoluminescence (Lindner & Grzybowski 1982), so the Tomaszów Soil probably developed during oceanic stage 7 (table 1).

6. Climatic comparison between interglacials by using loess soils

The interglacial soils developed from periglacial loess (table 1) show some consistent differences between interglacials. (i) The soil of stage 5e is always equally or more strongly developed than that of the Holocene (stage 1) in the same area; this agrees with the evidence of unburied soils in Europe and the U.S.A. (§ 4), but could result partly from the effects of man in the Holocene. (ii) Soils of stages 7 and 9 are equal to or weaker than that of 5e, except for the more strongly developed soils of stage 9 at Tutrakan and stage 7 at Stari Slankamen. (iii) Apart from Paks (Hungary), Ukraine and Charvak (Tashkent), the soil of stage 11 is always equal to or stronger than that of 5e. (iv) With a few exceptions, the soils of stages 13–23 are more strongly developed than those of the later interglacials (1–11).

In the absence of any other numerical data common to all the interglacial loess soils, the numbers 1–7 for the weathering sequence of interglacial soils in central and eastern Europe can be used in simple numerical calculations to compare soils in different regions and different interglacials. The mean soil values for each interglacial in the three main areas (table 3) are more variable in central–eastern Europe (2.0–6.2) than in either northern Europe (3.9–6.0, which is the upper part of this range) or Asia (2.0–4.3, the lower part of the central–eastern European range). This suggests that central–eastern Europe experienced very variable interglacial climates as well as large differences between cold and warm stages.

Because we have eliminated the parent material and relief factors in soil variation, the differences between mean soil values for interglacials can be attributed to climate–organisms and time (relative length of soil-forming periods). Before we can isolate the effect of climatic differences between interglacials, we should therefore allow for differences in the duration of pedogenesis in each interglacial, but this is difficult because many of the isotopic stage boundaries in oceanic cores are imprecisely dated. The 'terminations' at the beginning of each odd-numbered stage are reasonably well dated, but the ends of interglacials are much less clear. Also the good resolution for interglacial stages 5 and 7 shows that these were complexes of two or more warm episodes with colder periods between; the isotope curves for earlier warm stages are less well resolved and could also be complexes containing cold periods of unknown length.

Table 3. Relative development of loess soils during successive interglacials, and suggested climatic ranking of interglacials

oceanic warm stages	mean soil values: central–eastern Europe	northern Europe	Asia	mean soil ranking[a]	mean climatic ranking[b]
1	2.0	3.9	2.0	11	11
5e	3.9	4.3	3.5	8	5
7	4.0	4.2	2.8	9	9
9	3.5	4.0	2.5	10	10
11	3.8	5.0	3.3	6	8
13	6.2	5.0	3.3	1	1
15	5.8	4.0	4.3	3	6
17	5.8	5.3	2.7	4	4
19	5.3	5.5	3.0	4	2
21	4.0	5.0	3.0	6	7
23	6.0	6.0	2.5	1	2

[a] Ranking by mean, worldwide soil values.
[b] Mean, worldwide climatic ranking (from $Y = a + (B \log x)$ (Bockheim 1980)).

Bockheim (1980) found that the general equation which best explains the relationship between soil properties (Y), time (X, in years) and climate (b) over a range of climatic regions and parent-material types is the logarithmic function $Y = a + (b \log X)$, and for relations between three soil properties relevant to interglacial pedogenesis in loess (oxidation depth, percentage clay in the B horizon, and total profile depth) and mean annual temperature, the ratio $a:b$ was approximately 20:1 for each property. When the lengths of interglacials can be established more clearly from the oceanic isotopic record, this or a similar equation might be used to rank interglacials according to climatic characteristics (temperature and rainfall). At present it can be used only by taking the lengths of odd-numbered oceanic stages from the revised stage boundary dates of Imbrie *et al.* (1984). If these values are used for X, and the mean worldwide soil values (table 3) are used for Y, with $a:b = 20:1$ in Bockheim's equation, the climatic ranking of interglacials obtained is in fact little different from the ranking based on the unmodified mean worldwide soil values (table 3). This result suggests that, using Bockheim's equation, the differences in lengths of interglacials likely to emerge in future from better resolution of the oceanic oxygen-isotope record will probably make little difference to the climatic ranking of interglacials (table 3) based on the loess soils. However, Bockheim's equation may make too little allowance for time as a soil-forming factor, and should also be checked and refined. Present knowledge of the relations between climate, time and any simple soil properties likely to survive burial is scarcely adequate for the reconstruction of past climates from these soils. The tentative climatic ranking of interglacials given in table 3 is therefore probably the best that can be suggested from palaeopedological evidence until more quantitative information is available on processes of soil development in loess under different climatic régimes.

7. Conclusions

If we are unable to draw very definite conclusions at present from the long sequences of buried loess soils about climatic differences between interglacials, other soils are likely to be even less useful. They are more difficult to date, especially in terms of the exact length of the soil-forming period, and few deposits are lithologically as uniform as loess over large areas, so the effects of parent-material differences on soil properties must be considered. Also, other soils are less likely to form sequences representing all interglacials within areas small enough to have shown no geographic variation of climate during each interglacial.

Because of these additional problems with non-loessial soils, future work on palaeoclimatic interpretation of buried soils should concentrate on those in the thick loess sequences. But before these can be used effectively, it is important that reliable climofunctions for properties likely to survive burial are established in Holocene loess soils developed over known periods and in regions which have not suffered major climatic change during the Holocene. Suitable properties are depth of oxidation, humus incorporation and occurrence of secondary carbonate, thickness of Bt horizon and percentage of illuvial clay, amounts of sesquioxides and layer-silicate clays formed by weathering, and the extent of weathering of heavy minerals in the fine sand and coarse silt fractions.

References

An Zhisheng, Lu Yanchou & Wei Lanying 1982 A preliminary study of soil stratigraphy in the Lochuan loess section. In *Quaternary dust mantles of China, New Zealand and Australia; proceedings of a workshop* (ed. R. Wasson), pp. 31–44. Canberra: Australian National University.

Avery, B. W. 1980 Soil classification for England and Wales (higher categories). *Soil Surv. tech. Monogr.* no. 14. (67 pages.)

Barron, V. & Torrent, J. 1986 Use of the Kubelka–Munk theory to study the influence of iron oxides on soil colour. *J. Soil Sci.* **37**, 499–510.

Biquand, D. & Lautridou, J. P. 1979 Détermination de la polarité magnétique des loess et sables pléistocènes de Haute-Normandie: premiers résultats. *Bull. Ass. fr. Étude Quat.* **58–59**, 75–81.

Bockheim, J. G. 1980 Solution and use of chronofunctions in studying soil development. *Geoderma* **24**, 71–85.

Bresson, L. M. 1974 *Rubéfaction récente des sols en climat tempéré humide. Séquence évolutive sur fluvio-glaciaire calcaire dans le Jura méridional (étude de microscopie intégrée)*. Thèse, 3e cycle, University of Paris 7. (197 pages.)

Bronger, A. 1966 Lösse, ihre Verbraunungszonen und fossilen Böden. Ein Beitrag zur Stratigraphie des oberen Pleistozäns in Südbaden. *Schr. geogr. Inst. Univ. Kiel* **24**, (2), 1–114.

Bronger, A. 1969 Zur Mikromorphogenese und zum Tonmineralbestand Quartärer Lössböden in Südbaden. *Geoderma* **3**, 281–320.

Bronger, A. 1976 Zurquartären Klima- und Landschaftsentwicklung des Karpatenbeckens auf (paläo-) pedologischer und bödengeographischer Grundlage. *Kieler geogr. Schr.*, no. 45. (xiv+268 pages.)

Bronger, A. 1979 The value of mineralogical and clay mineralogical analyses of loess soils for the investigations of Pleistocene stratigraphy and paleoclimate. *Acta geol. hung.* **22**, 141–152.

Bronger, A., Kalk, E. & Schroeder, D. 1976 Über Glimmer- und Feldspatverwitterung sowie Entstehung und Umwandlung von Tonmineralen in Rezenten und fossilen Lössböden. *Geoderma* **16**, 21–54.

Brown, G. 1984 Crystal structures of clay minerals and related phyllosilicates. *Phil. Trans. R. Soc. Lond.* A **311**, 221–240.

Bruins, H. J. & Yaalon, D. H. 1979 Stratigraphy of the Netivot section in the desert loess of the Negev (Israel). *Acta geol. hung.* **22**, 161–169.

Brunnacker, K. 1978 Gliederung und Stratigraphie der Quartär-Terrassen am Niederrhein. *Kölner geogr. Arb.* **36**, 37–58.

Clayden, B. 1977 Palaeosols. *Cambria* **4**, 84–97.

Coope, G. R. 1977 Quaternary Coleoptera as aids in the interpretation of environmental history. In *British Quaternary studies; recent advances* (ed. F. W. Shotton), pp. 55–68. Oxford: Clarendon Press.

Dan, J., Yaalon, D. H. & Koyumdjisky, H. 1968 Catenary soil relationships in Israel. 1. The Netanya catena on coastal dunes of the Sharon. *Geoderma* **2**, 95–120.

Dodonov, A. E. 1979 Stratigraphy of the Upper Pliocene–Quaternary deposits of Tajikstan (Soviet Central Asia). *Acta geol. hung.* **22**, 63–73.

Frenzel, B. 1964 Zur Pollenanalyse von Lössen. *Eiszeitalter Gegenw.* **15**, 5–39.

Goudie, A. S., Cooke, R. U. & Doornkamp, J. C. 1979 The formation of silt from quartz dune sand by salt-weathering processes in deserts. *J. Arid Env.* **2**, 107–112.

Guillet, B. & Souchier, B. 1982 Amorphous and crystalline oxhydroxides and oxides in soils (iron, aluminium, manganese, silicon). In *Constituents and properties of soils* (ed. M. Bonneau & B. Souchier; trans. V. C. Farmer), pp. 21–42. London: Academic Press.

Harris, S. A. 1968 Comments on the validity of the law of soil zonality. *Trans. 9th Int. Congr. Soil Sci., Adelaide, Australia* **4**, 585–593. International Society of Soil Science; Angus and Robertson.

Hays, J. D., Imbrie, J. & Shackleton, N. J. 1976 Variations in the earth's orbit: pacemaker of the ice ages. *Science, Wash.* **194**, 1121–1132.
Heim, J., Lautridou, J. P. Maucorps, J., Puisségur, J. J., Sommé, J. & Thévenin, A. 1982 Achenheim: une séquence-type des loess du Pléistocène moyen et supérieur. *Bull. Ass. Fr. etude Quat.*, ser. 2, **10–11**, 147–159.
Heller, F. & Liu Tungsheng 1982 Magnetostratigraphical dating of loess deposits in China. *Nature, Lond.* **300**, 431–433.
Heller, F. & Liu Tungsheng 1984 Magnetism of Chinese loess deposits. *Geophys. J. R. astr. Soc.* **77**, 125–141.
Imbrie, J., Hays, J. D., Martinson, D. G., McIntyre, A., Mix, A. C., Morley, J. J., Pisias, N. G., Prell, W. L. & Shackleton, N. J. 1984 The orbital theory of Pleistocene climate: support from a revised chronology of the Δ ^{18}O record. In *Milankovitch and climate* (ed. A. Berger, J. Imbrie, J. Hays, G. Kukla & B. Saltzman), pp. 269–305. Dordrecht: Riedel.
Jenny, H. 1941 *Factors of soil formation: A system of quantitative pedology*. New York: McGraw Hill.
Jenny, H. 1961 Derivation of state factor equations of soils and ecosystems. *Soil Sci. Soc. Am. Proc.* **25**, 385–388.
Jersak, J. 1977 Cyclic development of the loess cover in Poland. *Biul. Inst. Geologicznego* **20**, 83–96.
Kemp, R. A. 1985*a* The Valley Farm Soil in southern East Anglia. In *Soils and Quaternary landscape evolution* (ed. J. Boardman), pp. 179–196. Chichester: Wiley.
Kemp, R. A. 1985*b* The cause of redness in some buried and non-buried soils in eastern England. *J. Soil Sci.* **36**, 329–334.
Ku, T., Bull, W. B., Freeman, S. T. & Knauss, G. 1979 Th^{230}–U^{234} dating of pedogenic carbonates in gravelly desert soils of Vidal Valley, southeastern California. *Bull. geol. Soc. Am.* **90**, 1063–1073.
Kukla, G. J. 1977 Pleistocene land-sea correlations. 1. Europe. *Earth-Sci. Rev.* **13**, 307–374.
Kukla, J. 1969 Die zyklische Entwicklung und die absolute Datierung der Löss-Serien. In *Periglazialzone, Löss und Paläolithikum der Tschechoslowakei* (ed. J. Demek & J. Kukla), pp. 75–96. Brno: Tschechoslowakische Akademie der Wissenschaften, Geographisches Institut.
Kukla, J. 1970 Correlations between loesses and deep-sea sediments. *Geol. För. Stockh. Förh.* **92**, 148–180.
Lautridou, J. P. 1977 Loess et sables de la Basse-Seine; lithostratigraphie et chronostratigraphie. *Bull. trimest. Soc. geol. Normandie* **64**, 81–91.
Lautridou, J. P., Monnier, J. L., Morzadec, M. T., Sommé, J. & Tuffreau, A. 1986 The Pleistocene of northern France. *Quat. Sci. Rev.* **5**, 387–393.
Lazarenko, A. A. 1980 Buried soils of the loess formation in Middle Asia and their paleogeographic significance. *Dokl. Akad. Nauk S.S.S.R.* **252** (1), 181–185. (In Russian.)
Lazarenko, A. A., Bolikhovskaya, N. S. & Semenov, V. V. 1981 An attempt at a detailed stratigraphic subdivision of the loess association of the Tashkent region. *Int. Geol. Rev.* **23**, 1335–1346.
Limmer, A. W. & Wilson, A. T. 1980 Amino acids in buried paleosols. *J. Soil Sci.* **31**, 147–153.
Lindner, L. & Grzybowski, K. 1982 Middle-Polish glaciations (Odranian, Wartanian) in southern central Poland. *Acta geol. pol.* **32**, 162–178.
Liu Tungsheng, An Zhisheng & Yuan Baoyin 1982 Aeolian processes and dust mantles (loess) in China. In *Quaternary dust mantles of China, New Zealand and Australia, proceedings of a workshop* (ed. R. J. Wasson), pp. 1–17. Canberra: Australian National University.
Liu Tungsheng, Zhang Shouxin & Han Jiaomao 1986 Stratigraphy and paleo-environmental changes in the loess of central China. *Quat. Sci. Rev.* **5**, 489–495.
Lozek, V. 1969 Paläontologische Charakteristik der Löss-Serien. In *Periglazialzone, Löss und Paläolithikum der Tschechoslowakei* (ed. J. Demek & J. Kukla), pp. 43–60. Brno: Tschechoslowakische Akademie der Wissenschaften, Geographisches Institut.
McKeague, J. A. 1983 Clay skins and argillic horizons. In *Soil micromorphology* (ed. P. Bullock & C. P. Murphy), pp. 367–387. Berkhamsted: A. B. Academic Publishers.
Paas, W. 1982 Fossile Böden auf den Rhein-Terrassen und deren Deckschichten in der Niederrheinischen Bucht. *Geol. Jb.* **F14**, 228–239.
Plass, W., Scheer, H. D. & Semmel, A. 1977 Löss-Sedimente und rote Böden im Altpleistozän Rheinhessens. *Catena* **4**, 181–188.
Pye, K. 1984 Loess. *Progr. phys. Geogr.* **8**, 176–217.
Rohdenburg, H. & Sabelberg, U. 1973 Quartäre Klimazyklen in westlichen Mediterrangebiet und ihre Auswirklungen auf die Relief und Bödenentwicklung. *Catena* **1**, 71–180.
Ruddiman, W. F. & McIntyre, A. 1976 Northeast Atlantic paleoclimatic changes over the past 600000 years. In *Investigation of late Quaternary paleoceanography and paleoclimatology* (ed. R. M. Cline & J. D. Hays), pp. 111–146. Geological Society of America Memoir no. 145.
Sabelberg, U. 1977 The stratigraphic record of late Quaternary accumulation series in south west Morocco and its consequences concerning the pluvial hypothesis. *Catena* **4**, 209–214.
Schwertmann, U., Fischer, W. R. & Taylor, R. M. 1974 New aspects of iron oxide formation in soils. *Proc. Xth I.S.S.S. Congr. Moscow* 1974 **6** (1), 237–247.
Schwertmann, U., Murad, E. & Schulze, D. G. 1982 Is there Holocene reddening (haematite formation) in soils of axeric temperate areas? *Geoderma* **27**, 209–223.

Semmel, A. & Fromm, K. 1976 Ergebnisse paläomagnetischer Untersuchungen an quartären Sedimenten des Rhein-Main-Gebietes. *Eiszeitalter Gegenw.* **27**, 18–25.

Shackleton, N. J. & Opdyke, N. D. 1973 Oxygen isotope and paleomagnetic stratigraphy of equatorial Pacific core V28–238: oxygen isotope temperatures and ice volumes on a 10^5 year and 10^6 year scale. *Quat. Res.* **3**, 39–55.

Shackleton, N. J. & Opdyke, N. D. 1976 Oxygen-isotope and paleomagnetic stratigraphy of Pacific core V28–239 late Pliocene to latest Pleistocene. In *Investigation of late Quaternary paleoceanography and paleoclimatology* (ed. R. M. Cline & J. D. Hays), pp. 449–464. Geological Society of America Memoir no. 145.

Soil Survey Staff 1975 *Soil taxonomy. A basic system of soil classification for making and interpreting soil surveys. U.S. Dept. Agric. Handbook* no. 436.

Torrent, J., Schwertmann, U. & Schulze, D. G. 1980 Iron oxide mineralogy of some soils of two river terrace sequences in Spain. *Geoderma* **23**, 191–208.

Urban, B. 1984 Palynology of central European loess-soil sequences. In *Lithology and stratigraphy of loess and paleosols* (ed. M. Pécsi), pp. 229–248. *Proc. Symp. INQUA Comm. Loess Palaeoped., XIth INQUA Congr., Moscow.* Budapest: Geographical Research Institute, Hungarian Academy of Sciences.

Van der Hammen, T., Wijmstra, T. A. & Zagwijn, W. H. 1971 The floral record of the late Cenozoic in Europe. In *The late Cenozoic glacial ages* (ed. K. K. Turekian), pp. 391–424. New Haven: Yale University Press.

Veklich, M. F. 1979 Pleistocene loesses and fossil soils of the Ukraine. *Acta geol. hung.* **22**, 35–62.

Vreeken, W. J. 1975 Principal kinds of chronosequences and their significance in soil history. *J. Soil Sci.* **26**, 378–394.

Wintle, A. G., Shackleton, N. J. & Lautridou, J. P. 1984 Thermoluminescence dating of periods of loess deposition and soil formation in Normandy. *Nature, Lond.* **310**, 491–493.

Yaalon, D. H. 1975 Conceptual models in pedogenesis: can soil-forming functions be solved? *Geoderma* **14**, 189–205.

Discussion

R. Paepe (*Belgian Geological Survey, Brussels, Belgium*). What is done if palaeosols might be polycyclic or double-buried?

J. A. Catt. Polycyclic soils, as defined in the paper, were omitted, but some of the double-buried soils were attributed to successive interglacials.

Phil. Trans. R. Soc. Lond. B **318**, 559–602 (1988)
Printed in Great Britain

The history of the great northwest European rivers during the past three million years

By P. L. Gibbard

Subdepartment of Quaternary Research, Botany School, University of Cambridge, Downing Street, Cambridge CB2 3EA, U.K.

This paper is based on a review of the histories of the Rivers Elbe, Saale, Weser, Rhine, Meuse, Scheldt, Thames, Somme and Seine. Two further rivers no longer in existence, the Baltic and Channel rivers, are also included. The histories of these rivers illustrate how the interplay of tectonics and climate have influenced the northwest European drainage system through the late Cainozoic.

The foundations of the modern drainage system were laid in the Miocene when earth movements associated with Alpine orogenesis and the opening of the North Atlantic were at their height. In general, these early rivers occupied shallow valleys and transported only chemically resistant minerals and lithologies.

The Pleistocene was marked by the appearance of cold climates. These climates resulted in fluvial dissection of the landscape, which stripped first regolith, then fresh material derived by periglacial processes. This material accumulated in the river valleys as gravel and sand deposits, which make up the overwhelming bulk of Pleistocene fluvial sediments. The rivers generally adopted braided courses during cold stages. The deeply incised modern valley system has developed largely as a result of rapid climatic changes over the past 2.4 Ma or so.

Throughout this period the river system has undergone repeated adjustments in response to continental glaciation. These responses are discussed. Particular attention is paid to the impact of the Anglian–Elsterian glaciation that blocked the southern North Sea to produce a vast ice-dammed lake, the overspill from which initiated the Dover Straits.

By contrast, interglacial sedimentation comprises predominantly fine, often fossiliferous sediments with rivers normally adopting single-thread channels, while estuarine sediments were deposited in areas invaded by high eustatic sea levels. The impact of sea-level change on the length of rivers and their courses is considered.

Introduction

Northwest Europe can be regarded as an unified geographical region which has undergone a broadly consistent geological and climatic history over the past few million years. Today the region lies in the temperate climatic and vegetational zone but during the recent past it has been subjected to climatic change that has given rise to long periods of periglacial and even glacial conditions. In coastal areas, sea-level change, largely driven by glacial eustasy, has caused intermittent regressions and transgressions; in and adjacent to glaciated regions, isostatic and forebulge effects have also caused major land- and sea-level variations. All these variations are superimposed on longer-term trends in climate and tectonic evolution of the continent.

The drainage system of northwest Europe has evolved in response to the development of the continent during the late Cainozoic. The position of the region at the margin of the Eurasian continent has been crucial in determining the events that have influenced its geological evolution. By their nature, rivers owe their existence to, and are strongly influenced by, tectonic activity and climatic variability. The interplay of these two major factors is responsible for determining the form of the modern river system, which should be considered as the product of continual remodelling.

Because river valleys are major zones of terrestrial deposition, an understanding of their detailed geological record will provide a calibration of the geological evolution of the North Atlantic region as a whole throughout the past three million years. In addition, the intercalation of marine or estuarine sediments in the river sequences allows correlation with the offshore and potentially with the deep-sea facies. As will be shown below, there is considerable evidence of palaeoclimatic significance in river-sediment sequences; from this evidence it is clear that the northwest European rivers have functioned as an integrated system.

For the purpose of this paper the great northwest European rivers are taken to include the drainage systems of the Elbe, Weser, Rhine, Meuse, Scheldt, Thames, Somme and Seine. In addition, two further rivers which are no longer in existence, the Baltic River and the Channel River, will also be considered. These rivers have been selected because their individual histories illustrate the impact and interplay of the major influencing factors. They also provide an important transect across the continent from north to south, i.e. from the zone of maximum impact of glaciation, particularly during the Middle and Upper Pleistocene, to a zone which was apparently never glaciated.

In order to appreciate the evolution of the northwest European drainage system it is vital to review the history of each basin. To provide a uniform stratigraphical timetable for sequences discussed, The Netherlands' Neogene stratigraphical terminology has been adopted (Zagwijn & Doppert 1978; Zagwijn 1985). In this scheme the Plio-Pleistocene boundary is placed at the base of the Praetiglian Stage (*ca.* 2.4 Ma BP). This widely applicable boundary is thought to be 'natural' in northwest Europe because it marks the first arrival of true cold climates. It has also been used extensively throughout the region in stratigraphical schemes. Recent redefinition of the Plio-Pleistocene boundary at a younger level (cf. Aguirre & Pasini 1985) is not easily applied to already established successions and therefore, for the sake of simplicity, has not been used here.

Throughout the paper, discussions are restricted to those concerning fluvial deposits and events. Although deposition by other agencies is mentioned in passing where relevant, it is not considered in detail. This particularly concerns loess, thick accumulations of which, together with associated palaeosols, are found in many of the river valleys discussed. Loess is not fluvial in origin and therefore, although its importance to the understanding of climatic evolution is appreciated (Catt, this symposium), it has only been discussed when required for stratigraphical purposes.

Tectonic background

Throughout the Cainozoic era, northwest Europe has been affected by two major tectonic influences: fragmentation of the Eurasian–North American plate, and Alpine orogenesis. The interplay of these two factors has resulted in great complexity of both structural and depositional patterns. In essence these two processes work in opposing ways, i.e. the break-up of the northern hemisphere plate produces tensional or extensional features whereas Alpine mountain building, arising from continent-to-continent collision between the Eurasian and

African plates, produces compressional features (cf. Ziegler 1978). The forces have led to activity along pre-existing tectonic lines that have been prevalent for long periods and in many instances have been inherited from previous phases (figure 1).

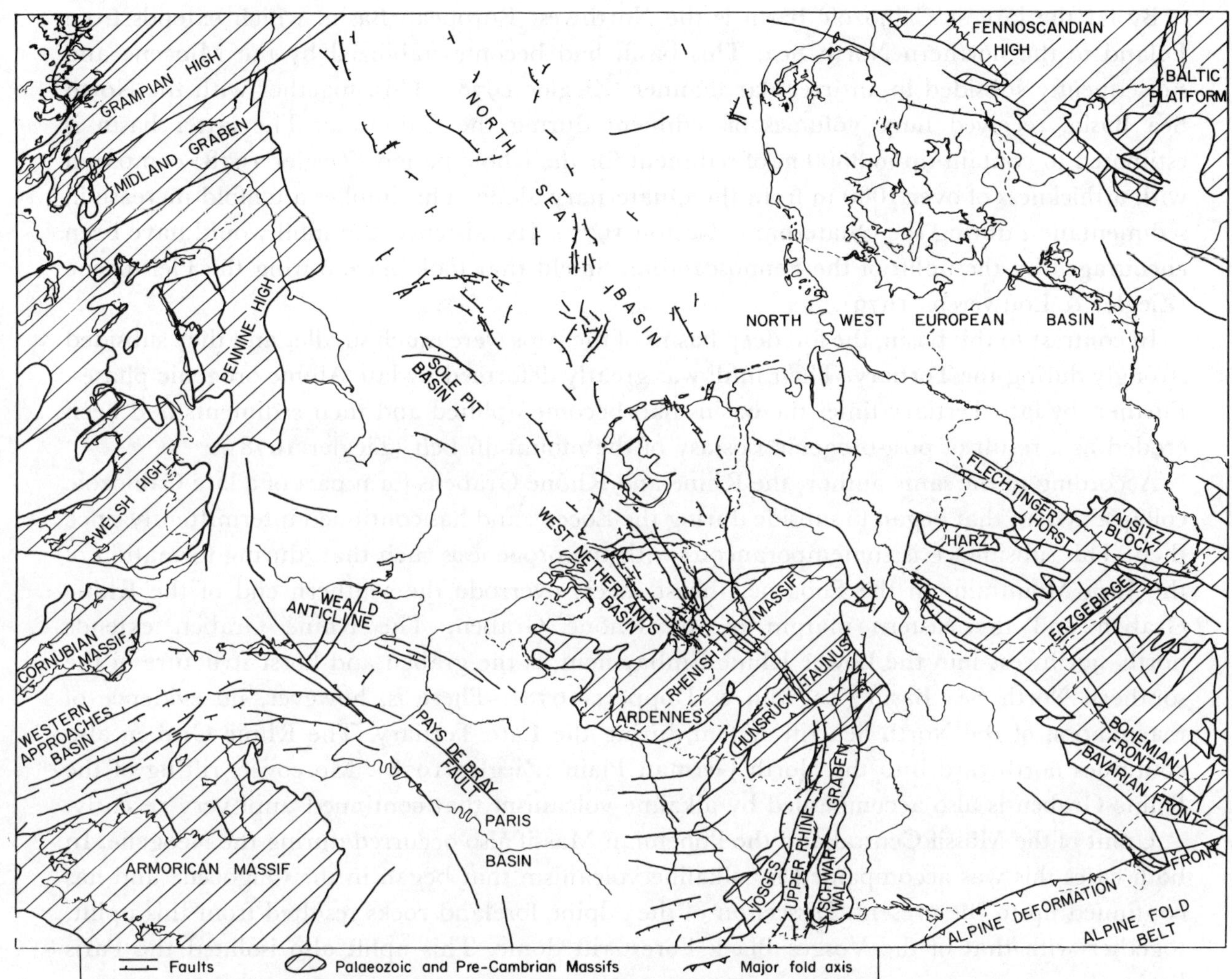

FIGURE 1. Tectonic map of the sub-Tertiary surface, modified from Ziegler (1978) with additions from Caston (1977), Zagwijn & Doppert (1978), Smith & Curry (1975), Quitzow (1974) and Ziegler & Louwerens (1979).

After the early Tertiary fragmentation of the Eurasian–North American crustal plate, the rift system of the North Sea Basin became inactive. This rift system was formed in the Mesozoic as a secondary rifting line in response to extensional stresses. Ziegler & Louwerens (1979) state that after the continental separation the crustal relaxation resulted in continued uniform basinal subsidence, which followed from a pre-existing pattern of differential subsidence initiated in the Jurassic (Ziegler 1978; Ziegler & Louwerens 1979).

The Channel may have had a similar origin, according to Smith & Curry (1975), in that ocean crust may possibly have been formed here, but that events affecting the northwest European continent may have deflected this zone southeastwards beneath the land mass. Later widespread uplift in post-Eocene times reached a peak in the Miocene resulting in inversion of the Hampshire and Channel Basins and the Bristol Channel and Western Approaches Trough, and updoming of the Pays de Bray Anticline in the Paris Basin (Ziegler 1978). This uplift was related to late phases of the Alpine orogeny. Local minor activity has continued into the Pleistocene (Bourdier & Lautridou 1974*a*).

By far the largest Cainozoic basin is the Northwest European Basin, which extends from Poland to the northern North Sea. This basin had become stabilized by the Miocene and subsequently subsided in an irregular manner (Ziegler 1978). This, together with the North Sea Basin, received huge volumes of sediment during the Cainozoic. The latter basin is estimated to contain up to 3500 m of sediment for the whole period (Ziegler 1978) compared with a thickness of over 1000 m from the Quaternary alone. This implies a tenfold increase in sedimentation during the Quaternary (Caston 1977). Its existence and infill would have been encouraged by the uplift of the Fennoscandian Shield that took place during the Pleistocene (Ziegler & Louwerens 1979).

In contrast to this basin, the foredeep basins of the Alps were much smaller and they subsided strongly during the Tertiary. Their infill was greatly deformed by late Alpine orogenic phases. Further, by late Tertiary times the basins had become uplifted and their sediments had been eroded as a result of post-orogenic isostasy of the mountain belt (Ziegler 1978).

According to the same author, the Rhine and Rhône Grabens form part of a late Cainozoic collapse system that began to subside during the Eocene and has continued intermittently since then. This subsidence is contemporaneous with the orogenesis such that, during formation of the Jura Mountains in the Pliocene, thrust sheets overrode the southern end of the Rhine Graben and the eastern margin of the Rhône Graben. The Rhine Graben extends north–northwest into the Lower Rhine Embayment as the graben and horst structure of the southern North Sea Basin (Zagwijn & Doppert 1978). There is, however, no evidence of reactivation of the North Sea rift system during the Late Tertiary. The Rhine Graben also continues northward into the North German Plain (Ziegler 1978). Moreover, rifting of the Rhine Graben is also accompanied by alkaline volcanism that continued until very recently.

Uplift of the Massif Central and the Bohemian Massif also occurred during the Neogene. In both cases this was accompanied by alkaline volcanism that began in the Oligocene and has continued up to the present. Dissection of the Alpine foreland rocks resulted from this uplift, together with that of the Vosges–Black Forest rift dome. This uplift also isolated the Paris Basin.

These important events have had profound effects on the fluvial system, which has undergone considerable adjustment and modification to adapt to the changes. As can be seen from a comparison of the tectonic and late Pliocene palaeogeography maps (figures 1 and 2) there is a close correspondence of major drainage lines and distribution of structural elements. For example, the Baltic River is aligned along the long axis of the Northwest European Basin, the central German rivers drain northwards towards this basin, the rivers Thames, Meuse and Rhine drain towards the North Sea Basin and the Seine drains towards the Channel and follows a major fault zone. In addition, the Rhine occupied the Central Graben region of the southern North Sea, Lower Rhine Embayment and the Upper Rhine Graben. Subsequent disruption of this pattern by later glaciation, sea-level change and river migration and capture has occurred, but nevertheless the tectonic effects remain an important controlling element.

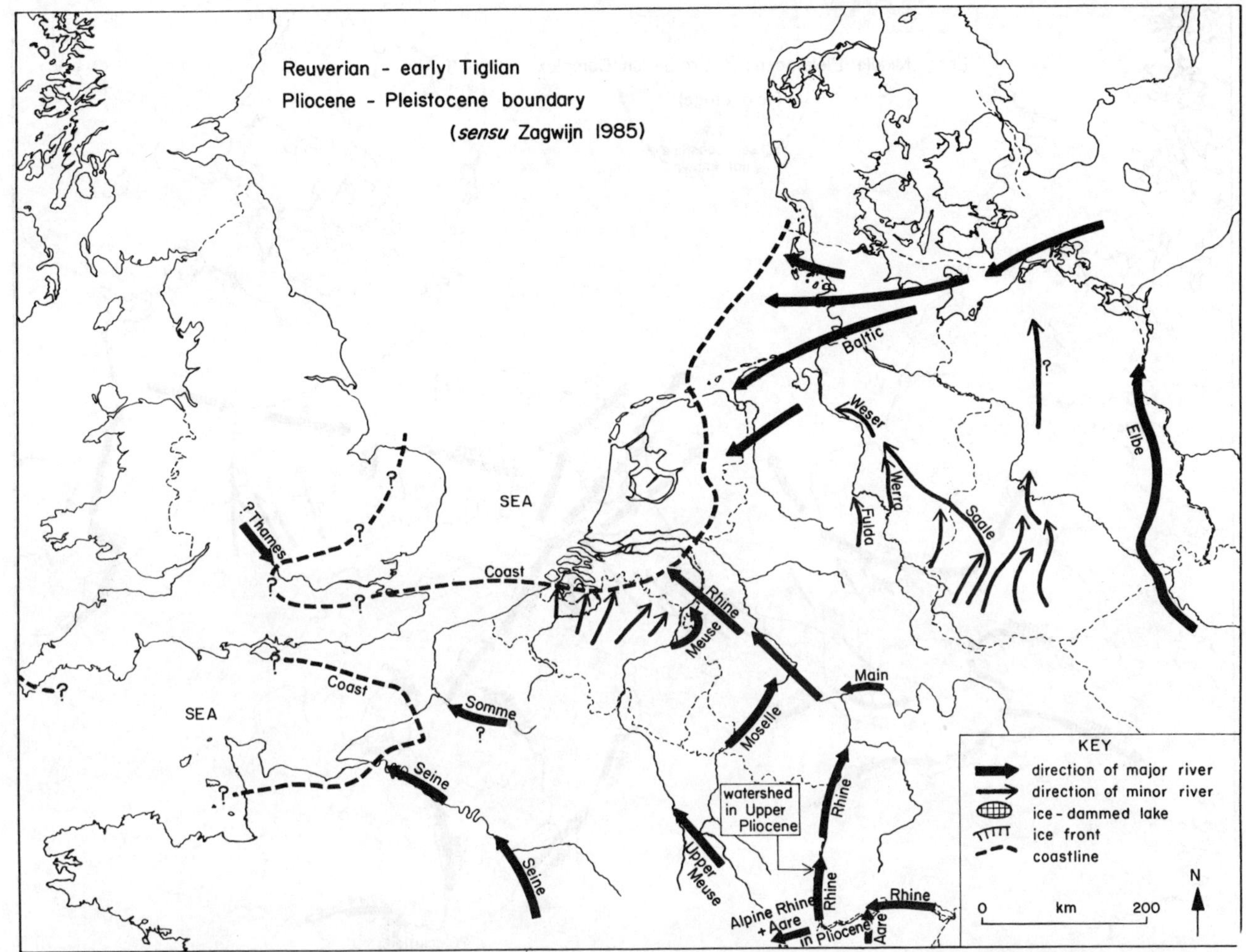

FIGURE 2. Schematic palaeogeographical reconstruction of the major drainage lines during the Reuverian (late Pliocene) to early Tiglian (early Pleistocene) Stages (*sensu* Zagwijn (1985)) based on sources quoted in the text.

THE HISTORY OF SELECTED NORTHWEST EUROPEAN RIVERS

This section comprises a series of reviews of the major river systems selected to illustrate evolution of the northwest European drainage system over the past 3–4 million years. The rivers are described in order, beginning in the north. For these reviews palaeogeographical summaries are provided in figures 2–6 for those periods for which it was practical to do so. In these reconstructions, periods of low eustatic sea level have been selected, to show the rivers at their maximum extent. During high sea-level events (i.e. interglacials), problems of location of the regional coastlines are acute, particularly for the older stages. Reconstructions for these periods have therefore been omitted.

(*a*) *Baltic River system*

Fluvial deposition became predominant in the Northwest European Basin in the Miocene (Quitzow 1953; Zagwijn & Doppert 1978) after infill of the North Sea Basin by marine sediments in the early Palaeogene. According to Bijlsma (1981) rivers from the Fennoscandian

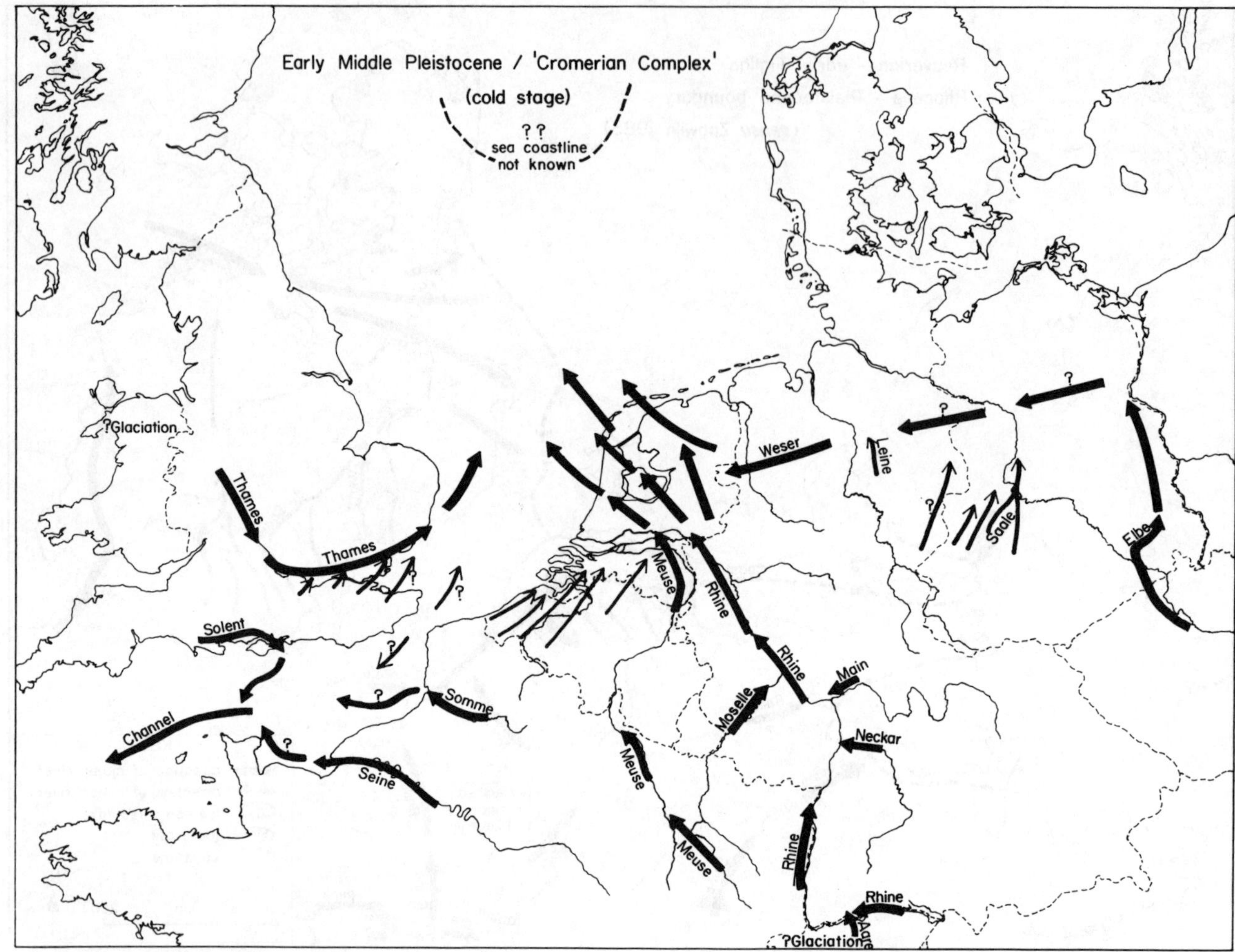

FIGURE 3. Schematic palaeogeographical reconstruction of the major drainage lines during the early Middle Pleistocene 'Cromerian Complex' at low sea-level stand. The courses are based on sources quoted in the text. (For key to symbols, see figure 2.)

Shield and the Baltic Platform in the north and the Variscan Massifs to the south (see Elbe and Weser systems, below) brought detritus into the Northwest European and East German–Polish Basins (cf. Ziegler 1978) from early in the Miocene. This sedimentary input was encouraged by uplift of areas marginal to the North Sea and East German–Polish Basins.

The deposits comprise white to grey–white quartz-rich sands intercalated with clay, gravel and brown coals (Crommelin 1954; Friis 1974). The gravel is predominantly formed of rounded quartz with subsidiary quartzites and silicified sediments, the latter mainly limestones derived from the eastern Fennoscandian Shield and southern Baltic Platform (Zandstra 1971). The river was therefore aligned through the area now occupied by the Baltic Sea and is therefore termed the Baltic River (Bijlsma 1981).

The first sediments of the Baltic River system occur in the Early to Middle Miocene of the G.D.R. and Poland, and are associated with brown coal deposition. According to Quitzow (1953) they can be subdivided into three sedimentary cycles, which may reflect periodic uplift of the source region. In the southwestern part of the North Sea Basin, fluvial and intercalated

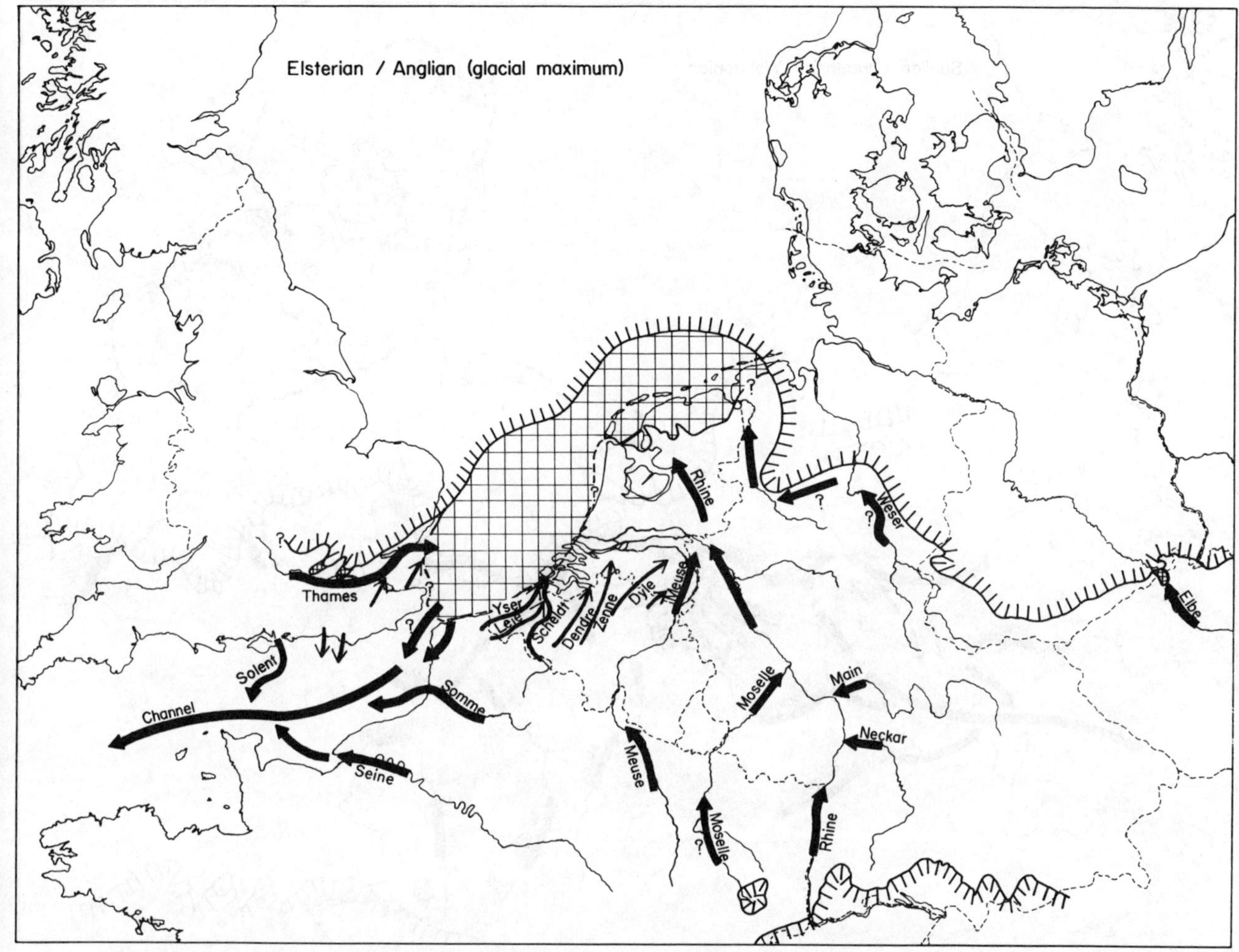

FIGURE 4. Schematic palaeogeographical reconstruction of the Elsterian–Anglian Stage at the glacial maximum. The ice-front positions and courses are based on sources quoted in the text, with additions from Kaiser (1960), Cepek (in Erd 1970) and Woldstedt (1967). (For key to symbols see figure 2.)

coastal sediments were laid down during the Miocene. In the F.R.G. and Denmark they are termed the Brown Coal Sands (Rasmussen 1961; Hinsch & Ortlam 1974).

Pliocene deposits of the Baltic River system are known from a narrow zone in the F.R.G. and The Netherlands. In the F.R.G. these Kaolin Sands have been described by Hinsch (1974). Their distribution is much affected by halokinetic movements: the greatest thicknesses are found between salt diapirs. The Kaolin Sands are exposed on the Isle of Sylt in Schleswig-Holstein, where the coarse to fine sands are of coastal and fluvial origin. They have been referred to the Brunssumian Stage by Weyl *et al.* (1955) and Averdieck (1971). Recent reinvestigation of these sands by Ehlers (1987) has shown that they contain evidence of presumed seasonal frost. Further south in Germany, brown coal layers interbedded in the sands have been found, for example at Oldenswort (Menke 1975) and Hamburg (Hallik, in Koch 1954).

The full areal extent of these deposits in northern Germany is not known. However, the course of the river seems to have been determined by subsidence in the North Sea Basin.

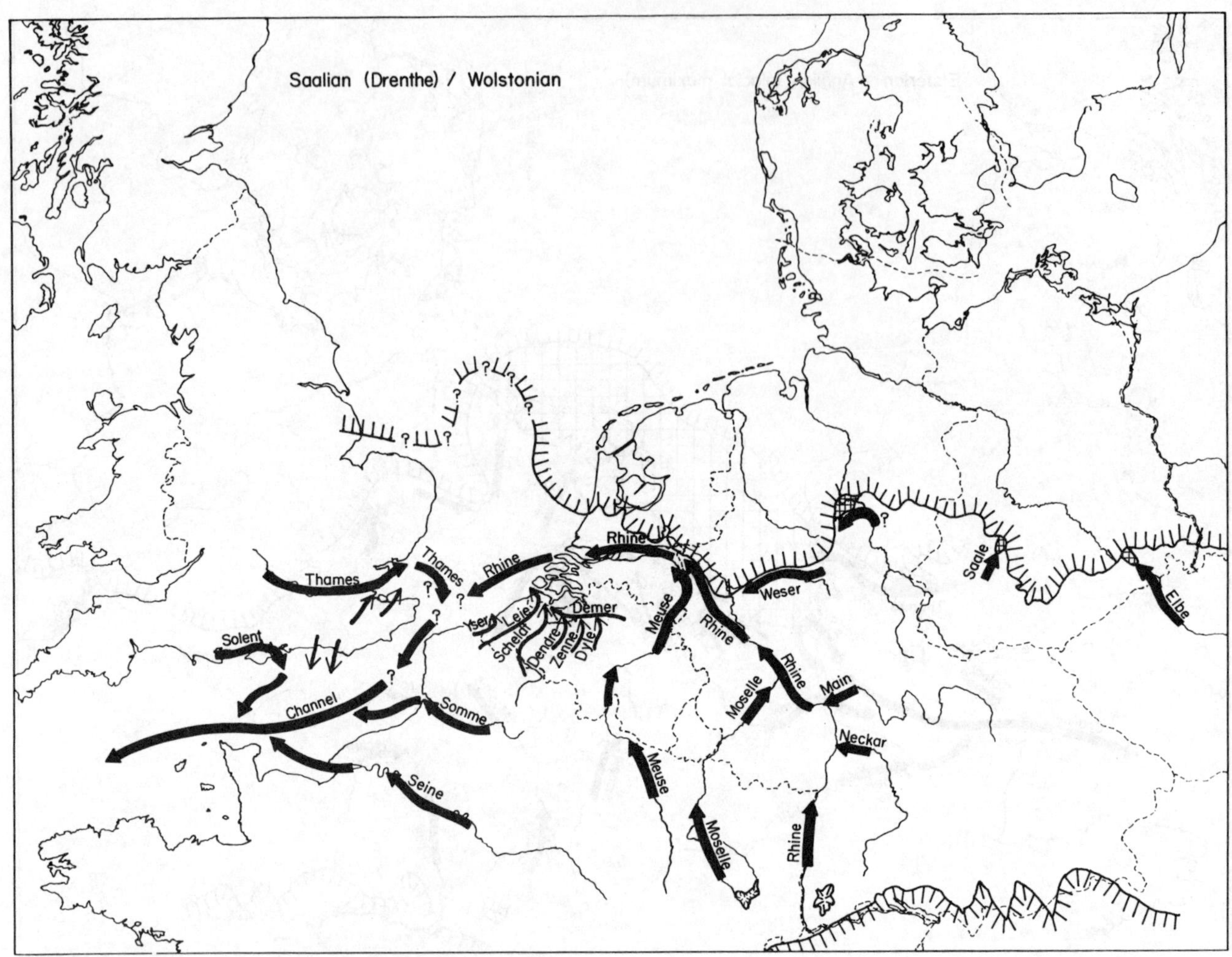

FIGURE 5. Schematic palaeogeographical reconstruction of the major drainage lines during the Saalian–Wolstonian Stage at the Drenthe substage glacial maximum. The ice-front positions and courses are based on sources quoted in the text, with additions from Kaiser (1960), Oele & Schüttenhelm (1979), Cepek (in Erd 1970) and Woldstedt (1967). (For key to symbols, see figure 2.)

In the northeastern Netherlands, deposits, predominantly fine sands but containing the characteristic pebble assemblage, are included in the Scheemda Formation of Doppert *et al.* (1975). This ranges from Brunssumian to Reuverian or possibly Early Pleistocene.

Baltic River system sediments are only known in the southern parts of the North Sea Basin in the Pleistocene, from northern Germany near Bremen to the Dutch–German border (Duphorn *et al.* 1973). They contain materials such as the characteristic lydite of the central German Uplands, as well as some large boulders that may have been ice-rafted from the eastern Baltic (Gripp 1964). In The Netherlands these coarse sands and fine gravels of the Harderwijk Formation (Doppert *et al.* 1975) range from early Tiglian to Waalian age. The earliest part of these braided river sediments is restricted to the northeast and eastern part of the Netherlands; however, by late Tiglian times and subsequently they had expanded both southwest and westwards to cover a large part of the country (Zagwijn & Doppert 1978). They also extend offshore to form a vast clastic sediment wedge (Zagwijn 1985; Balson & Cameron 1985). There are no younger sediments of this Baltic River. However, in the Enschede

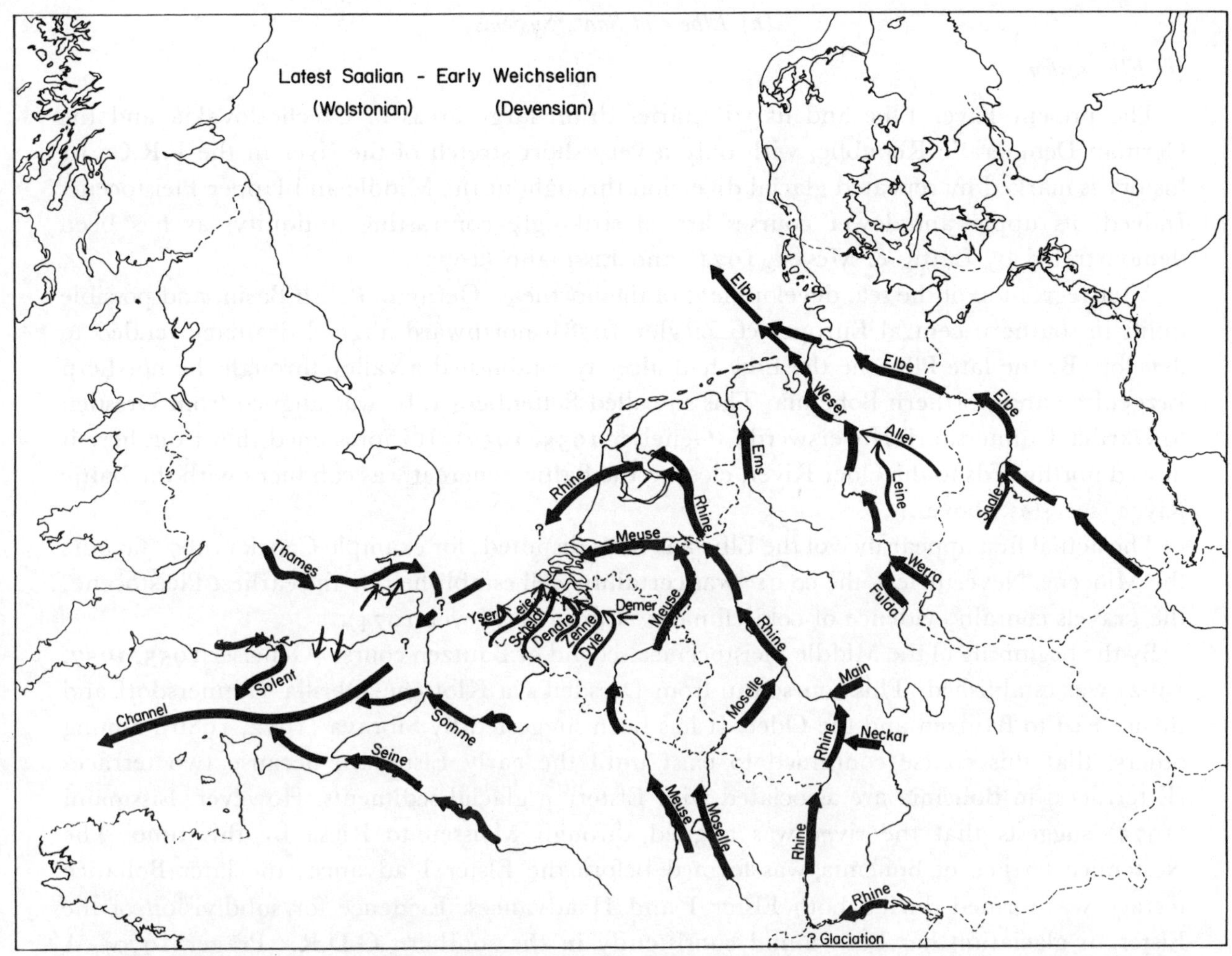

FIGURE 6. Schematic palaeogeographical reconstruction of the major drainage lines during the interval late Saalian–Wolstonian to early Weichselian–Devensian Stages at low sea-level stand. The courses are based on sources quoted in the text. (For key to symbols, see figure 2.)

Formation, which is largely of Menapian age (Doppert *et al.* 1975), crystalline rocks of Fennoscandian origin occur as large boulders or cobbles in the so-called 'Hattem Layers' of Lüttig & Maarleveld (1961) and Zandstra (1971, 1983). However, here they are intermixed with large volumes of gravel derived from central Germany via the Weser, etc. (i.e. north German rivers) and, near their southern margin, with Rhine and Meuse sediments. These 'Hattem Layers' indicate a change in depositional style and have therefore been interpreted by Bijlsma (1981) to result from contemporary glaciation of much of Fennoscandia in the Menapian. The meltwater streams from the ice sheet are thought to have discharged through the Baltic River and been confluent with the periglacial streams from central Germany. There is, however, no direct contemporary evidence of the Baltic River itself, nor is there evidence for its existence subsequently. Its destruction is attributed to glacial erosion, which resulted in formation of a proto-Baltic Basin.

After disappearance of the Baltic River, the north German rivers continued to flow into the eastern Netherlands until the Middle Pleistocene (cf. Maarleveld 1954) (see Weser (§*c*), figure 3).

(b) *Elbe and Saale Systems*

(i) *Elbe system*

The present River Elbe and its tributaries drain large areas of Czechoslovakia and the German Democratic Republic, with only a very short stretch of the river in the F.R.G. Its history is marked by repeated glacial diversion throughout the Middle and Upper Pleistocene. Indeed, its upper and lower courses are of strikingly contrasting antiquity, as has been demonstrated by Lüttig & Meyer (1974) and Eissmann (1975).

After regression of the sea, development of the northeast German–Polish Basin, and possible uplift in southern central Europe (cf. Ziegler 1978), northward-aligned drainage tended to develop. By the late Pliocene the Elbe had already established a valley through the northern Erzgebirge and northern Bohemia. This so-called Seftenberg Elbe was aligned from Dresden to Okrilla–Cunnersdorf–Hoyerswerda (Geneiser 1955, 1957). It is presumed that from here it flowed northwards to the Oder River, close to the Baltic, where it was confluent with the Baltic River (see § (*a*) above).

The actual first appearance of the Elbe has been disputed; for example Cepek (1967) favours the Miocene. Nevertheless, the course was certainly well established by the earliest Pleistocene; the gravels contain evidence of cold climates (Lüttig & Meyer 1974).

By the beginning of the Middle Pleistocene a second or Bautzen course (Geneiser 1955, 1957, 1962) was established. This course ran from Dresden via Klotsche–Okrilla–Cunnersdorf and thence east to Bautzen and the Oder. It has been suggested by Šibrava (1972, 1986), among others, that this course continued to exist until the early Elsterian, because two terraces (E terraces) in Bohemia are associated with Elsterian glacial sediments. However, Eissmann (1975) suggests that the river was aligned through Meissen to Riesa by this time. The Nestemice terrace of Bohemia was formed before the Elster I advance; the later Bohatice terrace was formed during both Elster I and II advances. Evidence for subdivision of the Elsterian glaciation has been found consistently in the southern G.D.R. (Präger 1970). A massive ice-dammed lake was formed in the Elbe and other valleys in front of the ice margin south of Dresden (Eissmann 1975). In this area the Elbe took up a course through Schmiedeberg, late in the Elsterian. However, this course must pre-date the glacial retreat, because Elsterian till rests on gravels of this phase. After the retreat of the ice sheet the river apparently established a course towards Berlin, where typical southern gravel has been found, in the Rathenow–Zossen–Wietstock area, filling valleys excavated into Elsterian glacial sediments. Mixing of reworked glacial and Elbe sediments has given rise to a characteristic pebble assemblage. Intercalated in these gravels are the so-called '*Paludina*' beds of Holsteinian age (Genieser 1962). According to the same author, the upper Elbe course through Meissen was not established before the Holsteinian, but this conflicts with the views of Eissmann (1975).

Advance of continental ice into eastern Germany during the Drenthe Stadial of the Saalian Stage caused the northerly course to be overridden as far south as Meissen, where extensive ice-dammed lakes were formed in the Dresden area (Eissmann 1975). The glacial lakes are recorded by sands and glaciolacustrine clays overlying fluvial gravel and sand of the so-called 'Main Terrace'. This 'Main Terrace' suffered later dissection so that upstream in Czechoslovakia (Šibrava 1972) it forms an important morphostratigraphical marker.

Renewed glaciation of the region during the Warthe Stadial resulted in profound changes in the upper Elbe drainage. The river was forced to flow towards the northwest by ice reaching the Fläming line. This diversion caused the Elbe to become linked for the first time with the Saale and Mulde rivers, which lost their lower courses (Woldstedt 1956). The continuation of the Elbe to the north took place late in the Warthe Stadial after ice recession, when the river cut through the Warthe Moraines at Magdeburg. It flowed through a series of glacial channels and topographic lows and established the present Elbe course through Hamburg into the North Sea Basin (Ehlers 1978; Meyer 1983).

By the earliest Weichselian the Bohemian–Saxonian drainage was operating as an integrated system. In the Late Weichselian, when ice advanced into eastern Schleswig-Holstein, meltwater from the ice sheet greatly enlarged the Elbe Valley in the Hamburg region (Meyer 1983). Moreover, its northwest extension can be recognized to over 150 km south of Helgoland on the North Sea floor (Figge 1980, 1983), and demonstrates that the river was graded to low eustatic sea level. The Elbe was also confluent with the Weser in the offshore area.

During the postglacial (Holocene) period this valley has been flooded as a result of high interglacial sea level in the estuary and fine sediments have accumulated on the late Weichselian fluvial gravels and sands.

(ii) Saale River

During the Pliocene, the German Central Uplands, Thüringia and the Weisse Elster basin were drained by northward-flowing streams. This probably resulted from uplift of the region (cf. Ziegler 1978). Before the Elsterian, rivers in the western Thüringia Wald transported crystalline rock pebbles to the Werra and Weser Rivers. Whereas rivers in the eastern Harz followed courses close to those at present, those from central Germany flowed northwards, to judge from discoveries of the typical heavy-mineral association in the Braunschweig area (Lüttig (1974, in Lüttig & Meyer (1974)). The latter, which includes topaz from Vogtland, is found in The Netherlands (Crommelin 1953, 1954; Maarleveld 1954; Bijlsma 1981) and is interpreted as evidence for inflow by rivers from the north German area before the Elsterian.

With the advance of Elsterian ice a series of ice-dammed lakes was formed in the river valleys in the Halle–Leipzig area. These may have coalesced into a single massive lake 450 km^2 in area (Eissmann 1975). The advance during the Elsterian brought ice to the Thüringia Basin and the Eastern Harz. In Elster I, glaciolacustrine clays were laid down on river gravels and till was deposited above. Subsequent ice retreat allowed laminated clays to accumulate once more, but these were overridden by the Elster II advance. The substantial moraines remaining after retreat of the ice caused many of the North German rivers to adopt new courses, some following glacial valleys; for example, the Saale, Mulde and Elster rivers formed a single river north of Zörbig (Knoth 1964; Ruske 1964).

In the Holsteinian, fine sediments were laid down in river valleys, for example, at Edderitz in the Riesdorf Saale Valley. A period of downcutting seems to have followed the Holsteinian and an Early Saalian interstadial (Dömnitz). Subsequent aggradation gave rise to a 'Main Terrace' found in each of the river valleys. For instance, this early Drenthe aggradation can be followed along the Saale river as far as the confluence with the present Elbe Valley. The lobate advance of ice in the Drenthe caused ice-dammed lakes to be formed in the river valleys. On retreat of this ice the rivers were again forced to adopt new courses; for example, the Saale changed its course to the west downstream of Merseburg (Schulz 1962; Lüttig & Meyer 1974).

In contrast to the Elbe, the Saale River was not overwhelmed by ice in the subsequent

Warthe Stadial. Instead, the river deposited gravels and sands downstream as far as Magdeberg. In the Gerwisch area an Acheulian industry is found in these sediments. The Saale joined the Elbe, as mentioned above, after diversion of the latter by the Fläming ice margin of the Warthe Stadial. Subsequent development took place under interglacial conditions in the Eemian and Holocene (Flandrian) with the accumulation of 'lower terrace' gravels during the Weichselian.

(c) *Weser system*

The River Weser is formed by confluence of two important rivers, the Fulda and Werra, and has been investigated by Lüttig (1974). Both these rivers drain the uplands underlain by Palaeozoic to Mesozoic rocks of the Rhenisches Schiefergebirge and the Thüringia Wald. However, below Porta Westfalica the Weser flows across the unconsolidated Quaternary rocks of the North German Plain. Near Verden, the broadly northwest to westerly flow direction of the Weser becomes more northwesterly where it is joined by its major tributary, the Aller.

In the Weser catchment there were two major tectonic influences during the Tertiary. Firstly, uplift of the Palaeozoic–Mesozoic rocks was associated with Saxonian tectonics. These late Cretaceous–early Tertiary movements reflect compression from the south during late Alpine orogenesis. This uplift is accompanied by contemporary downwarping of the Northwest European Basin (Ziegler 1978).

The earliest evidence of northward drainage from the Thüringia Wald is found in the Miocene lignite sands of the Kassel area, in which abundant flint and crystalline rocks occur. In the Neogene, uplift of the Lower Saxony uplands became more intense associated with faulting and basaltic volcanism. A series of gravels and sands derived from these upland areas are found but are restricted to local tectonic basins and depressions between salt domes. For example, post-basaltic flint and crystalline rock gravels and sands occur in association with clays of Reuverian (Late Pliocene) age in the Uslar Basin.

In the Pliocene, uplift continued south of a line from the North Harz to the Wiehengebirge of Lower Saxony. During this period the sea retreated to beyond the present coast. At the same time the rivers seem to have flowed towards the north and northwest, to judge from the occurrence of 'southern-type' quartz-bearing gravel spreads from areas between the present Weser and the Ems. Although these spreads are recognized, the source of the material and the actual river courses are not identified. Indeed Lüttig (1974) has postulated that the distribution suggests that the rivers may have been migrating across much of northwest Germany at this time.

Late in the Pliocene the rivers may have established more fixed courses, to judge from evidence in the upper Werra, upper Weser and Fulda valleys. Here gravels are found as elevated remnants at scattered localities. Pebbles from the present source area of the Weser are also found in a broad strip that extends generally westwards from the Minden–Nienburg area to the Netherlands and the Ems river. Pebbles of granite and porphyry from the Thüringia Wald are present. Similar material is also found at Braunschweig, deposited by the early Saale river which also flowed north of the Harz and followed the present Leine course. This westward-flowing stream was part of the Baltic River system (Maarleveld 1956; Bijlsma 1981).

Because of the structural relations in this region, i.e. uplift of the uplands and subsidence of the lowlands, older deposits progressively dip downvalley and pass below the modern floodplain level one after another in a transition zone. North of this zone the fluvial deposits occur in superposition with the oldest at depth. The transition zone at the upland front is not simple but is separated into structural strips marking the boundary between the two areas.

The so-called 'Upper Terrace' is much better known than the higher earlier deposits. Sediments of this complex unit are widespread in the upper Weser, Leine and other tributaries. The presence of well preserved 'Upper Terrace' remnants in the Leine–Aller system has been established by Lüttig (1960) on the basis of Thüringia Wald pebbles. At this time the Leine flowed westwards from Elze towards the Netherlands where the gravels are assigned to the Enschede Formation (Zagwijn 1985).

With the onset of continental glaciation in the Elsterian of north Germany, stratigraphical control is much improved. The lower courses of the rivers are still poorly known from this period; however, the upper courses have provided important evidence. The gravels and sands relating to this period are still included in the 'Upper Terrace' complex. In the Harz rivers, gravels and sands derived from the devegetated ground surfaces were laid down at the beginning of the Elsterian when the ice sheet entered the region and overrode river valleys. It brought about a complex series of drainage changes. A short recession was followed by a major readvance that took the ice across large areas of Lower Saxony and may have dammed the Weser at Hamelin, although this has recently been disputed by J. P. Groetzner *et al.* (unpublished). Water from the upper river courses, as well as the meltwater, must have been directed south of the Wiehengebirge into the Osnabrück area from where it may have continued into the Ems valley at this time. The contemporary ice margin extended from north of Minden west into the Emden area at Salzbergen (Meyer *et al.* 1977), and eastwards into Saxony and Silesia. East of the Weser Valley this margin marks the maximum extent of Pleistocene glaciation (figure 4 and 5). However, to the west the Saalian glaciation was the most extensive.

After glacial recession, proglacial lacustrine clays of the Lauenburg Clay Formation were laid down in basinal areas in the lower Weser area. This was accompanied by incision of the upper sections of the river valleys. With the climatic amelioration of the Holsteinian Stage, widespread interglacial sedimentation took place. These predominantly fine sediments have yielded much fossil evidence (see Lüttig (1974) for references). Reworked faunal material derived from interglacial sediment is found in the next unit of gravels and sand of the 'Middle Terrace' complex. This unit, which contains evidence for cold-climate braided river deposition, also includes Palaeolithic artefacts. The gravels and sands comprise local products of periglacial weathering. These early Drenthe Stadial deposits were overridden by continental Drenthe ice in the lower courses but remain undisturbed in the upstream areas. During the Rehburger phase the ice penetrated as far south as the central Netherlands, and to Braunschweig. The deposits of the 'Middle terrace' can be shown to be contemporary with this ice advance and ice-dammed lake sediments were laid down at Porta Westfalica (J. P. Groetzner *et al.*, unpublished). After an ice recession, downcutting occurred in the valleys; it was into the newly deepened valleys that ice flowed and deposited till. Drainage during this phase was aligned marginally to the ice and entered the Rhine in the Ruhr area (Thome 1980). This is supported by finds of exceptionally 'eastern' material in The Netherlands (Zandstra 1983). This second, Heisterberg-phase advance caused the ice to bulldoze preexisting fluvial and glacial deposits into a series of extensive marginal ridges.

The rivers, after having flowed westwards since the early Pleistocene, breached ice-pushed ridges and flowed northwards. The ice recession allowed the Weser to follow a series of glacially overdeepened areas to reach its present northwest course in the late Drenthe Stadial; this course may already have been exploited, before the glaciation, by the Aller River.

The subsequent interstadial and Warthe Stadial periods are not represented in the middle and upper Weser System except possibly by sands in the northern tributaries of the Aller.

By Eemian Stage times, however, the rivers had almost cut down to present levels and aggradation of fossiliferous sediments of argillaceous and/or calcareous lithologies took place (see Lüttig (1974) for references). The glacioeustatic rise in sea level caused the lower reaches of the river to be drowned by the sea. The river course was then almost as today.

Glaciation during the Late Weichselian did not reach the Weser system. Instead the area was subjected to intense periglacial conditions, which provided large volumes of coarse material by slope processes. The gravels and sands derived from these materials, the 'Lower Terrace' complex, reach substantial thicknesses and were deposited by the river in a braided mode. In some places the 'Lower Terrace' can be subdivided into lower and upper subunits; radiocarbon dates on material from these units indicate that sedimentation continued into the Late Glacial (Lüttig 1974).

In the Holocene fine sediments were once more laid down, dominated by the so-called 'flood loams' of overbank origin. Lowered humidity in the Atlantic period gave rise locally to sand-dune formation on the north and east banks of the Aller, the Leine below Hanover and the Weser below Minden. Today, however, these dunes are generally stable.

(*d*) *Rhine system*

The Rhine System is the longest and therefore the most important member of the present northwest European drainage system. At present it rises in the central Swiss Alps, where it is fed by glacial meltwater, and flows into the northern Alpine Foreland. In northern Switzerland it enters the Bodensee (Lake Constance) which marks the lower limit of the so-called Alpine Rhine. In this section the river, aligned northeast–southwest, has a very steep gradient and falls over 2000 m. Downstream of the lake, as far as Basle, the river passes through the Upper Rhine Valley. The gradient in this stretch, although much shallower than previously, is still rather steep and is broken at intervals by rapids and a large waterfall. Midway between Basle and Schaffhausen the first major tributary, the Aare, which drains much of the Alpine Bernese Oberland region, joins the Rhine. At Basel, the river turns northwards and enters the Upper Rhine Graben in which it flows to Oppenheim. No change of gradient occurs, however, even though this trough is one of active subsidence in which the river is confined for about 250 km. The Upper Rhine Graben is a major structural element that originated in the Eocene as a result of the late Alpine orogenic phases (Ziegler 1978). At Mannheim the second major tributary, the Neckar, joins the Rhine. The Neckar system drains much of the Black Forest.

Downstream of Oppenheim the Rhine passes into the Middle Rhine Valley by entering the Mainz Basin. The latter extends as far as Koblenz, an area of net uplift. At Mainz, the Main is confluent with the river; at Koblenz, it is joined by the Moselle. The former drains much of central Germany, whereas the latter brings water from the Vosges and Hunsrück regions. Throughout the Middle Rhine the river passes through the tectonically active uplifting Rhenish Massif (Rheinisches Schiefergebirge). The resistant lithologies in this region cause the river once more to steepen its gradient and occasionally to give rise to rapids (subsequently removed by man). This course has been occupied by the Rhine at least since the Pliocene (Bibus 1980). At Koblenz the river passes through the Neuwied Basin, which extends approximately to Andernach. Downstream the river crosses the northern boundary faults of the Rhenish Massif and enters the Lower Rhine Embayment, from where it has a gentle gradient into the Netherlands and the sea. However, in The Netherlands and southern North Sea Basin, subsidence has affected the Rhine downstream of Nijmegen (Quitzow 1974; Brunnacker 1975; Zagwijn & Doppert 1978). In The Netherlands the Rhine and Meuse (Maas) share a common delta and their history has been linked throughout the period considered here.

The tectonic complexity of the region through which it flows has had great impact on the type and preservation of deposits of the Rhine system. In essence this means that, in areas of net uplift like the Rhenish Massif, the vertical distance between terrace accumulations is increased relative to stable regions. By constrast, in areas of crustal depression, such as downstream of the Nijmegen–Cologne hinge zone and in the Upper Rhine Graben, thick sequences of younger sediments overlying older sediments are laid down (Brunnacker 1975). By its nature the Rhine has a long and detailed history, which is based here on Quitzow (1974), Boenigk (1978), Brunnacker (1975, 1978, 1986), Semmel (1973), De Jong (1965), Zagwijn (1974, 1979, 1985) and Zonneveld (1974).

The River Rhine seems to have formed in the middle Miocene, so that by the Pliocene it had already evolved considerably. According to Boenigk (1978) the early Pliocene Rhine deposited large quantities of coarse clastics in the Lower Rhine. The materials comprise highly stable minerals and rounded pebbles, mainly of quartz and quartzite derived from the Moselle area. These 'Hauptschotter' or Waubach Gravel sediments are included in the Kieseloolite Formation. At this time, the Rhine was only a local stream in the Rhenish Massif area.

In the Upper Rhine Graben, the southern Vosges and southern Germany, water flowed southwards to join a pre-Alp river aligned towards the southwest. This river almost certainly entered the Rhône system. Uplift of the Swiss Jura during the Middle Pliocene forced the Aare to flow to the northeast to join the Rhône tributary: the Doubs and the pre-Alp river southwest of Basle (Quitzow 1974). The Danube (Donau) was also created at this time. By the Middle–Late Pliocene the Rhine had cut back to a major watershed in the Upper Rhine Graben at Kaiserstuhl near Freiburg (figure 2). Depression of the Graben caused considerable expansion of the rivers Neckar and Main by capture of Danube tributaries in Swabia and north Bavaria (Quitzow 1974).

In the Lower Rhine the deposition of clastics decreased into the middle Pliocene; Waubach Gravel was replaced by widespread lagoonal clays and organic sediments (Red or Brunssum Clay) deposited under a warm-temperate climate. The contemporary river system seems to have adopted a meandering pattern with fine clastics restricted to narrow channels. Zagwijn (1963) has interpreted the couplet of Waubach Gravel and Brunssum Clay as a major sedimentary cycle. This he compared with stratigraphically higher levels deposited by the Rhine (such as the Late Pliocene Schinveld Sand and Reuver Clay; Early Pleistocene Tegelen Gravel and Tegelen Clay) and interpreted them as related to intermittent tectonic uplift of the central European uplands. Further coarse clastics of the Schinveld Sand and Gravels overlie the fine sediments and contain higher frequencies of less durable lithologies than the previous unit. These rocks, derived from the Rhenish Massif, mark the breakthrough of the Rhine into the Upper Rhine Graben. The gravels show evidence of deposition in an oscillating-discharge 'braided-type' system under a cool dry climate (Boenigk 1978). Deposition by the Rhine in The Netherlands was largely restricted to a delta in the Central Graben area (Zagwijn 1974).

In the Upper Reuverian the coarse clastics are replaced by widespread deposition of clay, sand and peat under a broadly humid, warm-temperate climate (Zagwijn 1960). According to Boenigk (1978), contemporary with this period there is a major change in the sediments in the basin, which he attributed to the expansion of the Rhine into the Alsace molasse region. This resulted in a great increase in the rivers' catchment.

In the Upper Pliocene–Lower Pleistocene, the southwest-flowing Aare filled a wide valley that developed between Basle and the present Doubs Valley with very coarse gravels termed the Sundgau Gravels. Quitzow (1974) equated these gravels with the earliest known deposits of the Upper Rhine Valley, the Older Deckenschotter, which he thinks were laid down by meltwater discharging into the Aare–Rhône system. By the time that the Younger Deckenschotter were deposited in the Upper Rhine, the Aare had broken through into the Upper Rhine Graben and established a connection with the Rhine. Exactly when this happened is not agreed; Quitzow (1974) equates it with the 'Hauptterrasse 3' (HT III; cf. Brunnacker (1975)) of the Lower Rhine on the basis of alpine rocks in Rhine gravels, whereas Zonneveld (1974) states it was not earlier than the Plio-Pleistocene boundary.

The Pliocene–Pleistocene boundary is represented in the Lower Rhine and The Netherlands by a widespread hiatus (Zagwijn & Doppert 1978), a possible consequence of local uplift or erosion. The southeast Netherlands and the Lower Rhine Embayment are marked by the development of a series of northwest–southeast aligned tectonic blocks at this time, differential movement of which gave rise to a series of basins and highs (cf. Zagwijn & Doppert 1978). The greatest thickness of Pleistocene sediments is found in the Central Graben near Eindhoven and its upstream extension, the Erft Basin; sediments also accumulated in the Venlo Graben and covered the intervening Peel Horst–Ville Block. In the earliest Pleistocene The Netherlands was beneath the sea, the Rhine discharging in the Central Graben region (Zagwijn 1974, 1979). Upstream in the Lower Rhine Embayment the Rhine deposited a series of gravels and sands which are interbedded with clays and organic deposits of limnofluvial origin. This complex of sediments forms the 'Hauptterasse 1' (HT I) unit of Brunnacker (1975, 1978, 1986). The basal Pleistocene gravels at Ville (Gravel b, ?Praetiglian) are of cold-climate origin, as are the later gravel units. This is overlain by temperate fossiliferous clays and intercalated sands referred to the Tiglian interglacial stage (Zagwijn 1960, 1985; Brunnacker 1975, 1978, 1986; Boenigk 1978). The next gravel deposits (gravel D), representing the Eburonian Stage, are the first to be deposited in thick units rather than sheets and for the first time include large ice-rafted blocks. However, according to Semmel (1973), drift blocks were already present in Praetiglian gravels in the Rhine–Main area. The river adopted a braided pattern during this and all subsequent gravel and sand depositional periods (Brunnacker 1975, 1978, 1986; Boenigk 1978). The remaining gravel and clay interbedded units of HT I are referred to the Waalian and Menapian Stages. The upper part of the HT I gravels and sands are probably the equivalent of the combined Rhine–Meuse Kedichem Formation in The Netherlands. The latter is overlain by the important Sterksel Formation, a unit of crossbedded coarse gravel-rich sands, about 50 m in thickness, which contain materials of Alpine provenance (De Jong 1965). It spans several interglacial and glacial periods (Zonneveld 1974). The uppermost part of the formation, the Weert Zone, which is equivalent to 'Hauptterrasse 2' (HT II) in the Lower Rhine (*sensu* Brunnacker 1975, 1978, 1986) dates from Glacial B of the 'Cromerian Complex' (Zagwijn 1985). The Weert Zone, a garnet-poor assemblage, is an important market horizon in The Netherlands. During the intervening period the Rhine–Meuse system, together with the North German rivers, had greatly expanded their deltas to the west and northwest into the southern North Sea Basin (Zagwijn 1974); these sediments are almost certainly the Yarmouth Roads Formation of the offshore area (Balson & Cameron 1985; Bowen *et al.* 1986).

In the Upper Rhine Graben, subsidence has resulted in the preservation of very thick Pleistocene successions; the thickest sequence (over 328 m) occurs near Heidelberg. The depression has also caused the tributaries to incise very deeply into the the valleys adjacent to the Graben (Quitzow 1974). In the Mainz Basin parts of the uplifted Kelsterbach Terrace may be equivalent to the Lower Pleistocene (von der Brelie, in Semmel (1973)).

During the deposition of the 'Hauptterrasse 3' (HT III) of the Lower Rhine, a general eastward shift of the river was in progress. This prevented deposition of the gravels and sands across the whole basin (Boenigk 1978). The gravels are rich in Eifel volcanic material (Brunnacker 1975). An important climatic development is also recorded in the sediments; the local appearance of widespread permafrost with ice-wedge casts and cryoturbation occurs for the first time in HT III (Brunnacker 1975; Boenigk 1978). The deposits of HT III are equated in The Netherlands with the Veghel Formation and are therefore of late 'Cromerian Complex' age. The continued eastward shift of the Rhine begun in HT III was completed by HT IV, when the river finally left the western basin because of uplift and infill by the Meuse.

Contemporary subsidiary deposition upstream continued in the Neuwied Basin, where equivalents of the Main Terraces can also be recognized. The early Middle Pleistocene sequence can be equated with the Upper Rhine Graben. The gravels and sands preserved in the Graben are intercalated with widespread fossiliferous sands of the Mosbachian, the equivalence of which is debated, but is probably of early Middle Pleistocene age. This deposit is also found in the Rhine tributaries, the Main and Neckar (Quitzow 1974). It is possible that the Mosbach deposits may also be in part equivalent to the Kelsterbach Terrace of the Mainz Basin.

Once the Lower Rhine was established east of the Ville Block, a series of terraces was formed, the so-called 'Middle Terraces' (*sensu* Brunnacker 1975, 1978, 1986). In spite of a thick loess cover, four morphologically defined gravel and sand units with intercalated fine sediments can be recognized. The earliest Middle Terrace 1 (MT I) gravels include acidic volcanic minerals from the Eifel in the Middle Rhine. The influx of volcanic minerals from the Eifel marks the change in The Netherlands from the Veghel Formation to the Urk Formation. This unit, marked by a lack of permafrost structures which Brunnacker (1975) relates to a mild climate, can also be equated with tectonic activity. Uplift of the Rhenish Massif during late deposition of the 'Hauptterrasse' caused massive aggradation in the Main and Neckar Valleys upstream, according to Brunnacker (1975). This may be a time-equivalent of the Günz period of the Alpine region (Quitzow 1974). Evidence of glaciation in northern Europe is indicated at this time ('Cromerian glacial C') in the northern Netherlands by the occurrence of Scandinavian erratics in the Weerdinge Member, basal Urk Formation (Ruegg & Zandstra 1977; Zandstra 1983).

The Lower Rhine Middle Terrace II of Brunnacker (1975) shows the return of periglacial conditions but with an intercalated temperate deposit of probable 'Cromerian IV' age (*sensu* Zagwijn 1985). The subsequent terrace MT III is underlain by coarse gravels, the so-called 'Channel Gravels', which, because they are overlain by the fossiliferous Krefeld Beds of Holsteinian age, are referred to the Elsterian Stage (Brunnacker 1975, 1978, 1986).

In the Middle Rhine Valley deepening continued into the Elsterian (Quitzow 1974), and similarly in the Upper Rhine. However, advance of ice in the Mindel (?Elsterian) certainly overran the Alpine and Upper Rhine. However, in The Netherlands, continental ice did not impinge on the Rhine in the onshore area. Nevertheless, ponding of the southern North Sea by ice caused the deposition of the glaciolacustrine Peelo Formation clays in the northern Netherlands (Ter Wee 1983*a*; Zagwijn 1974, 1979). The river discharged into this lake (see Channel, §*j*).

After deglaciation the North Sea transgressed into the Lower Rhine Valley in The Netherlands (Zagwijn 1974). Upstream fossiliferous sediments were laid down such as at Krefeld and Karlsruhe (Semmel 1973; Brunnacker 1975).

In the Saalian (?Riss), erosion of the Upper Rhine Valley reached a greater depth than subsequently, so that these earlier channel gravels underlie Weichselian deposits. The lower Middle Terrace can be equated with the outermost Alpine Riss endmoraines (Quitzow 1974). In the Lower Rhine the gravels of MT III, overlying the Krefeld Beds, are equated with ice advance in the Drenthe Substage (Brunnacker 1975, 1986; Zagwijn 1985) although Zagwijn equates this with MT IV. This advance caused bulldozing of Rhine sediments near Krefeld and in the Central Netherlands (Zagwijn 1974). The advance also deflected the Rhine from its northwest course in The Netherlands towards the west. However, after ice retreat the river partly readopted a northerly course by flowing through the Deventer basin towards Noordoostpolder (Van der Meene & Zagwijn 1978). This northern course, now occupied by the IJssel river, was abandoned by the Rhine in favour of a westerly course during the Middle Weichselian (Van der Meene & Zagwijn 1978).

A fourth Middle Terrace, MT IV, according to Brunnacker (1975) post-dates the Drenthe Substage. The age of this gravel unit can be defined by reference to the next lower accumulation, the Lower Terraces, which in The Netherlands are equivalent to the Kreftenheye Formation (Zagwijn 1985). These braided river gravels and sands of Rhine and Meuse origin include peats and organic material of Eemian age. The underlying gravels are therefore, by implication, of Saalian age (De Jong 1965; Zonneveld 1974; Zagwijn 1985). The Kreftenheye Formation closely follows the modern river course in The Netherlands and can be traced out beneath the southern North Sea to at least 50 km offshore (Jelgersma *et al.* 1979). According to these authors, the Rhine discharged through the Straits of Dover at this time (figure 6).

Deposition of the Kreftenheye Formation resumed in the Weichselian, a twofold subdivision of which is possible, as the upper horizons contain pumice from the Laachersee eruption (*ca.* 11 ka BP) in the Late Glacial Interstadial. Upstream two Lower Terraces can be identified by using the same criterion, according to Brunnacker (1975). Indeed equivalents of the Lower Terraces are known throughout the Rhine system and can be related to Würm endmoraines in the Alps (Quitzow 1974).

Since the last glaciation flood silts have been deposited on the underlying gravels in the Rhine Valley. In the Alpine Rhine the glacially overdeepened basin of Bodensee is being infilled by massive accumulations of deltaic deposits, which extend upstream to Chur.

(*e*) *Meuse* (*Maas*) *system*

Over the past few decades the history of the Meuse has received attention from Dutch, Belgian and French workers, most notably Zonneveld (1974), Paulissen (1973), Maarleveld (1956) and Pissart (1974).

The present River Meuse system drains the eastern part of the Paris Basin, Lorraine and the Ardennes in southeastern Belgium, from where it flows northeastwards into the Lower Rhine

Graben. After crossing a series of northwest–southeast-trending major faults, the river turns first northeast then eastwards to parallel the Rhine through the southern Netherlands to the sea. In their lower reaches the Meuse and Rhine have closely linked histories; they share the same general course and the Meuse has been a Rhine tributary for part of its Neogene existence.

According to Pissart (1974) a primitive Meuse system was already in existence in the Miocene and was represented by northwest flowing streams. However, by the late Pliocene–early Pleistocene, a series of captures in the region north of Mézières had resulted in the formation of northeastward flowing drainage. The Meuse system has lost a considerable part of its basin during the Quaternary as a result of river capture: for example, the Aire was lost to the westward-flowing Oise in northern France and the Upper Meuse was lost to the Moselle and thus to the Rhine.

The drainage-system changes must have been encouraged by differential local tectonic movements such as subsidence of The Netherlands' Basin region and uplift of the Paris Basin. These differential movements have given rise to local deviations of the terrace deposits and varying presentation of remains; for example, on the Mesozoic rocks of the Lorraine basin the Meuse deposits are very poorly preserved, whereas between Mézières and Liège they are better preserved on the more resistant Palaeozoic lithologies.

Divergence of older aggradations upstream of Namur indicates that the movement axis migrated northwards progressively during the Quaternary from Givet to Anseremme; local uplift of the Liège region may account for strongly inclined downstream aggradations. In South Limburg, where the river is no longer restricted to the valley, very wide accumulations are found. According to Zonneveld (1974) the quartz content of the aggradations decreases with age; this result indicates progressive incision of the Ardennes Massif through the Quaternary.

In Plio-Pleistocene times the river had adopted a wide, shallow valley trending in a southwest–northeast direction. Several levels of braided river sand and gravel aggradations have been distinguished. A peat bed within sediments of the Sempelveld Terrace is probably of late Tiglian age (Zagwijn 1985). The oldest deposits in the Lower Meuse, often referred to as 'Traînée Mosane' and comprising only very resistant quartzite and flint, occupy a very shallow valley, represented by gravel on the Ubachsberg Massif and in the Lower Rhine Basin (Boenigk 1978). It is referred to the late Pliocene. The oldest terraces are aligned south and east of this massif.

The next lowest aggradation in the South Limburg area, the Margraten Terrace, includes the first evidence of detritus derived from the Upper Meuse Vosges catchment into the Lower Meuse. This material indicates that by this time the drainage link from the Vosges region had been established, although Bustamente (discussion in Pissart (1974)) believes that this link was already established in Plio-Pleistocene times. Also at this time the Dutch–German border area was greatly influenced by tectonic movements along the Rhine Graben boundary faults. These led to the Meuse's adopting a more northerly course by periodic northwest migration; the resulting thick gravel aggradation forced the Rhine to flow further to the east in the Lower Rhine Valley. Organic sediments resting on gravels beneath the Valkenburg–Sibbe Terrace (Boenigk 1978) have been referred to the Bavel interglacial Substage of the Bavelian by Zagwijn (1985). Equivalent terrace aggradations of these older terraces exist upstream of Liège at least as far as Mézières according to Pissart (1974), although this conclusion is almost entirely based on altitudinal correlations, which provide a well-developed 'stairway'.

The so-called 'Middle Terraces' in southern Limburg, and their downstream equivalent, the

Veghel Formation in the buried valley of Middle and North Limburg and northern Brabant, span the Middle Pleistocene. The 'Middle Terrace' sequence was studied by Paulissen (1973) who noted that almost all the deposits accumulated under periglacial conditions with the river adopting a braided mode. The only possible exception is the Lanaken Terrace, which may be of interglacial origin. These deposits form part of the Veghel Formation of the southern Netherlands (Zonneveld 1958, 1974). The lowest part of the Veghel Formation contains clay deposits at Rosmalen equated with 'Cromerian Complex Interglacial III' (De Ridder & Zagwijn 1962). Deposits of this formation can be traced into the central Netherlands. According to Bustamente (1974) by the time of deposition of the Caberg Terrace gravels, heavy-mineral analyses indicate that supply of Vosges material had stopped entering the Meuse. This cessation reflects capture of the Upper Meuse by the Moselle. Because the Caberg Terrace is thought to be of early Saalian age (Zagwijn 1985) this capture most probably took place during the preceding Elsterian Stage (figures 4 and 5). At Maastricht–Belvédère a fossiliferous temperate deposit overlies gravels and sands of the Caberg Terrace. Detailed investigation of the sediments indicates deposition early in the Saalian Stage (Vandenberghe *et al.* 1985; van Kolfschoten & Roebroeks 1985).

From late in the Saalian the Kreftenheye Formation sediments were laid down by broad braided streams in the southern Netherlands (De Jong 1965). These Rhine–Meuse sediments include interglacial organic sediments of Eemian age. According to Van der Meene & Zagwijn (1978) diversion of the Rhine by the Drenthe ice caused confluence with the Meuse. This was severed after the ice retreat, but a westerly Rhine course was re-established in the Middle Weichselian.

Aggradation of the youngest of the Saalian terraces in Limburg, the Eisden–Lanklaar Terrace (Paulissen 1973) was disturbed by movement on the northwest–southeast-trending Feldbiss Fault, a major line from The Netherlands (Paulissen *et al.* 1985). Minor faulting also occurred in the early Weichselian. This gives rise to local thickening of the gravels north of the fault scarp.

The gravels and sands of the Kreftenheye Formation of The Netherlands have an upstream equivalent termed the Mechelen–Grosveld Terraces (Paulissen 1973; Zonneveld 1974). According to Juvigné (in Pissart (1974)) the Mechelen Terrace contains volcanic minerals derived from the Eifel area, eruptions from which are known to be of post-Eemian age.

The youngest or lowest terraces of the Meuse have been assigned to the Late Glacial and earliest Flandrian (Holocene) by Van den Broek & Maarleveld (1963) and Paulissen (1973). Consequently, formation of the present Meuse floodplain is thought to post-date the Preboreal.

(*f*) *Scheldt system*

The Scheldt system comprises a high-density network draining a series of small, low-altitude basins. It is surrounded on three sides, north, south and east, by the Meuse catchment and on the fourth by the North Sea where the valleys of Yser, Aa and Hem, etc., are partly drowned. The most important members of the Scheldt system include the Lys (Leie), Scheldt (Escaut), Dender (Dendre), Zenne (Senne), Dijle (Dyle) and Demer rivers. They drain the northern border of the Dinant Basin, the upper table country of the Namur Basin, and the northernmost part of France. The main streams, the Yser, Dendre, Zenne and Scheldt, are generally aligned parallel to the present coast but this alignment is abruptly interrupted by the east–west-trending Scheldt–Demer valley towards the Scheldt estuary. In general the rivers are flowing over Tertiary rocks, except for the upper headwaters of the Leie and Scheldt which drain the Mesozoic rocks of north France. In these cases the sole source of coarse detrital material is the Cretaceous rocks which provide flint, although silicified sandstones (sarsen) are also known from the Tertiary.

The primary consequent fluvial system seems to have developed on a marine regression surface towards a west-southwest–east-northeast coastline north of the present Scheldt–Demer valley during the late Neogene (Tavernier & De Moor 1974). This regression appears to be associated with general lowering of the North Sea Basin. By the early Pleistocene the coastline was aligned southwest–northeast and subsidence of the Lower Rhine Embayment and Central Graben caused the development of a bay (Zagwijn & Doppert 1978). This, accompanied by slight uplift of the Brabant Massif, encouraged the streams to flow northwards (Vandenberghe *et al.* 1986) (figure 2). According to Tavernier & De Moor (1974) the deposits of the pre-Quaternary streams are intensely chemically altered and suggest that warm, moist conditions prevailed during their deposition.

The detailed pattern of early Pleistocene drainage in Belgium is not known but it is presumed that deltaic deposits were laid down during the Tiglian under a perimarine environment (cf. Paepe & Vanhoorne 1970) in the Campine area. Regression of the sea allowed expansion of the rivers northwards into the southern Netherlands. Here a 'pre-Scheldt' river has been recognized by Vanderberghe *et al.* (1986) at Galder, where it was confluent with the early Meuse (see above). This river, flowing under a cold climate, had a braided form.

The onset of rapidly changing climates seems to have initiated alternating periods of downcutting and aggradation in the rivers. Early high-level terrace accumulations seem to be restricted to interfluve areas; this observation is interpreted by Tavernier & De Moor (1974) to reflect lateral displacement of streams. This is best seen in the trunk streams Leie and Scheldt. In the northernmost part of the area, during the Eburonian and Menapian, an alternation of fluvioperiglacial and perimarine deposits is found. This alternation reflects sea level oscillations, the former being formed by rivers during cold periods and the latter being deposited during the higher sea levels of the temperate periods when the sea invaded the lower reaches of river valleys, a pattern that continues to the present day. The northward-directed drainage system seems to have persisted throughout the Early Pleistocene and much of the Middle Pleistocene as well. For example, braided river deposits of the Scheldt can be traced to Dordrecht during the 'Cromerian Complex'. By the Elsterian the drainage system, particularly the upper parts of the southern valleys, were established with fluvioperiglacial terrace deposits, mostly gravels and sands, occurring on modern valley sides (e.g. Leie and Scheldt). The drainage probably continued to flow towards Breda early in this period. Later in the Elsterian the so-called Flemish Valley, an east–west-trending depression south of the present estuary occupied by the Scheldt and Leie, may have been excavated. Its formation may have been associated with the opening of the Dover Straits, which favoured east–west aligned drainage (Vandenberghe & De Smedt 1979) (figures 4 and 5).

Deep erosion of river valleys during the Elsterian enabled penetration of their lower courses by the sea after a eustatic sea-level rise during the Holsteinian. From this point onwards the Yser seems to have drained directly into the sea at the coast, whereas the Scheldt–Demer system entered the Flemish Valley. Meanwhile the lower courses of the rivers were accumulating estuarine or perimarine deposits, and fine freshwater sediments such as peats and organic clays were accumulating upstream.

The Saalian Stage is well represented in the Scheldt System by two marked periods of valley incision each followed by aggradation of fluvioperiglacial gravels and sands. The Flemish Valley and the upstream extension, the Scheldt–Demer Valley, were greatly enlarged by lowering of interfluves by periglacial processes. During this time the series of obsequent northern valleys draining into the Demer river were probably established. With the onset of temperate conditions in the Eemian, marine trangression again invaded the lower parts of the valleys and submerged the Flemish Valley. The latter became a large gulf, the floor of which was subjected to tidal scour. Upstream, organic sediments accumulated in the valleys; on interfluve surfaces pedogenesis was active and peat was formed in the obsequent northern valleys.

Regression of the sea in the Weichselian accompanied a return to deep incision, followed by rapid deposition throughout the drainage system, the form of which was virtually as at present (Vandenberghe & De Smedt 1979; Tavernier & De Moor 1974) (figure 6). The fluvioperiglacial gravels and sands were laid down as before with the rivers adopting a braided mode. Organic sediments are occasionally found intercalated in the sediments formed as abandoned channel or hollow fillings. The dates, according to Tavernier and De Moor (1974), group between 45600 and 21113 years BP. In the Late Weichselian, cover-sand deposition became dominant in the area. This sediment buried the gravels and in some areas the rivers became sluggish and took on a meandering form (Vandenberghe & De Smedt 1979). Contemporary organic sediments from the cover-sand period have yielded radiocarbon dates of 12856–10025 years BP.

In the Flandrian (Holocene) a return to interglacial conditions has led once more to drowning of the lower parts of valleys by the sea and accumulation of fine organic and inorganic sediments upstream, interspersed by minor periods of incision (Vandenberghe & De Smedt 1979).

(*g*) *Thames System*

The River Thames System is the largest drainage basin in Britain. The Middle and Lower Thames forms the axial stream of the London Basin syncline, which comprises Cretaceous chalk and overlying Tertiary clays and sands. However, at Reading the river is aligned northwards, passing through a deep chalk valley from the so-called Upper Thames region. Here the river crosses a series of gently dipping Mesozoic clay and limestone formations. Upstream of Oxford several tributaries converge to form the trunk Thames river. This part of the catchment includes the Cotswold Hills and the south Midlands.

In the late Pliocene to earliest Pleistocene, the London Basin was submerged beneath the sea. Fossiliferous sands with a lithology and fauna similar to that of the Red Crag Formation of Suffolk occur at heights of up to 180 m on both the north and south sides of the basin (Chatwin 1927; Dines & Chatwin 1930). After reinvestigation of the North Downs localities these sands were assigned by John (1980) to the Headley Formation; Moffat & Catt (1986) have shown that the sands and possibly some gravel are of marine origin. Contours on the sub-Red Crag surface indicate a relative tilting of *ca*. 180 m between the western London Basin and the Suffolk coast localities. Assuming a broadly uniform age (Pre-Ludhamian = ?Praetiglian (Zalasiewicz *et al*. 1988)), either the eastern part of the basin must have subsided (West 1972; Moffat & Catt 1986), or the western part been uplifted in the early Pleistocene. It is not known whether the Thames existed at this time.

The subsequent marine regression was followed by the development of an early Thames drainage system. This is indicated by fragmentary deposits, termed Pebble Gravel, which occur at 120–130 m OD on the south Hertfordshire plateau. The gravels are composed predominantly of flint, much of which is derived from the Tertiary rocks. However, in some areas they contain a characteristic chert derived from the Lower Greensand of the Weald. These finds indicate the existence of south-bank tributaries (Wooldridge & Linton 1955). The Pebble Gravels, part of which was termed Nettlebed Gravel by Gibbard (1985), fall in height towards the northeast. Their precise age cannot at present be determined.

After a period of incision, a marked change in gravel composition is recorded in the next lowest unit, the Stoke Row Gravel (Gibbard 1985). This is dominated by a dramatic increase in the quartz and quartzite clast frequency; this material is apparently derived from Triassic conglomerates in the west Midlands. The downstream equivalent of this unit, the Baylham Common Gravel of Allen (1984), can be recognized in southern East Anglia and forms the uppermost member of the Kesgrave Formation. This abrupt increase is probably the result of an upstream capture, but it has also been attributed to glaciation in the west Midlands and north Wales (Bowen *et al.* 1986). The first influx of quartz-rich gravel into northern East Anglia was found in sediments of pre-Pastonian age by Hey (1976, 1980), Funnell *et al.* (1979) and West (1980*a*).

Four further quartz-rich gravel units occur in the Upper and Middle Thames and can be followed both into the Upper Thames and northeast into East Anglia (see, for example, Clarke & Auton 1982), i.e. north of the present eastern course. In the latter area they are referred to (with the Stoke Row and equivalents) as the Kesgrave Sand and Gravel Formation (Rose & Allen 1977), whereas upstream they form part of the Middle Thames Gravel Formation (Gibbard 1985) in the Middle Thames and the Northern Drift Formation (Hey 1986) in the Upper Thames (see Bowen *et al.* (1986) for summary). The next youngest member, the Westland Green Gravel, can be traced throughout the area and contains evidence of deposition under a periglacial climate with the river adopting a braided mode. It contains ice-rafted blocks and smaller clasts of volcanic rocks derived from north Wales (Hey & Brenchley 1977; Hey 1980; Green *et al.* 1980). The same sedimentary structures and lithologies occur in the lower units and the latter have been interpreted as evidence of multiple glaciation in the north Welsh Snowdon and Berwyn districts (Whiteman 1983; Bowen *et al.* 1986). These deposits presumably accumulated during the later Lower and early Middle Pleistocene, because temperate fossiliferous sediments of 'Cromerian Complex' age have been found intercalated within the younger members of the Kesgrave Formation (P. L. G., unpublished observations) (figure 3). The quartz-rich gravels are also present in the offshore region east of East Anglia, where they form part of the Yarmouth Roads Formation (Balson & Cameron 1985; Bowen *et al.* 1986). At this time, therefore, the Thames was contributing to the greatly expanded deltas that occupied the southern North Sea (cf. Zagwijn 1974).

Advance of the continental ice sheet in the Anglian–Elsterian caused burial of the Kesgrave Formation in all but the southernmost part of East Anglia. A characteristic palaeosol, the Valley Farm Soil, formed on the terrace surfaces; this soil was also buried by Lowestoft Till deposited by the ice sheet and forms an important marker horizon (Kemp 1985; Rose & Allen 1977; Rose *et al.* 1985, 1976). In addition, the Anglian ice sheet advanced into and overrode the contemporary valley of the Thames in Hertfordshire and also advanced into southern tributary valleys. This dammed the rivers and the resulting lakes progressively overspilled until the water reached an unglaciated valley, the Medway, via which the river could return to its earlier course in easternmost Essex (Gibbard 1977, 1979, 1985; Bridgland 1980, 1985) (figure 4). During the course of this glaciation the connection with the source of quartz-rich material seems to have been severed; the late Anglian Black Park Gravel and all subsequent units are poor in quartz in the Middle and Lower Thames (Gibbard 1985). This accords well with the Upper Thames, where the post-Northern Drift gravels are similarly poor in quartz (Briggs & Gilbertson 1973, 1980).

During the Hoxnian–Holsteinian temperate Stage, the rise of the eustatic sea level caused inundation of the Lower Thames. This transgression deposited estuarine sands, silts and clays on organic freshwater sediments at Clacton (West 1972; Bridgland 1980, 1985). Upstream the brackish water extended to Swanscombe, east of London, where fossil-bearing fine channel and overbank sediments accumulated. This interglacial period is poorly represented further up the valley, although organic sediments at Sugworth, previously correlated with the Cromerian, may be of this age (Shotton *et al.* 1980; Gibbard 1985). The river seems to have adopted a meandering channel pattern at this time.

The Wolstonian–Saalian of the Thames Valley is represented by a series of gravel and sand deposits laid down under periglacial conditions (figures 5 and 6). In the Middle and Lower Thames three major units, the Boyn Hill, Lynch Hill and Taplow Gravels, accumulated. These can be correlated in the Upper Thames with the Hanborough Terrace gravels, equivalent to both the Boyn Hill and Lynch Hill members; the Wolvercote Terrace gravels are equivalent to the Taplow member. The Wolvercote gravels contain flint, thought to have been derived from till at Moreton-in-the-Marsh (Bishop 1958; Briggs & Gilbertson 1980). Because the till has been equated with the Wolstonian glaciation of the Midlands, the gravels must either be contemporary with or post-date the glacial event. The previous aggradations, Boyn Hill, Lynch Hill and their equivalents, characteristically include considerable quantities of Late Middle Acheulian artefacts throughout the valley, but particularly in the Middle Thames region (Wymer 1968; Gibbard 1985).

At the end of the Wolstonian, after deposition of the Taplow Gravel and downcutting, accumulation of a further thin gravel unit gave way to deposition of sands, silts, clays and organic deposits in both the Thames and its tributaries during the Ipswichian–Eemian Stage. Once again sediments from this period are poorly preserved upstream of the estuary area, although some sites exist in the Upper Thames (Briggs & Gilbertson 1980; cf. Briggs *et al.* 1985). However, from Central London downstream a series of localities such as Trafalgar Square, Peckham, Aveley, Purfleet, West Thurrock and Grays, correlated with the Ipswichian, record a drowning of the valley by a rise of sea level (Hollin 1977). Water level in the estuary rose to *ca.* 10 m OD and caused considerable aggradation of silts and clays.

Fall of sea level in the Devensian–Weichselian led to incision into these estuarine sediments and the climatic deterioration resulted in a return to gravel and sand deposition by the braided river. During cold stages solution of Chalk bedrock often associated with scour has considerably modified gravel and sand accumulations leading to collapse or local thickening of sequences (Gibbard 1985). One such feature occurs in the Thames tributary Kennet Valley at Brimpton where Bryant *et al.* (1983) have described an Early Devensian profile preserved beneath younger gravels. Evidence of two interstadial and three stadial events was preserved at this site. In the Upper Thames, probable Early Devensian gravels occur beneath the Summertown–Radley Terrace (Goudie & Hart 1975; Seddon & Holyoak 1985) and these can be followed into the Middle and possibly the Lower Thames Valleys (Gibbard 1985 and unpublished). They were laid down under periglacial conditions (figure 6).

By the Middle Devensian further downcutting, followed by aggradation, led to accumulation of gravel and sands. These occur below the modern floodplain in the Upper Thames, but downstream they emerge and are found above the valley floor. In the London area organic deposits interbedded in the Kempton Park Gravel contain herb-dominated floras indicating a full glacial environment and give radiocarbon dates of 45–30 ka BP. Further downcutting and aggradation of valley-bottom Shepperton Gravel and equivalents occurred during the Late Devensian. In the Middle and Upper Thames dates of 14.5–10 ka BP have been obtained for organic deposits intercalated in the gravels. The organic deposits again contain fossils indicating the prevalence of cold climates with treeless shrub- or herb-dominated vegetation. This evidence implies a period of downcutting or non-deposition from 30–*ca.* 15 ka BP in the Thames system. During this interval, clayey silt 'brickearth' (Langley Silt Complex) sediment accumulated in the Middle Thames. This sediment, largely of local origin, contains a loess component and has been dated by thermoluminescence to a period centred on 17 ka BP (Gibbard *et al.* 1987).

In the Flandrian, organic sediment, tufa, clay, silt and sand have been deposited on the gravels and sands underlying the modern floodplain. This deposition has led to an infilling of depressions and the adoption by the rivers of single meandering courses. Since the Neolithic period, land clearance has given rise to accumulation of overbank clay and silt by flooding.

(*h*) *Somme*

The Somme is not strictly one of the 'great European rivers'. It is, however, very important historically: it has been investigated for over 150 years because of the abundance of mammalian remains and Palaeolithic artefacts recovered from its deposits.

The course of the Somme is strongly related to the geological structure of the region. It occupies the axis of a major northwest–southeast-trending syncline in which smaller anticlinal structures separate neighbouring minor tributaries; the Yères, Bresle, Anthie and Cache parallel the Somme and similarly occupy the floors of small synclinal structures. The age of this structure is thought to be Miocene, but later earth movements in the late Tertiary and possibly early Quaternary are also thought to have effected the area (Bourdier 1974*a*; Bourdier & Lautridou 1974*a*).

The present Somme Valley is cut for its entire length in Chalk. It drains the Picardy region or the northern part of the Paris Basin and flows broadly east to west to Amiens where it turns towards the northwest. The interfluve areas represent the dissected remains of a gently undulating surface underlain by a thick spread of clay-with-flints similar to those found in southern Britain (Bourdier 1974*a*; Bourdier & Lautridou 1974*a*; cf. Catt 1986).

No late Pliocene and early Quaternary deposits are found in the Somme region, but close proximity to the Seine region suggests that the two areas have evolved similarly (see below). It is therefore unclear when downcutting began, but it is attributed by Bourdier (1974*a*) to the initiation of periglacial climates and possibly to marine regression. The highest and therefore the earliest deposits are plateau gravels, which occur as thin patches near Grâce and downstream of Abbeville. They are much disturbed by later movement, cryoturbation and solutional collapse (Bourdier & Lautridou 1974*b*). They contain no evidence of dating and can therefore only be assigned to a period before deposition of the subsequent Very High Terrace deposits. Between deposition of the plateau gravels and the gravels and sands of the so-called Very High Terrace, about 25–30 m of downcutting occurred. At Montières–Grâce they are calcareous, in contrast with those occurring at lower levels (Bourdier *et al.* 1974*a*). Moreover, the gravels and overlying loess have yielded an assemblage of small vertebrates which according to Chaline (1974) demonstrate a 'Cromerian Complex' age. They also pre-date the Brunhes–Matuyama palaeomagnetic epoch boundary (Biquand 1974). A possible find of Ardennes sandstone in this deposit by Commont (1911), which would indicate a link with the Oise Basin, has never been confirmed by recent finds (Bourdier 1974*a*).

The next lowest sequence is that of the High Terrace, a complex unit which mainly comprises periglacial fluviatile gravels and sands much disturbed by cryoturbation. Throughout the valley Palaeolithic artefacts have been found in the gravels. For example, the type site of the Acheulian, St Acheul near Amiens, occurs in this terrace (Bourdier 1974*b*). In the Abbeville area the gravels are overlain by white sands of temperate origin. These sands, which may be in part of estuarine origin, have yielded the remains of large vertebrates that suggest correlation with the Cromerian *sensu stricto* of Britain (Bourdier 1974*b*) (?'Cromerian Complex Interglacial IV'). However, the Abbeville sands are most important for their rich so-called Early Acheulian ('Abbevillian') artefact assemblages.

The Middle Terrace is the next complex unit and can be subdivided into three subunits, MT I–III (Tuffreau *et al.* 1982). Throughout the Somme and its tributary, the River Avre, they again record a return to the deposition of coarse sands and gravels thought to have originated under periglacial conditions. Each subunit has yielded Acheulian artefacts. In the Avre Valley, particularly at Cagny near Amiens, sands, clays and loess overlie the gravels of MT II. The floral and faunal evidence (Bourdier *et al.* 1974*b*) indicate a climatic amelioration; a soil in the upper part of the sediments is equated with the Holsteinian Stage. This implies that the gravels beneath and possibly those of MT I are Elsterian, whereas those of the lower MT III may be of early Saalian age.

During subsequent Saalian time, deep incision of the valley was resumed; indeed the total downcutting in the Middle Pleistocene is of the order of 42 m (cf. Bourdier 1974*a*). Deposition of coarse gravel and sand of the Low Terrace (Tuffreau *et al.* 1981) again took place under a periglacial climate, according to Bourdier *et al.* (1974*c*); loess also accumulated. These gravels have again yielded Acheulian artefacts. However, overlying clays and loess contain abundant Levalloisian and Mousterian material. The gravels are covered at Longpré by fluvial sands, clays and tufa of temperate origin (Tuffreau *et al.* 1981; Sommé *et al.* 1984). These sediments are presumably of Eemian age, although the latter authors equate a palaeosol in the overlying loess with this period, and therefore the deposits would have to be pre-Eemian. At Montières, silts containing hippopotamus remains have also been recorded from beneath loess (Bourdier *et al.* 1974*c*).

The Weichselian of the Somme Valley is represented by loess covering older sediments and by downcutting to 12 m in total below the pre-existing sediments. The deposits of the Very Low Terrace are coarse gravels and sands and can actually be subdivided into two aggradations, one with its surface above the modern floodplain and a second underlying the floodplain. The latter are overlain by Postglacial peat and in turn by alluvial silts and clays. The silts and clays are attributed to inwash of fine sediment from land clearance and agriculture (Bourdier 1974*a*).

(*i*) *Seine System*

The River Seine and its major tributaries, the Marne, Oise and Aube, drain much of the north, central and eastern Paris Basin. Throughout almost the whole catchment the bedrock comprises Mesozoic and Tertiary rocks in which flint is the major pebble-forming lithology, although Tertiary siliceous sandstone (sarsen) is also present. The drainage courses are greatly influenced by local geological structure. For example, the Lower Seine downstream of Paris closely follows a shallow northwest–southeast-trending synclinal structure, which turns towards the southwest downstream of Rouen. This structure is also penetrated by two major faults trending in the same direction (figure 1): the Seine fault, which crosses the whole Paris Basin from Rouen to Limagne, and the Fécamp–Bolbec Fault, which runs from near Rouen westwards and continues beneath the Channel (Bourdier & Lautriodou 1974*a*). Minor tectonic activity has continued in the region into the Pleistocene.

Similarly to the Somme, the present Lower Seine is incised deeply into the gently rolling chalk country of the Upper Normandy Plateau. The surface is likewise underlain by often thick accumulations of clay-with-flints (Bourdier & Lautridou 1974*a*).

The earliest evidence of the Seine is recorded in the Lozère Sands (Kuntz & Lautridou 1974; Cavelier 1980). The deposits comprise quartz-rich sands, which are found in widespread patches through the Lower Seine, south of Paris and as far south as the Bourbonnais country south of Nevers. In the Lower Seine, the Lozère Sands were deposited in the delta when the river was apparently draining the northern Massif Central, an area now within the catchment of the Loire (figure 2). According to Macaire (1984), epeiorgenic uplift of the south and southwest part of the Paris Basin in late Pliocene–early Pleistocene caused drainage to flow to the west to form the Loire, rather than towards the Seine. At La Londe, southwest of Rouen, a small tectonic downfaulted depression contains a sequence that spans the Pliocene to Lower Pleistocene. These sands occur at the base and are overlain by the marine St Eustache Sands, which have been assigned to the Upper Brunssumian or Lower Reuverian (Clet 1982*a*). The Lozère Sands must therefore be presumed to be of early Pliocene age.

Renewed marine incursion in the Late Pliocene is represented by the St Eustache Sands, but these give way to lagoonal deposits in their upper levels and are replaced by a series of estuarine–lacustrine black and brown silts and clays of the La Londe Formation. Cooling of climate towards the top of the silt and clays is recorded by the palynology, but is not paralleled by any sedimentary change. The clays are correlated by Clet (1982*a*) with the Reuverian to Praetiglian.

The return of fluviatile conditions is signalled at La Londe by deposition of the bedded sands of the Fourmetot Formation (Kuntz *et al.* 1974, 1979). They are overlain by fluviolacustrine sand and gravel of the Roumois Formation. Intermediate between these units are ice-wedge casts, which indicate that periglacial climates prevailed during the intervening phase (possibly equivalent to the Eburonian (Kuntz *et al.* 1979)). However, these deposits are not thought necessarily to represent the Seine itself, but record cutting of local stream valleys under a cold climate. This suggests that major river incision in the region followed regression from the relatively high Tiglian sea levels (recorded at Bosq d'Aubigny in Cotentin (cf. Clet 1982*b*)) and the development of periglacial conditions.

The Lower Seine is characterized at present by its spectacular series of deeply incised meanders. On the basis of the history of fluvial aggradations it appears that these meanders were formed very early in the river's history and have subsequently been progressively modified (cf. Chancerel 1985, 1986). Although some rare pebble spreads occur above it, the first true terrace-like alluvial deposits are the strongly weathered gravels and sands of the Madrillet Formation, which were laid down after *ca.* 35 m of valley incision. The meanders cut directly into bedrock were already fully formed (Chancerel 1986). Further incision is followed by deposition of the Bardouville Formation. These first two aggradations show very little downstream gradient and may therefore indicate uplift of the coastal region (Alduc *et al.* 1979; Chancerel 1986); all subsequent terraces show a normal gradient upstream to Paris (Lautridou 1982). Two younger gravel and sand accumulations separated by a period of incision follow: the Radicatel and Berville Formations of Chancerel (1986). These two units may be equivalent respectively to the Forêt de Bord and the Rond de France Terraces of the Elbeuf meander (Lefèbvre *et al.* 1986). By their relation to younger aggradations these units are assigned broadly to the early Middle Pleistocene (Alduc *et al.* 1979) (figure 3).

Considerable problems arise for longitudinal correlation of fluvial aggradations in the meanders of the Lower Seine because of the varying style of deposition and erosion. This is attributed to the orientation of the meanders and to the possible migration of a 'knick-point' between the Elbeuf and Mantes regions (Lautridou *et al.* 1984). The result is that from Elbeuf downstream there are clearly developed aggradational units separated by periods of incision. In contrast, upstream near Mantes an almost continuous overlapping sequence of small-scale sedimentary units separated by low step-like cliffs is found (Lautridou 1982). Correlation between the regions can only be achieved by using 'marker' units (Lautridou *et al.* 1984; Lécolle 1984; Chancerel 1985, 1986).

The next youngest aggradation is the St Pierre-les-Elbeuf Terrace of the Martot Formation or Middle Terrace (Alduc *et al.* 1979; Lautridou 1982; Lautridou *et al.* 1984). This unit is equivalent to the Bois Delamare Complex as seen at Anneville, where the sediments contain evidence for at least three erosional and depositional events (Chancerel 1986). The gravels are periglacial in origin (Lefèbvre *et al.* 1986). At the St Pierre type section, fluvial gravels and sands underlie an interglacial tufa and a very thick loess sequence which includes four palaeosol horizons. Correlation of the basal soil with the Holsteinian suggests that the terrace gravel and sands are of pre-Holsteinian, i.e. Elsterian or more probably late 'Cromerian Complex', age. Upstream the 'High Terrace' (P1, XII–XV) of Lécolle (1984) can be equated with these sediments on the basis of stratigraphical relations. The deposits are greatly disturbed by post-depositional bedrock solutional collapse and periglacial cryoturbation.

Possible estuarine deposits resulting from high eustatic sea level drowning the valley during the Holsteinian Stage may occur at Cléon near Tourville (Lautridou 1982) and possibly in part of the Tankerville sequence.

The next lowest aggradation is the Oissel Formation or Pont de l'Arche Terrace (Lefèbvre *et al.* 1986); this formation is equated with the Elsterian Stage (formations VII–XI of Lécolle 1984). The subsequent Tourville Formation and the Criquebeuf Terrace (Lefèbvre *et al.* 1986) are broadly of Saalian age (Lautridou 1982; Alduc *et al.* 1979). The base of the Tourville Formation comprises periglacial, fluvial gravels and sands that have yielded an abundant fauna of cold affinities. Within these gravels a silt horizon of warm climatic character is preserved. The gravels and sands are overlain by estuarine sands of the Tankerville Member, the latter also well developed at Tankerville itself and at Cléon (Alduc *et al.* 1979; Lautridou 1982). Part of this sequence is probably of Eemian age.

Post-Eemian dowcutting was followed by two further gravel and sand aggradations that, in the modern Lower Seine, are buried by Flandrian (Holocene) alluvium and estuarine sediments (Lefèbvre *et al.* 1974). These two units, the Rouen Formation (Porcher 1977) or Formations I and II of Lécolle (1984), are assigned to the Weichselian: the higher to the

Middle Weichselian, and the basal gravels flooring the valley to the Late Weichselian, according to Lefèbvre (1974), Alduc *et al.* (1979) and Lautridou *et al.* (1984).

Investigation of the offshore Seine Bay–Fosse Cotentin area by Alduc *et al.* (1979) has shown that it is possible to follow equivalents of Weichselian, Saalian and possibly pre-Saalian formations into the Fosse Centrale, north of Cotentin. At least two Weichselian accumulations, together with probable equivalents of the Tourville, Oissel and St Pierre Formations, i.e. six terraces, are represented (Lautridou 1982) (figures 4–6).

(*j*) *Channel River system and the Straits of Dover*

Detailed investigations over the past 30 years have revealed considerable evidence of the morphology and geology of shelf seas adjacent to northwest Europe. Investigations by French and British workers have shown that the Channel floor is generally smooth with its surface gently inclined from the continental shelf margin to the Dover Straits. Near the coasts, steeper, ramp-like features are present (Smith 1985). This smooth surface is underlain over most of the channel by a thin, often mobile sediment cover beneath which bedrock, much of which is of Mesozoic age, occurs (Larsonneur *et al.* 1979; Smith & Curry 1975). Cut into this bedrock in the central and eastern Channel is a complex series of narrow valleys or channels, most of which are infilled by unconsolidated sands, gravels and clays (Dingwall 1975). The best known of these, the Hurd Deep, has been shown by Hamilton & Smith (1972) to contain a multiple infilling and to have a composite origin by combined fluvial deposition and tidal scour. Dingwall (1975) and Smith (1985) have subsequently shown that the valleys are interlinked to form a drowned drainage system formed during periods of lowered sea level. The concept of greatly lowered sea levels during cold stages is well established; however, local evidence suggests sea-level minima of 90–130 m below present levels (for summary see Hamilton & Smith 1972). Fluvial downcutting and infilling during these periods would be followed by tidal scour by marine transgression during periods of rising and falling sea levels.

In detail, the valleys form a complex anastomosing system and many are overdeepened, particularly at confluence points, a feature also found in fluvial systems on land, e.g. the Thames (Berry 1979). Smith (1985) has demonstrated that they can be linked with the Seine, Somme, Béthune, Solent, Arun and other coastal rivers. The central or Lobourg valley can also be followed to the Dover Strait where it merges into the Fosse Dangeard of Destombes *et al.* (1975), a series of scour hollows formed on the Gault–Lower Chalk contact (Smith 1985).

There is little doubt therefore that the valley system is mainly of fluvial origin and that it functioned during periods of low eustatic sea level. This is contrary to suggestions that the system may have originated by glaciation, as proposed by Destombes *et al.* (1975) and Kellaway *et al.* (1975) (see Zagwijn (1979) and Oele & Schüttenhelm (1979) for further discussion).

Evidence for the age of this river system and therefore for the present form of the Channel is rather limited by problems of access. However, the system must certainly have been at least partly in existence by the time of deposition of the earliest river sediments in the neighbouring terrestrial river valleys. On the basis of the work by Alduc *et al.* (1979) this would indicate that valley formation pre-dates deposition, possibly of the Quillebeuf and certainly of the St Pierre-des-Elbeuf Terrace gravels of the Seine, i.e. it is at least pre-Elsterian or 'Cromerian Complex'. Earlier river terrace deposits are apparently not represented and the Plio-Pleistocene sequences found on the neighbouring land appear to be unrelated to the Channel palaeovalley system (cf. Smith 1985). Certainly the Channel existed in the Pliocene, to judge from the occurrence of marine sediments at St Erth, Cornwall (Jenkins *et al.* 1986), at La Londe, Normandy, and at several sites in Brittany (Lautridou *et al.* 1986) (figure 2). It would seem likely, therefore, that for much of the Lower Pleistocene the area was one of net erosion and, because there is some evidence for tectonic activity in the Western Approaches Basin during the early Pleistocene (Smith & Curry 1975), this may be partly the cause. It is conceivable that this uplift may have been contemporary with that experienced in the London Basin (§ (*g*) above).

River downcutting and deposition presumably occurred throughout the early Middle Pleistocene during periods of low sea level; contemporary periglacial fluvial deposits were laid down by the Somme, Seine and possibly some of the English South Coast rivers, e.g. the Solent River. A proto-Channel River system must have existed at the time (figure 3). However, the question of the existence of the Dover Straits is relevant here. Much debate has centred on the presence or absence of this passage and when precisely it was formed. In general it is assumed that the breach was established in the late Pleistocene (Destombes *et al.* 1975).

On the basis of Pliocene marine faunas in East Anglia, Funnell (1972) suggested that a link between the North Sea and the Atlantic through the Dover Straits was present at the time. However, Jenkins *et al.* (1986), after investigations of Pliocene foraminifera at St Erth, showed that the southern North Sea faunas were, by comparison, impoverished and indicated that no link was present. This view is supported by Zagwijn (1979).

Whether or not a Dover Strait existed in the Middle Pleistocene is also questioned, because sediments interpreted as indicating marine deposition occur in northern France at Herzeele (Sommé 1979; Paepe & Sommé 1975) and in southwest Belgium at Loo (Vanhoorne 1962; Paepe & Baeteman 1979). At least two marine transgressions have been recognized in these sediments, although part of this sequence, at least, is assigned to the Holsteinian (Vanhoorne 1962). This indeed would mean that the Straits may have been open before the Holsteinian; Zagwijn (1979) suggests possibly the Pastonian. Nevertheless, there is strong reason from indirect biostratigraphical evidence (West 1980*b*) and from subsequent events to believe that a substantial land barrier existed for much of the Pleistocene. This barrier was the extension of the Weald–Artois anticline in the Chalk (Smith 1985). Such a barrier may conceivably have been overtopped occasionally during periods of high sea level but for the prolonged cold stages it would have been dry land and drained by consequent streams analogous to those draining the neighbouring regions today (cf. Stamp 1927).

According to Zagwijn (1974, 1979), the southern North Sea Basin was greatly infilled during the Lower and early Middle Pleistocene by delta sediments, derived from the North German rivers, the Rhine and the Meuse, represented by the Yarmouth Roads and Winterton Shoal Formations offshore (Balson & Cameron 1985; Cameron *et al.* 1987). Smaller rivers such as the Thames also contributed to this huge delta complex (figure 3). Throughout this period and indeed into the Elsterian these rivers drained northwards. The advance of continental ice sheets from the north in the Elsterian–Anglian Stage would therefore be likely to have blocked the North Sea Basin, preventing water from escaping to the Atlantic, thus causing an immense ice-dammed lake to form in the remaining ice-free part of the basin (figure 4). Clearly such a lake would only have formed if the British and Scandinavian ice sheets coalesced; there is, however, much evidence to support this, such as the occurrence of Norwegian erratics (e.g. rhomb porphyry and larvikite) in East Anglia. In spite of this, until the ice sheets did form a barrier, the drastically lowered eustatic sea level and isostatic crustal depression would be expected to have caused major incision of river valleys. Just such an incision is indicated by deep valleys beneath the widespread freshwater glaciolacustrine sediments of the Peelo Formation, preserved extensively in the northern Netherlands (Ter Wee 1962, 1983*a*). Indeed, these valleys are so deep that they have been compared to those demonstrably formed beneath the ice sheet in Germany and East Anglia, even though they contain river sediments of southern derivation and were apparently never overridden by ice (Ter Wee 1983*a*; Zagwijn 1974, 1979). Nevertheless the considerable depth of the valleys (some reach to over 350 m locally (Ter Wee 1983*a*)) suggests that processes other than solely fluvial downcutting may also have been involved in their formation. Large areas of the North Sea adjacent to The Netherlands are also underlain by the Peelo Formation, according to Oele (1969, 1971); however, at least some of these deposits in the Dogger Bank area have more recently been correlated with the Saalian by Zagwijn (1979). Nevertheless, Balson & Cameron (1985) and Cameron *et al.* (1987) have identified widely distributed glaciolacustrine sediments of the Svarte Bank Formation in the British offshore sector.

Evidence from the British side of the North Sea has not previously been linked with that in The Netherlands. However, it has long been recognized that the Corton Sands, which occur intermediate between the Cromer Till and Lowestoft Till Formations in eastern East Anglia, were deposited during local ice recession (Corton Interstadial) in a substantial ice-dammed lake (West & Wilson 1968; Pointon 1978; Bridge & Hopson 1985). More recently, re-examination of Cromer Till sequences in Norfolk by the author and colleagues (P.L.G., unpublished results) has indicated that large parts of these sediments originated from the ice sheet's advancing into a large waterbody, because much of the till is of waterlain origin (cf. Kazi & Knill 1969; Gibbard 1980). Associated sand members within this formation appear to represent meltwater discharge events.

There is therefore widespread evidence for deposition in a glaciolacustrine environment during the Elsterian–Anglian Stage in the southern North Sea Region. The presence of such sediments above present sea level implies either isostatic depression of the area or elevation of lake level or else an interplay of both factors.

The water level in such a lake, formed by inwash from the rivers draining into the southern North Sea and from ice-sheet meltwater derived from much of western Europe, would undoubtedly rise until a suitable drainage outlet was reached. It seems highly likely that this outlet occurred at what is now the Straits of Dover (cf. Smith 1985). Indeed, a thick channel fill of gravels, clays and sands, deposited by a large southward-flowing river, has been identified near Wissant on the northwest French coast (Roep *et al.* 1975). However, it is probable that more than one channel functioned at this time and that the second became the Straits (figure 4).

When discussing this possibility Smith (1985) proposed that outflow from an ice-dammed southern North Sea lake would be catastrophic, comparable to the outburst from Lake Missoula in the northwest United States in the last glaciation. There is, however, a major difference between the latter and the Dover Straits examples. The drainage of Lake Missoula resulted from collapse of an ice-dam that blocked a valley, whereas a substantial bedrock Chalk ridge was present at the Dover Straits at this time. Overspill of an ice-dammed lake of the latter type would be expected to be balanced only by input, so that when a bedrock col was reached, the resulting erosional capacity of the water would be equivalent to the water entering the lake.

Given that chalk is highly susceptible to dissolution by cold water, downcutting by the outflow stream may have been relatively rapid, but certainly not catastrophic. Such an overspill may have followed pre-existing stream valleys, joined the Channel River and discharged through this to the sea.

Progressive enlargement of the outflow col would have followed, the southward flow possibly causing rivers in the adjacent areas of the southernmost North Sea to take up courses towards the outflow as the col was lowered. Although realignment of the Scheldt System streams has been attributed to this process by Vandenberghe & De Smedt (1979) and Vandenberghe *et al.* (1985), there is as yet no unequivocal evidence for a similar adjustment of the Thames. On the other hand, Thames deposits do show a southward tendency in the offshore Thames Estuary area (D'Olier 1975; D. R. Bridgland & B. D'Olier, personal communication). In addition, no Thames deposits of post-Elsterian age have been found in the offshore region so far investigated by the British Geological Survey (T. D. J. Cameron, personal communication).

From the distribution of laminated (varved) clays of the Peelo Formation and the remarkably similar Lauenburg Clay deposits found in northwest Germany and as far north as Denmark (Ehlers 1983), as well as laminated clays filling deep glacial valleys in East Anglia, such as the Nar Valley (Ventris 1985), and offshore (Balson & Cameron 1985) it is clear that glaciolacustrine conditions prevailed in the North Sea Basin until very late in the Stage. It is not certain whether all the finds of laminated clays represent a single lake or not, particularly during the deglaciation period. Nevertheless, the widespread occurrence of these sediments strongly suggests that the North Sea lake persisted as a major feature during the Late Anglian–Elsterian period. This implies that the Dover Straits outlet may have been subject to isostatic uplift, which prevented complete drainage of the lake. Such uplift might be expected to have favoured adoption of northerly courses by the rivers. However, because the lake seems to have persisted very late, downcutting of the Dover Straits col could have continued, keeping pace with uplift, enlarging the trough and forcing those rivers near the col to adopt southerly courses.

Similar events to those in the Elsterian may have been repeated in the Saalian Stage; glaciolocustrine sediments, albeit of smaller scale, are again found associated with the Drenthe ice advance off the Dutch coast (Oele & Schüttenhelm 1979). The Drenthe advance certainly forced the Rhine and Meuse to flow westwards (De Jong 1965; Zagwijn 1974; Ter Wee 1983*b*), so that, during this period of low eustatic sea level, the Rhine, Meuse, Thames and Scheldt could have become confluent and drained through the Dover Straits into the Channel River (figure 5). Such a pattern might, however, require an ice-dam across the North Sea during this time, a possibility that seems less likely from the latest investigations (cf. Cameron *et al.* 1987). Nevertheless, both Oele & Schüttenhelm (1979) and Jelgersma *et al.* (1979) have stated that the Rhine and Meuse drained southwards after the Drenthe Substage until the Late Weichselian, on the basis of the distribution of the Kreftenheye Formation sediments. If these rivers did so, then the Thames and Scheldt must surely also have adopted this course (figure 6) despite recent studies of the bedrock topography by B. D'Olier & D. R. Bridgland (personal communication) that suggest this might not be the case for the former. Clarification of this matter must await further research on the deposits.

It seems likely, however, that the formation of the Dover Straits has been polycyclic, so that once the breach was formed by fluvial action, it was enlarged by tidal scour after interglacial marine transgression, then fluvially dissected once more during a subsequent period of low sea level, and so on.

The progressive increase in size of the Dover Straits is reflected in the indirect evidence of impoverishment of British interglacial palaeontological assemblages through the later Pleistocene (cf. West 1980*b*). For example, during the Holsteinian Stage the Dover Strait may have been land for much of the interglacial, only being breached during the period of maximum sea level. However, by the Eemian (Ipswichian) a substantial barrier to plant and animal migration seems to have been present.

It is interesting to note that the origin of the Dover Straits and the existence of a substantial southern North Sea ice-dammed lake suggested above were proposed originally by Stamp (1927) and regularly ever since by Zonneveld (1958), Gullentops (1974), Dingwall (1975) and most recently by Smith (1985).

It is clear therefore that the Channel River has been a major drainage line throughout much of the later Pleistocene and for at least part of its history carried the runoff of a huge area of northwest Europe. If the correlations proposed here are correct, it last functioned during the Late Weichselian and earliest Postglacial (Alduc *et al.* 1979).

Discussion and conclusions

From the foregoing, it is apparent that the response of the northwest European drainage system to the external influences of the past three million years has been complex. However, certain patterns emerge and indicate the impact of climatically and tectonically controlled variables on the system as a whole. These patterns will be discussed below.

The foundations of the modern drainage system were laid in the Miocene, when earth movements associated with both the Alpine Orogeny and the opening of the North Atlantic were at their height. Although greatly modified subsequently, the precursors of the modern rivers can be identified. The form of this system was apparently closely linked to geological structure and took a broadly simple form, although this may be an artefact of the preservation of the sedimentary sequences.

Throughout the Miocene and Pliocene the rivers were generally transporting chemically resistant minerals and lithologies, such as quartz or flint, derived from long-term weathering under moist warm-temperate climates. Relatively little mechanical weathering is recorded except when associated with tectonic uplift, which resulted in rejuvenation of rivers such as the Rhine and gave rise to deposition of gravels. Otherwise, rivers seem to have occupied shallow channels and meandered across large regions with little evidence of deep valley incision. There is no indication of the development of permafrost, although some seasonal frost activity did possibly occur in the late Pliocene.

The onset of true cold climates in the Praetiglian (Pleistocene) Stage caused a marked change in depositional style with the input of gravel and sand derived by stripping of the deep regolith inherited from the Pliocene, and the adoption of a braided form by the rivers Rhine and Meuse. This however did not have great impact across the entire region, to judge from the apparently limited change in the Seine system. The return to temperate conditions in the Tiglian caused a readoption of the meandering river form and a renewed deposition of fine sediments.

In the Praetiglian, glaciation may already have been established in the Alps; ice-rafted blocks occur in Rhine sediments. By this time the Alpine Rhine may have been captured by the Upper Rhine system, having previously been flowing to the southwest to join the Rhône. Whether or not glaciation occurred elsewhere in northwest Europe at this time is not clear.

The return to cold climates in the Eburonian–Menapian period established considerable frost weathering and therefore the loading of rivers with freshly weathered detrital material. The resulting periodic incision and later deposition under braided river régimes is well marked. The great increase in sediment deposition is recorded by the vast expansion of deltas in The Netherlands (Zagwijn 1974). Glaciation seems to have become a widespread phenomenon, not only in the Alps, but also in upland Wales and in Scandinavia at this time. The result of the latter was the destruction of the Baltic River in the Menapian, after the possible formation of a proto-Baltic Basin by glacial scour.

The first records of the prevalence of true permafrost in lowland areas comes from the Dutch–Belgian border where cryoturbation of Tiglian T4c age has been observed (J. Vandenberghe, personal communication). Moreover, at La Londe in the Seine region, ice-wedge casts occur at the base of the Roumois Formation and are thus assigned to the Eburonian Stage. Nevertheless, it would seem very unlikely that permafrost did not occur before this in the Praetiglian, to judge from the evidence for persistent cold climates.

By the later Middle Pleistocene the pattern of repeated incision and sediment accumulation was well established. However, advance of continental ice sheets into lowland northwest Europe for the first time in the Elsterian Stage had widespread effects. The area overridden by the ice was subjected to total landscape remodelling, with old river courses destroyed or buried. An entirely new landscape was formed beneath the ice by glacial and glaciofluvial erosion and deposition; the sculpturing of deep glacial valleys was to have a striking palaeogeographic impact after the ice retreat. At the ice margin, major river valleys were dammed all across the region. The Thames and its tributaries were diverted southwards, the Elbe was dammed and the North German rivers were deflected westwards. However, the most striking feature was the development of a massive ice-dammed lake in the southern North Sea, into which the Thames, Rhine, Meuse, Scheldt and possibly the Ems all discharged. Overspill of this lake almost certainly initiated the Dover Straits and greatly enlarged the Channel River system.

Subsequent glaciation in the Saalian and Weichselian Stages had comparable effects, once again causing major drainage diversion and landscape remodelling. In the Saalian, major ice-pushed ridges in the Netherlands and Germany forced the Rhine to follow a more southerly course; after deglaciation the Elbe took up its present course through Hamburg. Glaciation in the Weichselian seems to have influenced the drainage system least, as it did not reach lowland northwest Europe except in the extreme north: where meltwater discharged through the Elbe, in central and northern Britain, and in the Alps where water discharged along the Rhine. The rest of the region outside that subjected to glaciation had a periglacial climate during this period. In periglacial areas, where both direct and indirect effects of glaciation have been minimal, rivers have maintained their courses except when river capture has taken place, e.g. the Seine and Somme.

Throughout the cold stages of the Pleistocene the rivers overwhelmingly deposited gravels and sands derived either from periglacial weathering or glacial sources. The result is that valley systems contain vast thicknesses of cold-climate sediments deposited by rivers that flowed in a braided or wandering, often multi-channelled, form. These sediment accumulations are generally separated by periods of non-deposition or incision outside subsiding areas. This incision was previously often attributed to interglacial events. It is now, however, clear from modern process studies that incision almost certainly also occurs when river runoff is highly seasonal but when limited supplies of detritus are available, i.e. predominantly under cold climates. Highly peaked discharges characterize modern rivers in periglacial regions. This pattern is caused by storage of water as snow on the land throughout the winter and subsequent rapid melting in spring, giving rise to the nival flood event. Such events may carry much of the year's precipitation in a few days. The result is high flow velocities and therefore strong erosion and transporting power of very short duration. In glacial meltwater streams these discharge peaks may be modified by cycles of glacial melt and discharge of stored water, such as jökulhlaups. Lack of vegetation and the development of permafrost ensure a rapid return of precipitation in catchments to rivers, so that storms may also produce marked flood events in such periglacial streams.

This situation contrasts markedly with that pertaining under temperate climates, where the dense vegetation cover and the development of ground-water storage causes a cushioning of storm or spring melting effects. Vegetation cover protects the land surface by fixing and stabilizing the mineral soil, and plants also take up water that is transpired directly into the atmosphere. Storage of water and slow flow in the ground reduces flood events by controlling the volume of water entering a drainage system. This results in river discharge becoming much less peaked and causes rivers to flow all year round. In general the lack of coarse material available, except that obtained by erosion of the channel itself, and the reduced flow velocities prevent lowland rivers from greatly altering their courses. They tend therefore to adopt single-thread meandering courses with floodplains, the hollows in which become filled with organic sediments or flood silts, clays or sands.

These observations are borne out by the sequences preserved throughout the region, in which interglacial sediments tend to be fine inorganic or organic materials intercalated in gravel and sand sequences. For this reason they have a low preservation potential and in general represent less than 10% of the total fluvial sequences. The remaining 90% of sediments are of cold-climate origin.

The comparison of modern, deeply incised river valleys with the apparently shallow features predominant during the earlier Neogene strongly suggests that alternating deposition and incision, which gives rise to these deep valleys, is a direct consequence of rapid climatic change. In particular it implies that frequent downcutting and aggradation in the region, where tectonic activity can be discounted, is a result of the occurrence of cold climates and the supply of abundant fresh materials by periglacial processes.

The effects of additional climatically controlled parameters are, with one exception, poorly understood. For example, the variation in amount of precipitation and its distribution through the year will certainly influence river flow and habit. However, relatively few studies have been undertaken in sufficient detail to determine this. Exceptions include those by De Gans (1981), Rose *et al.* (1980), Vandenberghe & De Smedt (1979) and Vandenberghe *et al.* (1984).

The one exception to this general lack of knowledge is sea-level change. Glacioeustatically and isostatically controlled sea level changes are characteristic of the Pleistocene. In northwest Europe low sea levels during the cold periods caused great expansion of the drainage system onto the surrounding continental shelves. Indeed, deposition on the huge river floodplains of the North Sea and Channel areas almost certainly provided the major source of silt that formed the widespread loess deposits present in adjacent land areas.

For much of the Lower and Middle Pleistocene the southern North Sea was occupied by the huge delta complex of the North German Rivers, the Rhine, Thames, Meuse and Scheldt. After the early Pleistocene, marine incursions during interglacial stages remained absent until the late Middle Pleistocene Cromerian Stage. A similar feature is found in the Channel Region. Marine transgression has subsequently been particularly marked during interglacial stages, and sea levels over 100 m below present are known from glacial maxima. Truncation of the greatly expanded fluvial system by marine transgression is very significant; the Channel River, for example, was approximately 800 km in length, 3.3 times longer than the present River Thames. Moreover, the drowned fluvial deposits are subjected to tidal scour, deposition of marine sediments and remobilization.

A further effect of marine transgression has been the infilling of the pre-existing river valleys by wedge-like accumulations of estuarine sediments. These sediments comprise silts, clays, and in some cases sands interbedded with peats and detrital organic sediments (for Flandrian (Holocene) examples see Devoy (1979); Jelgersma *et al.* (1979); Huault *et al.* (1974); Porcher (1977). Interglacial examples are also widespread; in areas of subsidence they are buried by later deposits, whereas in relatively stable areas they are dissected by subsequent fluvial incision and remain as eroded remnants on valley sides.

It therefore is apparent that the impact of the various forms of climatic change on the northwest European drainage system has been very great. These changes have been superimposed on long-term climatic and geological trends resulting from plate tectonics and world geographical evolution over the past three million years or so. These trends might reasonably be expected to continue into the future.

I thank Dr P. Balson, Dr. D. Bridgland, Dr T. Cameron, Dr J. Ehlers, Professor F. Gullentops, Dr J. De Jong, Dr J.-P. Lautridou, Professor J. Lewin, Mr J. Rose, Dr M. Sharp, Dr C. Turner, Mr C. Whiteman and Dr J. A. Zalasiewicz for invaluable discussions at various stages of this work. It would not have been possible without the patient drafting and typing by Sylvia Peglar, and the translation of two German papers by Jan Lettau. Lastly, I owe a debt of gratitude for unfaltering support to Professor R. G. West, F.R.S., and Ann Jennison.

References

Aguirre, E. & Pasini, G. 1985 The Plio-Pleistocene boundary. *Episodes* **8**, 116–120.

Alduc, D., Auffret, J.-P., Carpentier, G., Lautridou, J.-P., Lefèbvre, D. & Porcher, M. 1979 Nouvelles données sur le Pléistocène de la basse vallée de la Seine et son prolonguement sous-marin en Manche Orientale. *Bulletin d'information des géologues du Bassin de Paris* **16**, 27–34.

Allen, P. 1984 *Field guide to the Gipping and Waveney Valleys. May, 1982.* Cambridge: Quaternary Research Association.

Averdieck, F.-R. 1971 Palynologische Untersuchungen zum Tertiär auf Sylt. *Meyniana* **21**, 1–8.

Balson, P. S. & Cameron, T. D. J. 1985 Quaternary mapping offshore East Anglia. *Mod. Geol.* **9**, 221–239.

Berry, F. G. 1979 Late Quaternary scour hollows and related features in central London. *Q. Jl Engng Geol.* **12**, 9–29.

Bibus, E. 1980 Zur Relief, Boden- und Sedimententwicklung am unteren Mittelrhein. *Frankfurter geow. Arb.* D, band 1. (296 pages.)

Bijlsma, S. 1981 Fluvial sedimentation from the Fennoscandian area into the North-West European Basin during the Late Cenozoic. In *Quaternary geology: a farewell to A. J. Wiggers* (ed. A. J. van Loon). *Geologie Mijnb.* **60**, 337–345.

Biquand, D. 1974 Position chronologique de la très haute nappe alluviale de Grâce (vallée de la Somme) par rapport à la limite paléomagnétique Brunhes–Matuyama. *Bull. Ass. fr. Étude Quat.* **40–41**, 157–159.

Bishop, W. W. 1958 The Pleistocene geology and geomorphology of three gaps in the Midlands Jurassic escarpment. *Phil. Trans. R. Soc. Lond.* B **241**, 255–306.

Boenigk, W. 1978 Die Flussgeschichtliche Entwicklung der Niederrheinischen Bucht im Jungtertiär und Altquartär. *Eiszeitalter Gegenw.* **28**, 1–9.

Bourdier, F. 1974*a* Essai sur le creusement de la Vallée de la Somme au Quaternaire. In *L'évolution Quaternaire des bassins fluviaux de la Mer du Nord méridionale* (ed. P. Macar), pp. 241–252. Liège: Centre de Société Géologique de Belgique.

Bourdier, F. 1974*b* Le complexe Mindelien. I. La haute terrasse de la Somme. *Bull. Ass. fr. Étude Quat.* **40–41**, 165–168.

Bourdier, F. & Lautridou, J.-P. 1974*a* Les grands traits morphologiques et structuraux des régions de la Somme et de la Basse-Seine. *Bull. Ass. fr. Étude Quat.* **3–4**, 109–111.

Bourdier, F. & Lautridou, J.-P. 1974*b* Les dépôts du Quaternaire ancien. *Bull. Ass. fr. Étude Quat.* **40–41**, 129–135.

Bourdier, F., Chaline, J., Munaut, A. V. & Puissegur, J.-J. 1974*a* La très haute nappe alluviale de la Somme. *Bull. Ass. fr. Étude Quat.* **40–41**, 137–143.

Bourdier, F., Chaline, J., Munaut, A. V. & Puissegur, J.-J. 1974*b* Le Complex Mindelien. II. La moyenne terrasse de L'Avre. *Bull. Ass. fr. Étude Quat.* **40–41**, 168–180.

Bourdier, F., Munaut, A. V., Prat, F. & Puissegur, J.-J. 1974*c* Les dépôts du complexe Rissien de la Somme. *Bull. Ass. fr. Étude Quat.* **40–41**, 219–227.

Bowen, D. Q., Rose, J., McCabe, A. M. & Sutherland, D. 1986 Correlation of Quaternary glaciations in England, Ireland, Scotland and Wales. *Quat. Sci. Rev.* **5**, 299–340.

Bridge, D. McC. & Hopson, P. M. 1985 Fine gravel, heavy mineral and grain-size analyses of mid-Pleistocene glacial deposits in the Lower Waveney Valley, East Anglia. *Mod. Geol.* **9**, 129–144.

Bridgland, D. R. 1980 A reappraisal of Pleistocene stratigraphy in north Kent and east Essex and new evidence concerning the former course of the Thames and Medway. *Quat. Newsl.* **32**, 15–24.

Bridgland, D. R. 1985 The Quaternary fluvial deposits of north Kent and east Essex. Ph.D. thesis (C.N.A.A.)., City of London Polytechnic.

Briggs, D. J. & Gilbertson, D. D. 1973 The age of the Hanborough Terrace of the River Evenlode, Oxfordshire. *Proc. Geol. Ass.* **84**, 155–173.

Briggs, D. J. & Gilbertson, D. D. 1980 Quaternary processes and environments in the Upper Thames Valley. *Trans. Inst. Br. Geogr.* **5**, 53–65.

Briggs, D. J., Coope, G. R. & Gilbertson, D. D. 1985 The chronology and environmental framework of early man in the Upper Thames Valley: a new model. *Oxford: Br. Archaeol. Rep. Br.* Ser., no. 137. (176 pages.)

Brunnacker, K. 1975 The Mid-Pleistocene of the Rhine Basin. In *After the Australopithecines* (ed. K. W. Butzer & G. L. Isaac), pp. 189–224. The Hague: Mouton Publishers.

Brunnacker, K. 1978 Gliederung und Stratigraphie der Quartär-Terrassen am Niederrhein. *Kölner geogr. Arb.* **36**, 37–58.

Brunnacker, K. 1986 Quaternary stratigraphy in the Lower Rhine area and northern Alpine foothills. *Quat. Sci. Rev.* **5**, 373–379.

Bryant, I. D., Holyoak, D. T. & Moseley, K. A. 1983 Late Pleistocene deposits at Brimpton, Berkshire, England. *Proc. Geol. Ass.* **94**, 321–343.

Bustamente Santa-Cruz, L. 1974 Les minéraux lourds des alluvions du bassin de la Meuse. *C. r. hebd. Séanc. Acad. Sci., Paris* **278**, 561–564.

Cameron, T. D. J., Stoker, M. S. & Long, D. 1987 The history of Quaternary sedimentation in the UK sector of the North Sea Basin. *J. geol. Soc. Lond.* **144**, 43–58.

Caston, V. M. D. 1977 A new isopachyte map of the Quaternary of the North Sea. *Rep. Inst. geol. Sci.* no. 77/11, pp. 1–8.

Catt, J. A. 1986 The nature, origin and geomorphological significance of the clay-with-flints. In *The scientific study of flint and chert* (ed. G. de G. Sieveking & M. B. Hart), pp. 151–159. London: Cambridge University Press.

Cavelier, C. 1980 Miocène et Pliocène. In *Synthèse géologique du Bassin de Paris* (ed. C. Megnien & F. Megnien) *Mem. Bur. Rech. Gèol. Miner.* **101**, 416–437.

Cepek, A. G. 1967 Stand und Probleme der Quartärstratigraphie im Nordteil der DDR. *Ber. dt. Ges. geol. Wiss.* A **12**, 375–404.

Chaline, J. 1974 Les rongeurs, l'âge et l'environnement de la très haute terrasse de Grâce à Montières (Somme). *Bull. Ass. fr. Étude Quat.* **40–41**, 151–157.

Chancerel, A. 1985 *Le val de Seine D'Elbeuf à Caudebec-en-Caux. Evolution morphologique.* Thèse, 3e cycle, Université Paris 7.

Chancerel, A. 1986 Le système des nappes alluviales de la Seine en aval de Rouen. *Travaux Groupe Seine 2, Bull. Cent. Géomorph.* **31**, 73–82.

Chatwin, C. P. 1927 Fossils from the iron sands of Netley Heath (Surrey). *Mem. Geol. Surv. Summ. Prog. 1926*, pp. 154–157.

Clarke, M. R. & Auton, C. A. 1982 The Pleistocene depositional history of the Norfolk – Suffolk borderland. *Rep. Inst. geol. Sci.*, part 1, pp. 23–29.

Clet, M. 1982*a* The sections in the Forêt de la Londe: palynology. In *The Quaternary of Normandy. Field handbook* (ed. J.-P. Lautridou), pp. 47–49. Cambridge: Quaternary Research Association.

Clet, M. 1982*b* Bosq d'Aubigny Formation. In *The Quaternary of Normandy. Field handbook* (ed. J.-P. Lautridou), pp. 67–69. Cambridge: Quaternary Research Association.

Commont, V. 1911 La chronologie et la stratigraphie des dépôts quaternaires dans la vallée de la Somme. *Annls Soc. géol. Belg.* **39**, 156–180.
Crommelin, R. D. 1953 Over de stratigraphie en herkomst van de preglaciale afzettingen in Midden-Nederland. *Geologie Mijnb.* **15**, 305–321.
Crommelin, R. D. 1954 Über den Einfluss der nord- und mitteldeutschen Flüsse auf das ältere Pleistozän der Niederlande. *Mitt. geol. St Inst. Hamb.* **23**, 86–97.
Destombes, J.-P., Shephard-Thorn, E. R. & Redding, J. H. 1975 A buried valley system in the Strait of Dover. *Phil. Trans. R. Soc. Lond.* B **279**, 243–256.
Devoy, R. J. N. 1979 Flandrian sea level changes and vegetational history of the lower Thames Estuary. *Phil. Trans. R. Soc. Lond.* B **285**, 355–410.
De Gans, W, 1981 *The Drentsche Aa valley system.* Amsterdam: Vrije Universiteit te Amsterdam and Rodopi.
De Ridder, N. A. & Zagwijn, W. H. 1962 A mixed Rhine–Meuse deposit of Holsteinian age from the south-eastern part of the Netherlands. *Geologie Mijnb.* **41**, 125–130.
De Jong, J. 1965 Quaternary sedimentation in the Netherlands. *Geol. Soc. Am. Spec. Pap.* **84**, 95–123.
Dines, H. G. & Chatwin, C. P. 1930 Pliocene sandstone from Rothamstead (Hertfordshire). *Mem. Geol. Surv. Summ. Prog. 1929.*, pp. 1–7.
Dingwall, R. G. 1975 Sub-bottom infilled channels in an area of the eastern English Channel. *Phil. Trans. R. Soc. Lond.* A **279**, 233–241.
D'Olier, B. 1975 Some aspects of the Late Pleistocene–Holocene drainage of the River Thames in the eastern part of the London Basin. *Phil. Trans. R. Soc. Lond.* A **279**, 269–277.
Doppert, J. W. C., Ruegg, G. H. J., van Staalduinen, C. J., Zagwijn, W. H. & Zandstra, J. G. 1975 Formaties van het Kwartair en Boven-Tertiair in Nederland. In *Toelichtingen bij geologische overzichtskaarten van Nederland* (ed. W. H. Zagwijn & C. J. van Staalduinen), pp. 11–56. Haarlem: Rijks Geol. Dienst.
Duphorn, K., Grube, F., Meyer, K.-D., Streif, H. & Vinken, R. 1973 Area of the Scandinavian glaciation. 1. Pleistocene and Holocene. In *State of research on the Quaternary of the Federal Republic of Germany* (ed. E. Schönhals & R. Huckriede) (*Eiszeitalter Gegenw.* **23–24**), pp. 222–250.
Ehlers, J. 1978 Die quartäre Morphogenese der Harburger Berge und ihrer Umgebung. *Mitt. geogr. Ges. Hamb.* **68**, 1–181.
Ehlers, J. 1983 The glacial history of north-west Germany. In *Glacial deposits in north-west Europe* (ed. J. Ehlers), pp. 229–238. Rotterdam: Balkema.
Ehlers, J. 1987 Die Entstehung des Kaolinsandes auf Sylt. In *Fossilien von Sylt* (ed. U. von Hacht), Vol. 2, pp. 249–267. Hamburg: Von Hacht.
Eissmann, L. 1975 Das Quartär der Leipziger Tieflandsbucht und angrenzender Gebiete um Saale und Elbe. *Schriftenr. geol. Wiss.* **2**, 1–263.
Erd, K. 1970 Pollenanalytical classification of the Middle Pleistocene of the German Democratic Republic. *Palaeogeogr. Palaeoclimatol. Palaeoecol.* **8**, 129–145.
Figge, K. 1980 Das Elbe-Urstromtal im Bereich der Deutschen Bucht (Nordsee). *Eiszeitalter Gegenw.* **30**, 203–211.
Figge, K. 1983 Morainic deposits in the German Bight area of the North Sea. In *Glacial deposits in north-west Europe* (ed. J. Ehlers), pp. 299–304. Rotterdam: Balkema.
Friis, H. 1974 Weathered heavy-mineral associations from the young-Tertiary deposits of Jutland, Denmark. *Sediment. Geol.* **12**, 199–213.
Funnell, B. M. 1972 The history of the North Sea. *Bull. geol. Soc. Norfolk* **22**, 2–10.
Funnell, B. M., Norton, P. E. P. & West, R. G. 1979 The Crag at Bramerton, near Norwich, Norfolk. *Phil. Trans. R. Soc. Lond.* B **287**, 489–534.
Genieser, K. 1955 Ehemalige Elbeläufe in der Lausitz. *Geologie* **4**, 223–279.
Genieser, K. 1957 Ehemalige Elbeläufe im Raum zwischen Dresden, Görlitz und Berlin. *Hallesches Jb. mitteldt. Erdgesch.* **2**, 262–266.
Genieser, K. 1962 Neue Daten zur Flussgeschichte der Elbe. *Eiszeitalter Gegenw.* **13**, 141–156.
Gibbard, P. L. 1977 Pleistocene history of the Vale of St. Albans. *Phil. Trans. R. Soc. Lond.* B **280**, 445–483.
Gibbard, P. L. 1979 Middle Pleistocene drainage in the Thames Valley. *Geol. Mag.* **116**, 35–44.
Gibbard, P. L. 1980 The origin of stratified Catfish Creek Till by basal melting. *Boreas* **9**, 71–85.
Gibbard, P. L. 1985 *Pleistocene history of the Middle Thames Valley.* Cambridge University Press.
Gibbard, P. L., Wintle, A. G. & Catt., J. A. 1987 Age and origin of clayey silt 'brickearth' in West London, England. *J. Quat. Sci.* **2**, 3–9.
Goudie, A. S. & Hart, M. G. 1975 Pleistocene events and forms in the Oxford region. In *Oxford and its region* (ed. C. G. Smith & D. I. Scargill), pp. 3–13. Oxford University Press.
Green, C. P., Hey, R. W. & McGregor, D. F. M. 1980 Volcanic pebbles in Pleistocene gravels of the Thames in Buckinghamshire and Hertfordshire. *Geol. Mag.* **117**, 59–64.
Grimmel, E. 1973 Überlegungen zur Morphogenese des Norddeutschen Flachlandes, dargesteldt am Beispiel des unteren Elbtales. *Eiszeitalter Gegenw.* **23–24**, 76–88.
Gripp, K. 1964 *Erdgeschichte von Schleswig-Holstein.* Neumünster: K. Wachholtz Verlag.

Gullentops, F. 1974 The southern North Sea during the Quaternary. In *L'évolution Quaternaire des bassins fluviaux de la Mer du Nord méridionale*. Liège: Cent. Société Géologique de Belgique.

Hamilton, D. & Smith, A. J. 1972 The origin and sedimentary history of the Hurd Deep, English Channel, with additional notes on other deeps in the Western English Channel. *Mem. Bur. Rech. Géol. Miner.* **79**, 59–78.

Hey, R. W. 1976 Provenance of far-travelled pebbles in the pre-Anglian Pleistocene of East Anglia. *Proc. Geol. Ass.* **87**, 69–82.

Hey, R. W. 1980 Equivalents of the Westland Green Gravels in Essex and East Anglia. *Proc. Geol. Ass.* **91**, 279–290.

Hey, R. W. 1986 A re-examination of the Northern Drift of Oxfordshire. *Proc. Geol. Ass.* **97**, 291–301.

Hey, R. W. & Brenchley, P. J. 1977 Volcanic pebbles from Pleistocene gravels in Norfolk and Essex. *Geol. Mag.* **114**, 219–225.

Hinsch, W. 1974 Das Tertiär im Untergrund von Schleswig-Holstein. *Geol. Jb.* A **24**, 34.

Hinsch, W. & Ortlam, D. 1974 Stand und Probleme der Gliederung des Tertiärs in Nordwestdeutschland. *Geol. Jb.* A **16**, 3–25.

Hollin, J. T. 1977 Thames interglacial sites, Ipswichian sea levels and Antarctic ice surges. *Boreas* **6**, 33–52.

Huault, M. F. & Lefèbvre, D. 1974 Le Postglaciaire de la Basse-Seine: sediments et chronostratigraphie. *Bull. Ass. Fr. Étude Quatern.* **40–41**, 253–256.

Jelgersma, S., Oele, E. & Wiggers, A. J. 1979 Depositional history and coastal development in the Netherlands and the adjacent North Sea since the Eemian. In *The Quaternary history of the North Sea* (ed. E. Oele, R. T. E. Schüttenhelm & A. J. Wiggers) *Acta univ. ups. symp. univ. ups. a. quingent. celebr.* **2**, pp. 115–142.

Jenkins, D. G., Whittaker, J. E. & Carlton, R. 1986 On the age and correlation of the St Erth Beds, SW England, based on planktonic foraminifera. *J. Micropalaeont.* **5**, 93–105.

John, D. T. 1980 The soils and superficial deposits on the North Downs of Surrey. In *The shaping of southern England* (ed. D. K. C. Jones), pp. 101–130. London: Academic Press.

Kaiser, K. 1960 Klimazeugen des periglazialen Dauerfrostbodens in Mittel- und Westeuropa. *Eiszeitalter Gegenw.* **11**, 121–141.

Kazi, A. & Knill, J. L. 1969 The sedimentation and geotechnical properties of the Cromer Till between Happisburgh and Cromer, Norfolk. *Q. Jl Engng Geol.* **2**, 63–86.

Kellaway, G. A., Redding, J. H., Shephard-Thorn, E. R. & Destombes, J.-P. 1975 The Quaternary history of the English Channel. *Phil. Trans. R. Soc. Lond.* A **279**, 189–218.

Kemp, R. A. 1985 The Valley Farm Soil in southern East Anglia. In *Soils and Quaternary landscape evolution* (ed. J. Boardman), pp. 179–196. Chichester: John Wiley.

Knoth, W. 1964 Zur Kenntnis der pleistozänen Mittelterrassen der Saale und Mulde nördlich von Halle. *Geologie* **13**, 598–616.

Koche, E. 1954 Vom Untergrund Hamburgs. *Mitt. geol. St Inst. Hamb.* **23**, 10–17.

Kolfschoten, T. van & Roebroeks, W. 1985 The Maastricht–Belvédère project; an intermediate synthesis. *Meded. Rijks. geol. Dienst.* **39**, 119–121.

Kuntz, G. & Lautridou, J.-P. 1974 Contribution a l'étude du Pliocène et du passage Pliocène–Quaternaire dans les dépôts de la Forêt de la Londe près de Rouen. Corrélations possibles avec divers gisements de Haute Normandie. *Bull. Ass. fr. Étude Quat.* **3–4**, 117–128.

Kuntz, G., Lautridou, J.-P., Cavelier, C. & Clet, M. 1979 Le Plio-Quaternaire de Haute-Normandie. *Bull. Inf. géol. Bassin Paris* **16**, 93–126.

Larsonneur, C., Vaslet, D. & Auffret, J. P. 1979 Carte des sediments superficiel de la Manche. *Bur. Rech. Géol. miner.* 1:50000 map.

Lautridou, J.-P. (ed.) 1982 *The Quaternary of Normandy. Field handbook.* Cambridge: Quaternary Research Association.

Lautridou, J.-P., Monnier, J. L., Morzadec, M. T., Somme, J. & Tuffreau, A. 1986 The Pleistocene of northern France. *Quat. Sci. Rev.* **5**, 387–393.

Lautridou, J.-P., Lefèbvre, D., Lécolle, F., Carpentier, G., Descombes, J. C., Gaquerel, C. & Huault, M. F. 1984 Les terrasses de la Seine dans le méandre d'Elbeuf, corrélations avec celles dans la region Mantes. *Bull. Ass. Fr. Étude Quatern.* **1–2–3**, 27–32.

Lécolle, F. 1984 Phases érosives et cycles sédimentaires: les alluvions de la Seine au sud du Vexin. *Bull. Ass. fr. Étude Quat.* **1–2–3**, 33–36.

Lefèbvre, D. 1974 Le cailloutes würmien de fond de vallée, Basse Seine. *Bull. Ass. fr. Étude Quat.* **40–41**, 251–252.

Lefèbvre, D., Carpentier, E. & Evrard, H. 1986 Les terrasses de la Seine de Pont-de-l'Arche à Elbeuf. Travaux Groupe Seine 2. *Bull. Centre Géomorph.* **31**, 41–71.

Lefèbvre, D., Huault, M. F., Guyader, J., Giresse, P., Hommeril, P. & Larsonneur, C. 1974 Le prisme alluvial de l'estuaire de la Seine: synthèse sédimentologique, stratigraphique et paléogéographique. *Bull. Inf. Géol. Bassin Paris* **39**, 27–36.

Lüttig, G. 1960 Neue Ergebnisse quartärgeologischer Forschung im Raume Alfeld-Hameln-Elze. *Geol. Jb.* **77**, 337–390.

Lüttig, G. 1974 Geological history of the River Weser. In *L'évolution Quaternaire des bassins fluviaux de la Mer du Nord meridionale* (ed. P. Macar), pp. 21–34. Liège: Cent. Société Géologique de Belgique.
Lüttig, G. & Maarleveld, G. C. 1961 Nordische Geschiebe in Ablagerungen prä-Holstein in den Niederlanden (Komplex von Hattem). *Geologie Mijnb.* **40**, 163–174.
Lüttig, G. and Meyer, K.-D. 1974 Geological history of the River Elbe, mainly of its lower course. In *L'évolution Quaternaire des bassins fluviaux de la Mer du Nord méridionale* (ed. P. Macar), pp. 1–19. Liège: Cent. Société Géologique de Belgique.
Maarleveld, G. C. 1954 Über fluviatile Kiese in Nordwestdeutschland. *Eiszeitalter Gegenw.* **4–5**, 10–17.
Maarleveld, G. C. 1956 Grindhoudende Midden-Pleistoceen sedimenten. *Meded. geol. Sticht.* C**4** (6). (105 pages.)
Macaire, J. J. 1984 Les vallées et formations alluviales Plio-Quaternaires dans le sud et le sud-ouest du Bassin de Paris: genèse et signification dynamique. *Bull. Ass. fr. Étude Quat.* **1–2–3**, 37–40.
Menke, B. 1975 Vegetationsgeschichte und Florenstratigraphie Nordwestdeutschlands im Pliozän und Frühquartär. Mit einem Beitrag zur Biostratigraphie des Weichselfrühglazials. *Geol. Jb.* A **26**, 3–151.
Meyer, K.-D. 1983 Zur Anlage der Urstromtäler in Niedersachsen. *Z. Geomorph.* N. S. **27**, 147–160.
Meyer, K.-D., Schmid, F. & Wolburg, J. 1977 Erläuterungen zu Blatt Salzbergen Nr 3610. Geol. Karte Niedersachsen. (111 pages). Hannover: *Niedersächsisches Landesamt für Bodenforschung.*
Moffat, A. J. & Catt, J. A. 1986 A re-examination of the evidence for a Plio-Pleistocene marine transgression on the Chiltern Hills. III. Deposits. *Earth Surf. Process. Landf.* **11**, 233–247.
Oele, E. 1969 The Quaternary geology of the Dutch part of the North Sea, north of the Frisian Isles. *Geologie Mijnb.* **48**, 467–480.
Oele, E. 1971 The Quaternary geology of the southern area of the Dutch part of the North Sea. *Geologie Mijnb.* **50**, 461–474.
Oele, E. & Schüttenhelm, R. T. E. 1979 Development of the North Sea after the Saalian glaciation. In *The Quaternary history of the North Sea* (ed. E. Oele, R. T. E. Schüttenhelm & A. J. Wiggers) (*Acta univ. ups. symp. univ. ups. a. quingent. celebr.* **2**), pp. 191–215.
Paepe, R. & Sommé, J. 1975 Marine Pleistocene transgressions along the Flemish coast (Belgium and France). *North Sea* (ed. E. Oele, R. T. E. Schüttenhelm & A. J. Wiggers) (*Acta univ. ups. symp. univ. ups. a. quingent. celebr.* **2**), pp. 143–146.
Paepe, R. Sommé, J. 1975 Marine Pleistocene transgressions along the Flemish coast (Belgium and France). *Quaternary glaciations in the Northern Hemisphere. Int. Geol. Correl. Prog. project* **24**, no. 2, pp. 108–116.
Paepe, R. & Vanhoorne, R. 1970 Stratigraphical position of periglacial phenomena in the Campine Clay of Belgium, based on palaeobotanical analysis and palaeomagnetic dating. *Bull. Soc. Belge geol.* **79**, 201–211.
Paulissen, E. 1973 De morfologie en de Kwartairstratigrafie van de Maasvallei in Belgisch Limburg. *Verh. K. vlaam. Akad. Wet.* **35** (127). (266 pages.)
Paulissen, E., Vandenberghe, J. & Gullentops, F. 1985 The Feldbiss Fault in the Maas valley bottom (Limburg, Belgium). *Geologie Mijnb.* **64**, 79–87.
Pissart, A. 1974 La Meuse en France et en Belgique. Formation du bassin hydrographique. Les Terraces et leurs enseignements. In *L'évolution Quaternaire des bassins fluviaux de la Mer du Nord méridionale* (ed. P. Macar), pp. 105–132. Liège: Cent. Société Géologique de Belgique.
Pointon, W. K. 1978 The Pleistocene succession at Corton, Suffolk. *Bull geol. Soc. Norfolk* **30**, 55–76.
Porcher, M. 1977 Lithostratigraphie des alluvions fluviatiles holocènes de la vallée de la Seine. *Bull. Soc. géol. Norm.* **64**, 181–201.
Präger, F. 1970 Die Bedeutung stratigraphischer Untersuchungen im Quartär am Beispiel von angewandten Arbeiten im Gebiet von Nieschütz bei Meissen. *Wiss. Z. Univ. Halle*, **19** (1), 49–55.
Quitzow, H. W. 1953 Altersbeziehungen und Flözzusammenhänge in der jüngeren Braunkohlenformation nördlich der Mittelgebirge. *Geol. Jb.* **68**, 27–132.
Quitzow, H. W. 1974 Das Rheintal und seine Entstehung. Bestandsaufnahme und Versuch einer Synthese. In *L'évolution Quaternaire des bassins fluviaux de la Mer du Nord Meridionale* (ed. P. Macar), pp. 53–104. Liège: Cent. Sociéte Géologique de Belgique.
Rasmussen, L. B. 1961 Mittel- und Ober-Miocän von Dänemark. *Meyniana* **10**, 59–62.
Roep, T. B., Holst, H., Vissers, R. L. M., Pagnier, H. & Postma, D. 1975 Deposits of southward flowing Pleistocene rivers in the Channel region, near Wissant, NW France. *Palaeogeogr. Palaeoclimatol. Palaeoecol.* **17**, 289–308.
Rose, J. & Allen, P. 1977 Middle Pleistocene stratigraphy in southeastern Suffolk. *J. geol. Soc. Lond.* **133**, 83–102.
Rose, J., Allen, P. & Hey, R. W. 1976 Middle Pleistocene stratigraphy in southern East Anglia. *Nature, Lond.* **263**, 492–494.
Rose, J., Boardman, J., Kemp, R. A. & Whiteman, C. A. 1985 Palaeosols and the interpretation of the British Quaternary stratigraphy. In *Geomorphology and Soils* (ed. K. S. Richards, R. R. Arnett & S. Ellis), pp. 348–375. London: George Allen & Unwin.
Rose, J., Turner, C., Coope, G. R. & Bryan, M. D. 1980 Channel changes in a lowland river catchment over the last 13000 years. In *Timescales in geomorphology* (ed. R. A. Cullingford, D. A. Davidson & J. Lewin), pp. 159–175. Chichester: John Wiley.

Ruegg, G. H. J. & Zandstra, J. G. 1977 Pliozäne und pleistozäne gestauchte Ablagerungen bei Emmerschans (Drenthe, Niederlande). *Meded. Rijks geol. Dienst* **34**, 7–9.

Ruske, R. 1964 Das Pleistozän zwischen Halle (Saale), Bernburg und Dessau. *Geologie* **13**, 570–597.

Schulz, W. 1962 Gliederung des Pleistozäns in der Umgebung von Halle (Saale). *Geologie Beih.* **36**, 1–69.

Seddon, M. & Holyoak, D. T. 1985 Evidence for sustained regional permafrost during deposition of fossiliferous Late Pleistocene sediments at Stanton Harcourt (Oxfordshire, England). *Proc. Geol. Ass.* **96**, 53–71.

Semmel, A. 1973 Area between the Scandinavian and the Alpine Glaciation. In *State of research in the Quaternary of the Federal Republic of Germany* (ed. E. Schönhals & R. Huckriede) (*Eiszeitalter Gegenw.* **23/24**), pp. 293–305.

Shotton, F. W., Goudie, A. S., Briggs, D. J. & Osmaston, H. A. 1980 Cromerian interglacial deposits at Sugworth near Oxford, England, and their relation to the Plateau Drift of the Cotswolds and the terrace sequence of the Upper and Middle Thames. *Phil. Trans. R. Soc. Lond.* B **289**, 55–86.

Šibrava, V. 1972 Zur Stellung der Tschechoslowakei im Korrelierungssystem des Pleistozäns in Europa. *Sb. geol. věd.* (A) **8**, 1–288.

Šibrava, V. 1986 Scandinavian glaciations in the Bohemian Massif and Carpathian Foredeep and their relationship to the extra-glacial areas. *Quat. Sci. Rev.* **5**, 381–386.

Smith, A. J. 1985 A catastrophic origin for the palaeovalley system of the eastern English Channel. *Mar. Geol.* **64**, 65–75.

Smith, A. J. & Curry, D. 1975 The structure and geological evolution of the English Channel *Phil. Trans. R. Soc. Lond.* A **279**, 3–20.

Sommé, J. 1979 Quaternary coastlines in northern France. In *The Quaternary history of the North Sea* (ed. E. Oele, R. T. E. Schüttenhelm & A. J. Wiggers) (*Acta univ. ups. symp. univ. ups. a. Quingent. celebr.* **2**), pp. 147–158.

Sommé, J., Fagnart, J. P., Leger, M., Munaut, A. V., Puissegur, J. J. & Tuffreau, A. 1984 Terrasse fluviatiles du Pléistocène moyen en France septentrionale: signification dynamique et climatique. *Bull. Ass. fr. Étude Quat.* **1–2–3**, 52–58.

Stamp, L. D. 1927 The Thames drainage system and the age of the Strait of Dover. *Geogrl. J.* **70**, 386–390.

Tavernier, R. & De Moor, G. 1974 L'évolution du bassin de l'Escaut. In *L'évolution Quaternaire des bassins fluviaux de la Mer du Nord méridionale* (ed. P. Macar), pp. 273–280. Liège: Cent. Société Géologique de Belgique.

Ter Wee, M. W. 1962 The Saalian glaciation in the Netherlands. *Meded. Geol. Sticht.* (n.s) **15**, 57–77.

Ter Wee, M. W. 1983*a* The Elsterian glaciation in the Netherlands. In *Glacial deposits in north-west Europe* (ed. J. Ehlers), pp. 413–415. Rotterdam: Balkema.

Ter Wee, M. W. 1983*b* The Saalian glaciation in the northern Netherlands. In *Glacial deposits in north-west Europe* (ed. J. Ehlers), pp. 405–412. Rotterdam: Balkema.

Thome, K. N. 1980 Der Vorstoss des nordeuropäischen Inlandeises in das Münsterland in Elster- und Saale-Eiszeit. *Westf. geogr. Stud.* **36**, 21–40.

Tuffreau, A., Munaut, A. V., Puissegur, J. J. & Sommé, J. 1981 Les basse terrasse dans les vallées du Nord de la France et de la Picardie: stratigraphie et paléolithique. *Bull. Soc. prehist. Fr.* **78**, 291–305.

Tuffreau, A., Munaut, A. V., Puissegur, J. J. & Sommé, J. 1982 Stratigraphie et environnement des industries Acheuléennes de la moyenne terrasse du Bassin de la Somme (Région d'Amiens). *Bull. Ass. fr. Étude Quat.* **2–3**, 73–82.

Vandenberghe, J. and De Smedt, P. 1979 Palaeomorphology in the eastern Scheldt Basin (Central Belgium) – the Dijle-Demer-Grote Nete confluence area. *Catena* **6**, 73–105.

Vandenberghe, J., Krook, L. and van der Valk, L. 1986 On the provenance of the Early Pleistocene fluvial system in the southern Netherlands. *Geologie Mijnb.* **65**, 3–12.

Vandenberghe, J., Mücher, H. J., Roebroeks, W. & Gemke, D. 1985 Lithostratigraphy and palaeoenvironment of the Pleistocene deposits at Maastricht-Belvédère, Southern Limburg, the Netherlands. *Meded. Rijks. Geol. Dienst.* N.S. **39**, 7–18.

Vandenberghe, J., Paris, P., Kasse, C., Gouman, M. & Beyens, L. 1984 Paleomorphological and botanical evolution of small lowland valleys – a case study of the Mark valley in northern Belgium. *Catena* **11**, 229–238.

Van den Broek, J. M. M. & Maarleveld, G. C. 1963 The late Pleistocene terrace deposits of the Meuse. *Meded. Geol. Sticht.* N.S. **16**, 13–24.

Van der Meene, E. A. & Zagwijn, W. H. 1978 Die Rheinläufe im deutsch–niederländischen Grenzgebiet seit der Saale-Kaltzeit. Überblick neuer geologischer und pollenanalytischer Untersuchungen. *Fortschr. Geol. Rheinld Westf.* **28**, 345–359.

Vanhoorne, R. 1962 Het interglaciale veen te Loo (België). *Natuurw. Tijdschr.* **44**, 58–64.

Ventris, P. 1985 Pleistocene environmental history of the Nar Valley, Norfolk. Ph.D. thesis, University of Cambridge.

West, R. G. 1972 Relative land-sea level changes in southeastern England during the Pleistocene. *Phil. Trans. R. Soc. Lond.* A **272**, 87–98.

West, R. G. 1980*a* *Pre-glacial Pleistocene of the Norfolk and Suffolk coasts.* Cambridge University Press.

West, R. G. 1980*b* Pleistocene forest history in East Anglia. *New Phytol.* **85**, 571–622.

West, R. G. & Wilson, D. G. 1968 Plant remains from the Corton Beds, Lowestoft, Suffolk. *Geol. Mag.* **105**, 116–123.

Weyl, R., Rein, U. & Teichmüller, M. 1955 Das Alter des Sylter Kaolinsandes. *Eiszeitalter Gegenw.* **6**, 5–15.
Whiteman, C. A. 1983 Great Waltham. In *Diversion of the Thames. Field guide* (ed. J. Rose), pp. 163–169. Cambridge: Quaternary Research Association.
Woldstedt, P. 1956 Die Geschichte des Flussnetzes in Norddeutschland und angrenzenden Gebieten. *Eiszeitalter Gegenw.* **7**, 5–12.
Woldstedt, P. 1967 The Quaternary of Germany. In *The Quaternary* (ed. K. Rankama), vol. 2, pp. 000–000. New York: Interscience Publications.
Wooldridge, S. W. & Linton, D. L. 1955 *Structure, surface and drainage in South-east England.* London: G. Philip.
Wymer, J. J. 1968 *Lower Palaeolithic archaeology in Britain.* London: John Baker.
Zagwijn, W. H. 1960 Aspects of the Pliocene and Early Pleistocene vegetation in the Netherlands. *Meded. geol. Sticht.*, ser. C-III-I, **5**, 78.
Zagwijn, W. H. 1963 Pleistocene stratigraphy in the Netherlands, based on changes in vegetation and climate. *Verh. K. ned. geol.-mijnb. Genoot.*, geol. ser., **20**, 173–196.
Zagwijn, W. H. 1974 Palaeogeographic evolution of the Netherlands during the Quaternary. *Geologie Mijnb.* **53**, 369–385.
Zagwijn, W. H. 1979 Early and Middle Pleistocene coastlines in the southern North Sea Basin. In *The Quaternary history of the North Sea* (ed. E. Oele, R. T. E. Schüttenhelm & A. J. Wiggers) (*Acta univ. ups. symp. univ. ups. a. quingent. celebr.* **2**), pp. 31–42.
Zagwijn, W. H. 1985 An outline of the Quaternary stratigraphy of the Netherlands. *Geologie Mijnb.* **64**, 17–24.
Zagwijn, W. H. & Doppert, J. W. C. 1978 Upper Cenozoic of the southern North Sea Basin: palaeoclimatic and palaeogeographic evolution. In *Keynotes of the MEGS-II (Amsterdam 1978)* (ed. A. J. van Loon) (*Geologie Mijnb.* **57**), pp. 577–588.
Zalasiewicz, J. A., Mathers, S., Hughes, M. J., Gibbard, P. L., Peglar, S. M., Harland, R., Nicholson, R. A., Boulton, G. S., Cambridge, P. & Wealthall, G. P. 1988 Stratigraphy and palaeoenvironments of the Red and Norwich Crags between Aldeburgh and Sizewell, Suffolk, England. *Phil. Trans. R. Soc. Lond.* B. (In the press.)
Zandstra, J. G. 1971 Geologisch onderzoek in de stuwwal van de oostelijke Veluwe bij Hattem en Wapenveld. *Meded. Rijks. geol. Dienst.* **22**, 215–258.
Zandstra, J. G. 1983 Fine gravel, heavy mineral and grain-size analyses of Pleistocene, mainly glaciogenic deposits in the Netherlands. In *Glacial deposits in north-west Europe* (ed. J. Ehlers), pp. 361–377. Rotterdam: Balkema.
Ziegler, P. A. 1978 North-Western Europe: tectonics and basin development. In *Keynotes of the MEGS-II (Amsterdam 1978)* (ed. A. J. van Loon) (*Geologie Mijnb.* **57**), pp. 589–626.
Ziegler, P. A. & Louwerens, C. J. 1979 Tectonics of the North Sea. In *The Quaternary history of the North Sea* (ed. E. Oele, R. T. E. Schüttenhelm & A. J. Wiggers) (*Acta univ. ups. symp. univ. ups. a. quingent. celebr.* **2**), pp. 7–22.
Zonneveld, J. I. S. 1958 Litho-stratigraphische eenheden in het Nederlandse Pleistoceen. *Meded. geol. Sticht.* N.S. **12**, 31–64.
Zonneveld, J. I. S. 1974 The terraces of the Maas and the Rhine downstream of Maastricht. In *L'évolution Quaternaire des bassins fluviaux de la Mer du Nord méridionale* (ed. P. Macar), pp. 133–158. Liège: Cent. Société Géologique de Belgique.

Discussion

J. Rose (*Department of Geography, Birkbeck college, University of London, U.K.*) I should like to make two points concerning the topic considered by Dr Gibbard, and to ask him one question.

The evidence that Professor Smith presented for the catastrophic proglacial lake meltwater erosion which resulted in the formation of the Straits of Dover is based on the palaeochannel pattern and associated sea-bed sediments in the eastern part of the English Channel and not just on the geomorphology (Smith 1984).

The period of Quaternary time when the northwest European rivers had their greatest influence on the landscape is represented by the Sterksel Formation of The Netherlands, the Yarmouth Roads Formation of the southern North Sea and the Kesgrave Formation of East Anglia (Bowen *et al.* 1986). These bodies of clastic sediments represent the largest accumulations of Quaternary fluvial deposits within their respective areas. The vast quantities of sediment are the product of effective erosion in the headwater regions of the Alps in the case of the Rhine and of Wales in the case of the Thames, probably enhanced by the early episodes of glaciation in these upland regions. In terms of the Quaternary history of northwestern Europe these deposits were formed during the later part of the early Pleistocene (Bowen *et al.* 1986, table 2).

On all the palaeogeographical maps shown in the lecture the 'Solent River' in the region of Dorset, southeast Hampshire and the Isle of Wight is anomalous, either in terms of its relation with the coastline shown on the early reconstruction, or in terms of the river networks shown on the later reconstructions. Can Dr Gibbard explain this anomaly?

References

Bowen, D. Q., Rose, J., McCabe, A. M. & Sutherland, D. G. 1986 Correlation of Quaternary glaciations in England, Ireland, Scotland and Wales. *Quat. Sci. Rev.* **5**, 299–340.

Smith, A. G. 1984 Structural evolution of the English Channel region. *Annls Soc. geol. Nord* **103**, 253–264.

P. L. Gibbard. Mr Rose is correct in asserting that Professor Smith's interpretation of catastrophic drainage of the North Sea glacial lake was based partly on the palaeochannel pattern that superficially resembles that formed during the drainage of Lake Missoula. The palaeochannel system associated with the latter was formed in a very short, single event. This is, however, not the case for the Channel floor palaeochannels, which Smith himself concedes have a multiphase infilling (Hamilton & Smith 1972; A. G. Smith, personal communication). According to French workers the palaeochannels may in part pre-date drainage of the lake (cf. Alduc *et al.* 1979). It is possible that the palaeochannels may have been formed by superimposition of river courses following repeated marine regression.

I agree that the period before the arrival of widespread lowland continental glaciation in the Middle Pleistocene was dominated in the region by fluvial activity. The Kedichem and Sterksel Formations in the Rhine–Meuse system, and the Thames' Kesgrave Formation, mark the input of large quantities of detrital material. However, during the Early and early Middle Pleistocene, considerable quantities of sediment were deposited as the Harderwijk and Enschede Formations by the westward-flowing North German rivers. These combined sedimentary accumulations certainly indicate that contemporary periglacial erosion was very effective.

The evolution of the 'Solent River' system is fascinating and its course, at first sight, appears to be anomalous. This anomaly seems to arise from structural control. The course lies north of the Isle of Wight – Purbeck anticline, which trends west–east parallel to the present coast to beyond the east coast of the Isle of Wight. It then turns towards the southeast to cross the Channel towards the Normandy coast. I suspect that the Solent River was forced to flow eastwards by this structure, possibly as a consequence of regional uplift in the west, which may have been active in the Pleistocene (see text). The river presumably continued eastwards until it could break through into the Somme–Seine system.

D. R. Bridgland *Nature Conservancy Council, Peterborough, U.K.*). A comment was offered on behalf of myself and B. D'Olier, of the City of London Polytechnic, on the Straits of Dover.

The results of offshore reflective seismic profiling by B. D'Olier show that the Dover Strait incorporates the southern end of a submerged valley, some 50 km long, cut into the Chalk to a depth of *ca.* 55 m below sea level. This valley has no appreciable slope, but its highest point lies approximately 15 km from its southern end. At the same time, the submerged extension of the Thames valley can be traced offshore from the Essex coast towards the central North Sea, reaching 65 m below sea level by latitude 52 °N. Projection of the floor of the late Devensian

'Buried Channel' of the Thames offshore, by the route suggested by Dr Gibbard for the post-Elsterian Thames–Rhine, indicates that a depth of at least −70 m OD would be required for the river to have passed through the Dover Strait at this time. Given these facts it would seem that, unless differential subsidence or uplift is invoked, the Thames had returned to a North Sea exit during the Devensian.

P. L. GIBBARD. Erosional channels are known from the Channel and Dover Straits area and are referred to by authors quoted in the text. In the past these features have been variously interpreted to result from glacial erosion or more probably from considerable tidal scour (see discussion in Zagwijn (1979)). The presence of an extensive, sub-horizontal channel lacking a sedimentary fill in this region is not unexpected and may itself also be of tidal origin. Alignment of the Rhine, Thames and other tributaries through the Dover Straits during periods of low sea level from the late Middle Pleistocene to the end of the last cold stage is based on the distribution of sediments mapped by the Dutch and British Geological Surveys. I am assured by colleagues in both organizations that there is no evidence that the Thames adopted a northerly course in the southern North Sea after the Anglian–Elsterian. Similarly, there is no evidence of a northerly directed Rhine–Meuse from sediments preserved after the Saalian Drenthe Substage. Indeed, the distribution of the Rhine–Meuse Kreftenheye Formation sediments indicates that these rivers were aligned towards the south during the Late Pleistocene.

The answer to the altitudinal problem may lie in the fact that the gradient of the Late Devensian Thames would be expected to have been reduced offshore at its confluence with the Rhine–Meuse. Such a shallow gradient would have been required for the river to pass through the Dover Straits. If the river did pass through in the Late Devensian this could be confirmed by the distribution of its deposits. It would imply that the channel to which Dr Bridgland refers in its present form postdates the Late Devensian.

Phil. Trans. R. Soc. Lond. B **318**, 603–617 (1988)
Printed in Great Britain

Climatic variability during the past three million years, as indicated by vegetational evolution in northwest Europe and with emphasis on data from The Netherlands

By J. de Jong

Rijks Geologische Dienst, Postbox 157, 2000 *AD Haarlem, The Netherlands*

Temperate and cold stages comparable to those of the last interglacial–glacial have alternated for *ca.* 2.4 Ma, a time-level regarded as the base of the Quaternary. A curve showing climatic fluctuations according to a number of glacials, interglacials, and temperate oscillations of small amplitude or short duration (interstadials) is given, and the value of pollen records in this context is discussed. Because the position of the individual intervals with known vegetational development on the total timescale is controlled by superposition, the lithostratigraphic position of some of them is reviewed.

Basic differences in vegetational evolution between the Tiglian, Waalian, and later interglacials, as well as the extinction of certain trees at around the time of the transition of the Early–Middle Pleistocene, probably indicate lower temperatures during the glacials which have occurred in the past million years.

Although the established record shows some resemblance to the oxygen-isotope curve of the deep sea, precise correlation is not yet possible. A tentative correlation is discussed.

Introduction

The occurrence of climatic fluctuations in the past is no longer disputed, and discussion has shifted to the question of their number and intensity. Palaeontological data which can reflect such variations have been extensively used as a basis for time-stratigraphic subdivisions. Related to the geographical position, such data have been applied to larger areas, in our case northwest Europe. In The Netherlands, the system of subdivision described by van der Vlerk & Florschütz (1950, 1953) incorporated palaeoclimatic interpretation very early on, especially including information provided by pollen analysis as well as other biostratigraphic data. Initially, the number of cold and temperate stages, based mainly on lithostratigraphic data (cf. Penck & Brückner 1909), was restricted, but investigations, especially those based on palynology, of sediments of Early Pleistocene age (Zagwijn 1957) showed an increasing number of warm and cold phases. Since then the number has increased (van der Heide & Zagwijn 1967; Zagwijn 1974, 1985; Zagwijn & Doppert 1978; Zagwijn & de Jong 1984).

Limitations of the basic data

Pollen analysis has contributed substantially to the detection and reconstruction of vegetational variations in the past. This method, which reveals changes in the composition of the pollen content of sediment sequences, provides an impression of the behaviour of the vegetation during the period covering the sediment sequence under investigation. The information is not complete, however, owing mainly to discontinuity of the sedimentation and the fact that only organic material (peat, gyttja) and fine-grained clastic deposits (clay) are appropriate for this type of investigation. Because these materials are not frequently present in the sediment column (mainly composed of sand, gravel, etc.), continuous sections covering more than one interglacial are scarce for the northwestern part of Europe.

Although pollen diagrams may provide information about changes in past vegetation, the evidence is indirect. The same holds for the climate. In this connection mention can be made of the differences in pollen production between plant species, as a result of which the proportions of the pollen of certain plants in sediments are not the same as those of the same plants in the vegetation.

Other factors, too, complicate extrapolation of pollen data from sediments to vegetation and further to climate. For example, both low temperature and drought impede growth of forests. Interpretation of such open vegetation in pollen diagrams is often difficult, because it is not easy to determine whether they are the result of low temperature, a low rate of precipitation, or both. Another factor is the time needed for immigration. After a period of deforestation of an area, a quick rise in temperature will not lead immediately to the restoration of a forest, because immigration of trees takes time. This means that the re-establishment of forests, as indicated by pollen diagrams, occurred later than the actual climatic change. Inferences about climatic changes should therefore be based on increases in the plants already present in the area. From all this it may be concluded that estimates of temperatures during warm intervals give only minimum values.

Changes in soil conditions can also cause vegetational changes that might be erroneously interpreted as climatic changes. The changes in forest composition at the end of an interglacial in northwest Europe are of interest in this context. In this situation, after progressive leaching of the soil, acidiphilous vegetation expands (e.g. spruce in the forest) while moor vegetation and heath replace trees. Such changes can imitate those resulting from deterioration of the climate. However, only careful examination of other plants or other available palaeontological data can provide confirmation. Present northern limits of plant distribution can coincide with certain isotherms, and such data can be used in palaeoclimatic interpretation. A good example is the present northern timberline in northwest Europe, which coincides with the 10 °C mean summer isotherm. This coincidence is very useful for defining the main outlines of the vegetational history of northwest Europe. It remains to be seen, however, whether present limits are in equilibrium with the climate, whether immigration has been completed, or, for instance, whether man has contributed to recent distribution. The limits of distribution may also have been governed by different factors in different areas, and the coincidence with some isotherms may be accidental.

In spite of these major difficulties, vegetational changes reflected by pollen diagrams can, under certain conditions, be shown to indicate changes in temperature. It must be kept in mind, however, that caution is imperative and that the results are approximations: the values are estimates within a few degrees and are only valid for a given amplitude and latitude.

Climatic variations

The last interglacial–glacial (temperate to cold) cycle can serve as a model for vegetational and climatic variability during the Quaternary. Temperate climatic oscillations of small amplitude or short duration, or both, can be called interstadials. However, a clear-cut distinction between interstadials and interglacials cannot always be made. In principle, there can be a whole range of amplitudes of such fluctuations. The resolving power of both the method applied and the character of the sediment sequence under investigation determine whether, for instance, intervals of a very small amplitude will be detectable.

Not only the latitude but also the presence of a considerable area of lowlands make northwest Europe a favourable region for registration of vegetational changes in the past. Here the vegetation varies widely during glacials and interglacials, changing from a very open landscape (barren, tundra, park–tundra) during the glacials to complete afforestation with deciduous trees and conifers in the interglacials. Such changes favour the chance of recognizing climatic fluctuations in the pollen rain recorded in pollen diagrams. As an example, figure 1 shows the

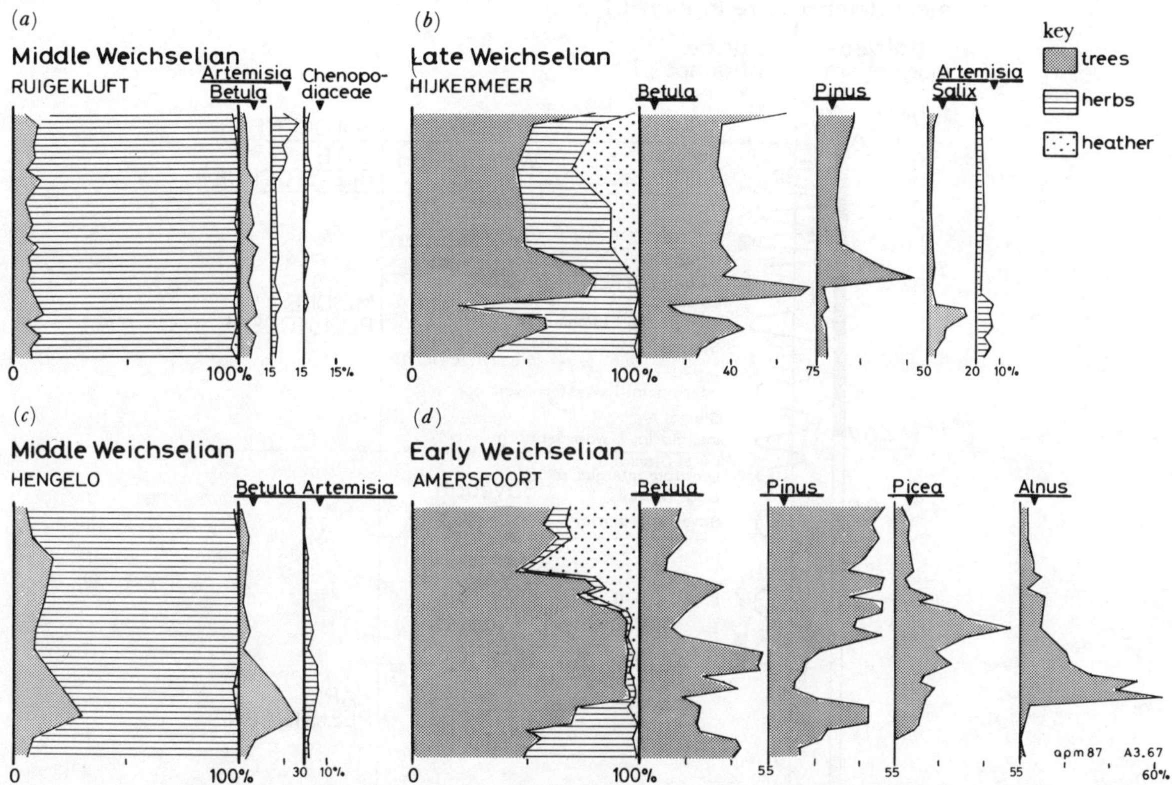

FIGURE 1. Interstadials of the Weichselian cold stage with differences in vegetational development. (*a*, *c*) Hengelo Interstadial of the Middle Weichselian (Zagwijn 1974); (*b*) late glacial with Bølling–Allerød Interstadial of the late Weichselian (van der Hammen 1949); (*d*) Brørup interstadial of the early Weichselian (Zagwijn 1961).

pollen records of four interstadials of the Weichselian, each with a different development of the vegetation. Ruigekluft and Hengelo (figure 1*a*, *c*) (Zagwijn 1974) both represent the Hengelo interstadial of the Middle Weichselian. In Ruigekluft only an increase of herbs already present in the vegetation indicates amelioration of the climate. In Hengelo, dwarf birch in particular increased. Both the diagrams refer roughly to the same period (*ca.* 39 ka BP). Although the differences in the pollen record in these two diagrams may have been caused by a slight difference in age, in which case they would represent different phases in the vegetational development, it is more likely that the variance can be ascribed to different environmental conditions.

The Hijkermeer diagram (figure 1*b*) (van der Hammen 1949) shows a more progressive stage in the vegetational development. In this diagram, characteristic for the late glacial, the vegetational development is demonstrated by the higher values of *Artemisia*, the presence of tree birches, and the appearance of *Pinus* (Bølling–Allerød interstadial). In the upper part of the diagram, further development of forest elements was impeded by the colder climatic conditions prevailing during the subsequent late Dryas stadial.

A still more progressive vegetational development is demonstrated in the Brørup interstadial of the early Weichselian in the Amersfoort region (Zagwijn 1961). Besides *Betula* and *Pinus*, *Picea* and *Alnus* are also represented, with significant values.

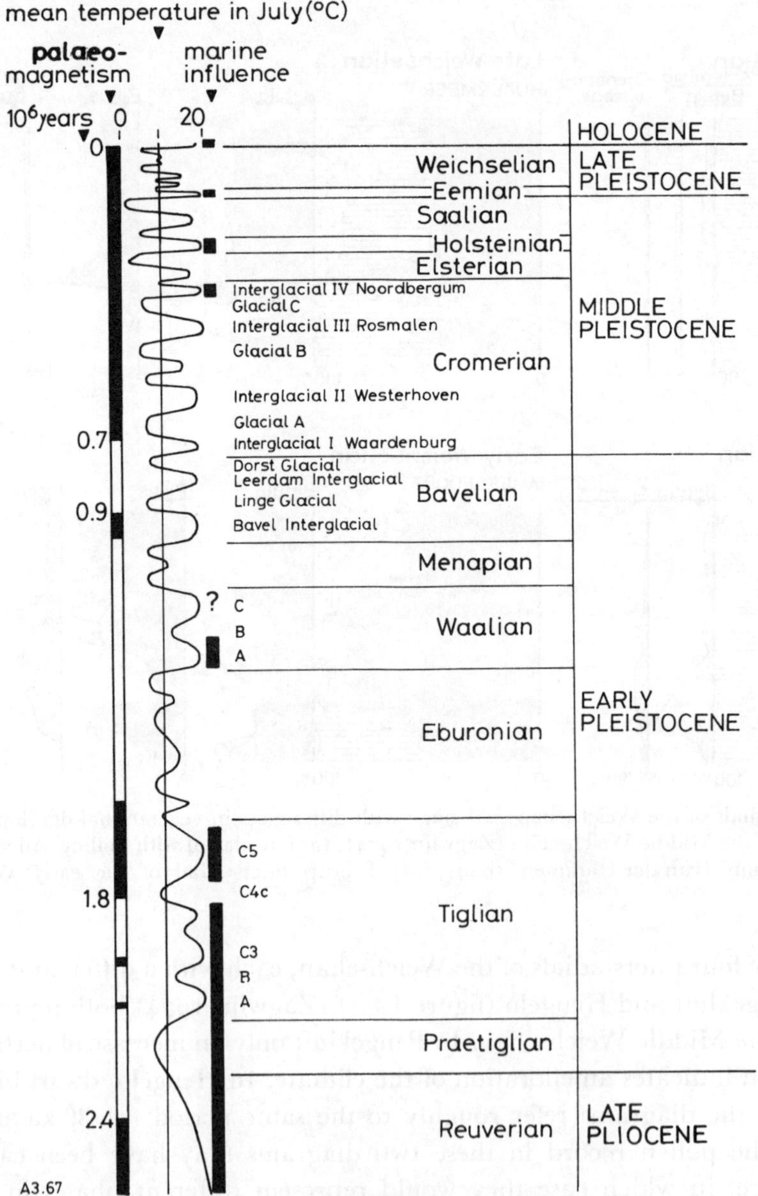

FIGURE 2. Climate curve and chronostratigraphy for the Quaternary of The Netherlands (mainly after Zagwijn 1985).

Stages: chronostratigraphical subdivisions

Cold phases with a vegetation comparable to that of the last glacial stage have occurred since about 2.4 Ma BP. This time-level is regarded as the base of the Quaternary. In the preceding Pliocene interval there were phases of cooling of the climate but this cooling was less intense than it was after 2.4 Ma BP. Originally there was a tendency to assign stage names, such as Tiglian, Eburonian, Waalian, Menapian and Cromerian, to individual interglacial and glacial climatic phases. However, as more palaeoclimatological data became available, particularly those provided by pollen analysis, it became clear that climatic variation had been even more complicated than formerly envisaged.

Predominantly cold stages, such as the Eburonian and the Menapian, have been found to include some intervals of less intensely cold climate. The Tiglian and Waalian interglacials have proved to be complex, because cool phases and even short cold phases were found to have occurred within these periods. Thus the problem arose as to whether each of these warm or cold fluctuations should be considered an interglacial or glacial stage.

This problem became even more critical when it was found that many more true interglacial and glacial phases must have occurred during the interval between the predominantly cold Menapian stage and the cold Elsterian stage, an interval that was originally thought to represent one interglacial called the Cromerian. As long as the true number of climatic variations of this kind was not known (and this may still be the case) it did not seem possible to introduce new stage-names for any of the already recognized interglacial intervals. Definition of the stage boundaries according to the model provided by Late Pleistocene interglacial–glacial change would only be meaningful if it were known with certainty that no other fluctuations had taken place between a given interglacial stage and a given glacial stage. Furthermore, it was evident that the nomenclature would become more and more complicated and difficult to memorize. This can be judged from the climatic curve for The Netherlands, shown in figure 2.

Stratigraphy

In considering the climatic curve of figure 2, it should be kept in mind that the available information about vegetational developments during the Pleistocene is rather fragmentary. The arrangement of the individual phases with known vegetational records in the overall timescale is determined by stratigraphic superposition. Therefore, data concerning the lithostratigraphic position are indispensable. Some cases for which the lithostratigraphic position is of particular interest for the sequence presented in figure 2 deserve discussion.

The present chronostratigraphic system for the Quaternary of The Netherlands is based on evidence from two different areas (figure 3), the first, for the lower part of the curve, being the Central Graben area in the south, starting with the Praetiglian as the base and ending with the Cromerian. For this area it has recently become clear that a series of deposits belonging pollen-analytically to the Early Pleistocene (and therefore thought to be older than the top of the Menapian) and hitherto partly assigned to the Waalian, are in reality younger because they overlie typical Menapian deposits in the stratotype region, also in the Central Graben. Further studies showed that at least two interglacials and two glacials had occurred within this interval, which is now termed Bavelian. In the Central Graben area the Bavelian beds underly the first interglacial of the Cromerian complex (Zagwijn & de Jong 1984).

FIGURE 3. Topographic map of The Netherlands. Geological features and localities of interest in the present context are indicated.

The second area, the source of the upper part of the curve (figure 4), derives from the northern part of The Netherlands, and the evidence is related to the features of glaciation during the Saalian and the Elsterian. In this area the position of an interglacial phase, called Cromerian IV, is of particular interest. The partly marine beds of this interglacial underly a suite of deposits belonging to the Peelo Formation (figure 5). The Peelo Formation partly fills very deep erosional depressions and often has a characteristic facies, called pottery clay (*potklei*). There can be no doubt that the Peelo Formation is in almost every respect the equivalent of the Lauenburg Clay of northwest Germany, which is Elsterian in age. In both areas these deposits underly Holsteinian interglacial beds as well as Saalian till. At first it was thought that the interglacial at the top of the sequence in the southern part of The Netherlands (Cromerian III; see also de Ridder & Zagwijn (1962) concerning the same interglacial), was identical to the interglacial near the base of the Middle Pleistocene sequence in the northern part of The Netherlands (Cromerian IV), but pollen analysis has shown that this is not the case. For instance, Cromerian IV (figure 6, Roswinkel) has *Abies*, whereas Cromerian III (figure 6, Het Zwinkel) does not; Cromerian III has *Carpinus* early in the interglacial, but Cromerian IV does not; and there are some minor differences as well.

For the northern part of The Netherlands, mention may be made of two interstadials known from the Early Saalian (Hoogeveen and Bantega interstadials) (figure 4). In Bantega (Zagwijn

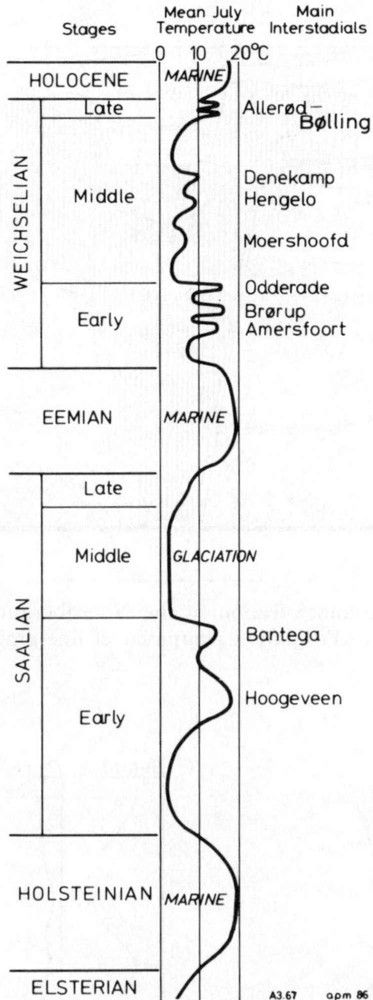

FIGURE 4. Climate curve and chronostratigraphy of the later part of the Quaternary of The Netherlands, based mainly on data for the northern part of The Netherlands.

1973) these interstadials are represented by partly organic beds intercalated in periglacial sediments (Eindhoven Formation), and are underlain by deposits of Holsteinian age and overlain by the Saalian till. The Hoogeveen interstadial in particular shows evidence of rather warm climatic conditions. It seems possible that the vegetational development is not yet completely known. Figure 7 gives a schematic representation of data obtained from various locations. The pollen diagram shows some features known from interglacials at our latitude.

In the southern part of The Netherlands, in the Central Graben area, periglacial deposits with intercalated loam and organic layers (Nuenen Group) overlie fluvial deposits from the river Meuse (Veghel Formation) (figure 8). The latter are partly of Cromerian III age (Het Zwinkel, figure 6).

So far, only fragmentary pollen-analytical information about the deposits of the Nuenen Group has become available. At shallow depths there are organic layers of Eemian age; at the base the deposits are presumably of Elsterian age. A considerable portion of the deposits must be of Saalian age. Although for this sedimentation area, outside the glaciated part of The

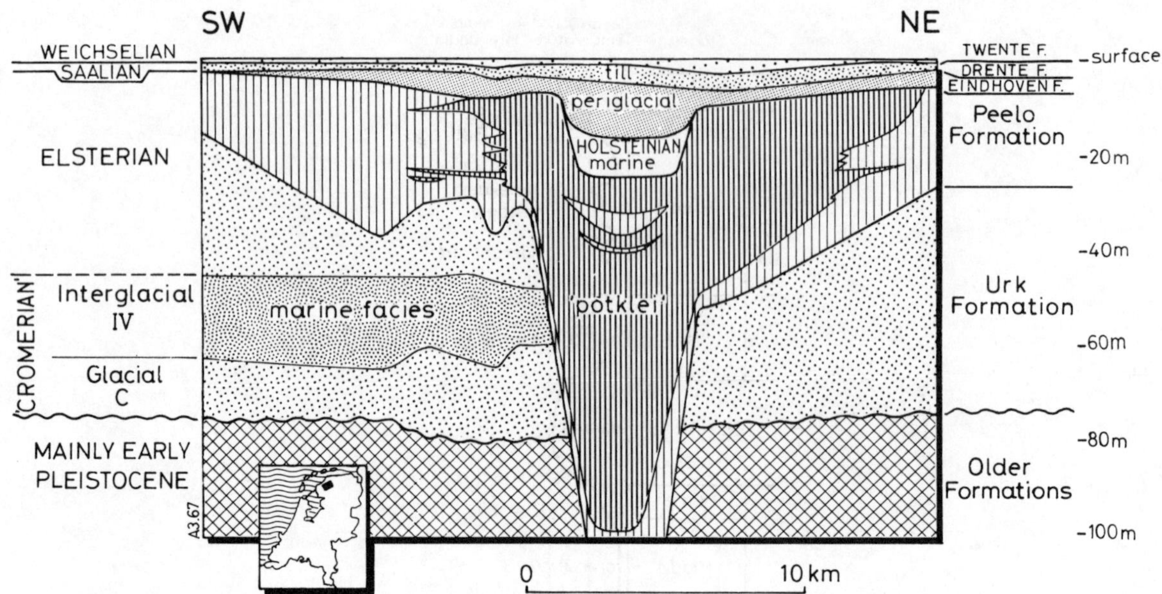

FIGURE 5. Northern part of The Netherlands: section in the Noordbergum area. Modified after Zagwijn and van Staalduinen (1975, p. 17). Peelo Formation composed of fine-grained *potklei* (fine hatching) and more sandy deposits (wide hatching).

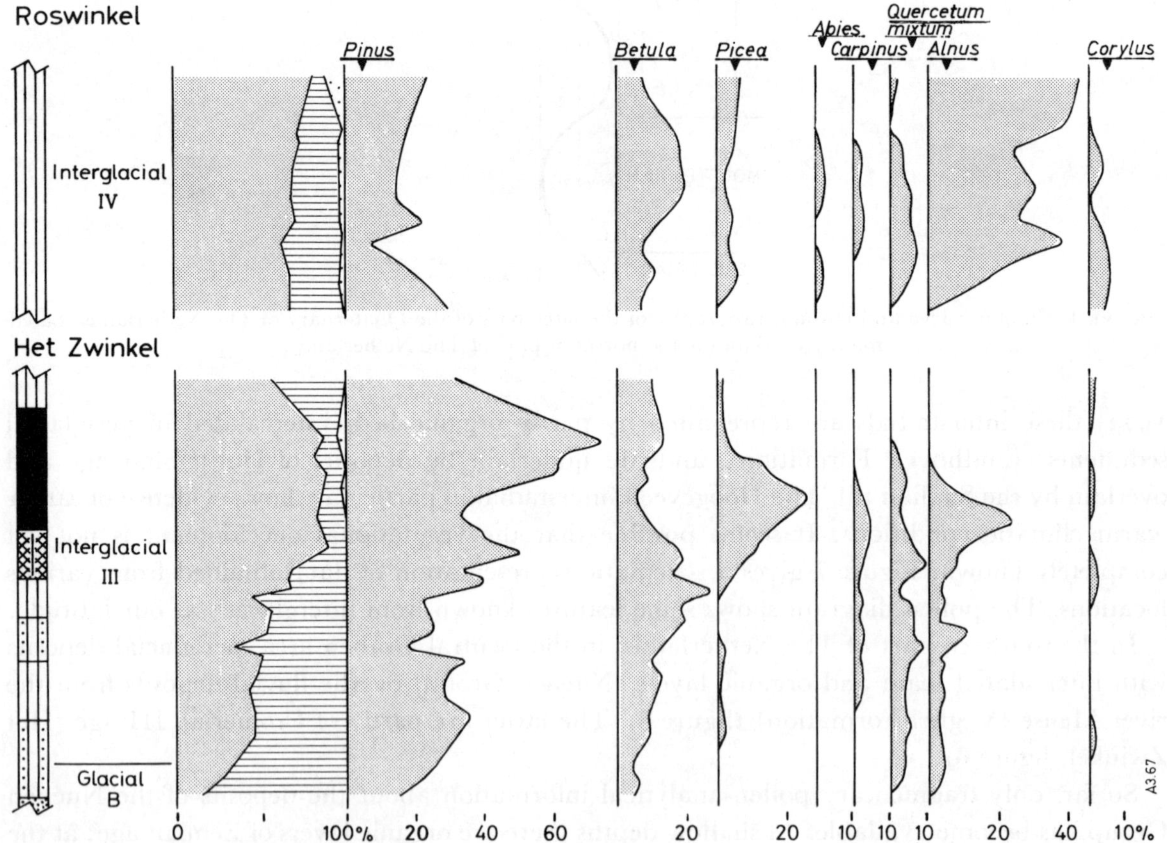

FIGURE 6. Pollen diagrams showing vegetational development in the Cromerian III (Het Zwinkel) and Cromerian IV (Roswinkel) interglacials. (For key to pollen diagrams, see figure 1.) Lithology: hatched, clay and loam; stippled, sand and sandy clay; black, peat with underlying gyttja.

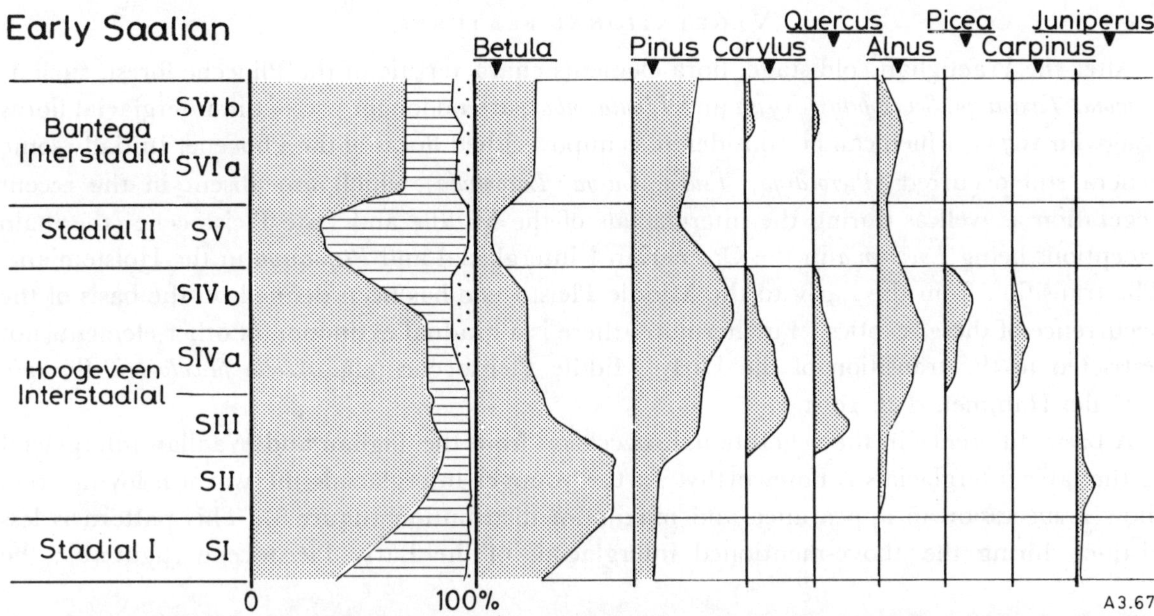

FIGURE 7. Schematic representation of vegetational succession in the Early Saalian, based on data obtained in three localities. (For key to pollen diagram, see figure 1.)

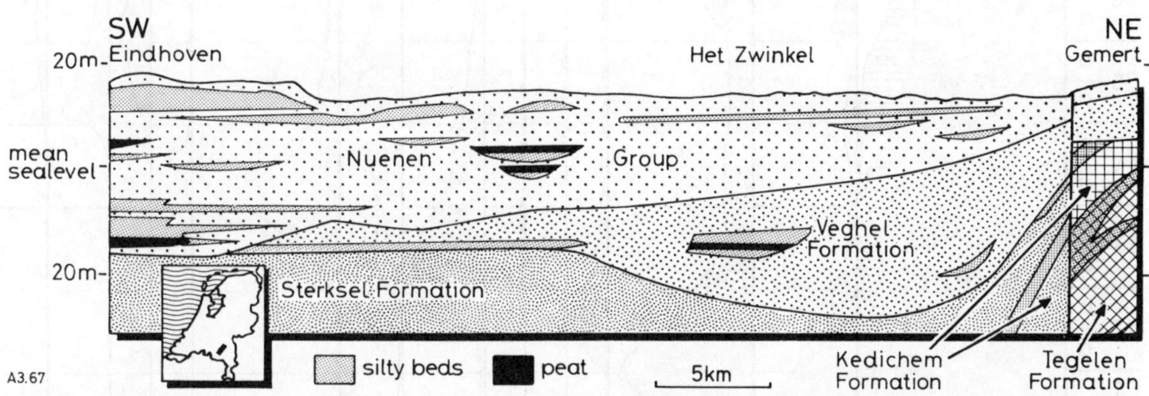

FIGURE 8. Stratigraphic section of the Middle and Late Pleistocene deposits in the Central Graben in the southern part of The Netherlands (modified after Bisschops 1973). Periglacial deposits of the Nuenen Group Overlying fluvial sediment of the Rhine (Sterksel Formation) and Meuse (Veghel Formation). The loamy and organic beds intercalated in the mainly sandy deposits of the latter indicate the position of the Cromerian III interglacial (Het Zwinkel).

Netherlands, rather complete information concerning the development of the Saalian could be expected, the pollen-analytical information is very incomplete. Besides sequences with very open vegetation, there are intervals with a pine-dominated interstadial (Bakel interstadial), presumably correlatable with the Bantega interstadial in the north. There are fragments representing a more temperate interval; these are considered to be parts of an interstadial (Hoogdonk interstadial) which presumably is correlatable with the Hoogeveen interstadial in the north. Just as in the northern part of the country, it is impossible or hardly possible to distinguish short intervals of this interstadial from those of Holsteinian age. In this area the position of the Holsteinian is not definitely known.

Vegetational features

After the Praetiglian cold stage, flora elements characteristic of the Pliocene forest, such as *Sequoia*, *Taxodium*, *Sciadopitys*, *Nyssa* and *Liquidambar*, are no longer present in interglacial floras (Zagwijn 1975), which can be considered as impoverished floras of the Pliocene. Initially some genera still occurred (*Pterocarya*, *Tsuga*, *Carya*, *Eucommia*) which are absent in the recent vegetation as well as during the interglacials of the Middle and Late Pleistocene (the main exceptions being *Eucommia* in the Cromerian I interglacial and *Pterocarya* in the Holsteinian). The transition from the Early to the Middle Pleistocene has been defined on the basis of the occurrence of these 'exotics'. Furthermore, there is a gradual extinction of other elements not restricted to the transition of the Early–Middle Pleistocene (e.g. *Azolla filiculoides*, *Brasenia* (van der Hammen *et al.* 1971)).

A basic difference in the vegetational succession from the Tiglian and Waalian interglacial to the later interglacials is noteworthy. In the younger interglacials the warmth-loving trees show a succession in appearance and maximum distribution (figure 8). This pattern is less distinct during the above-mentioned interglacials of the Early Pleistocene (figure 9); the

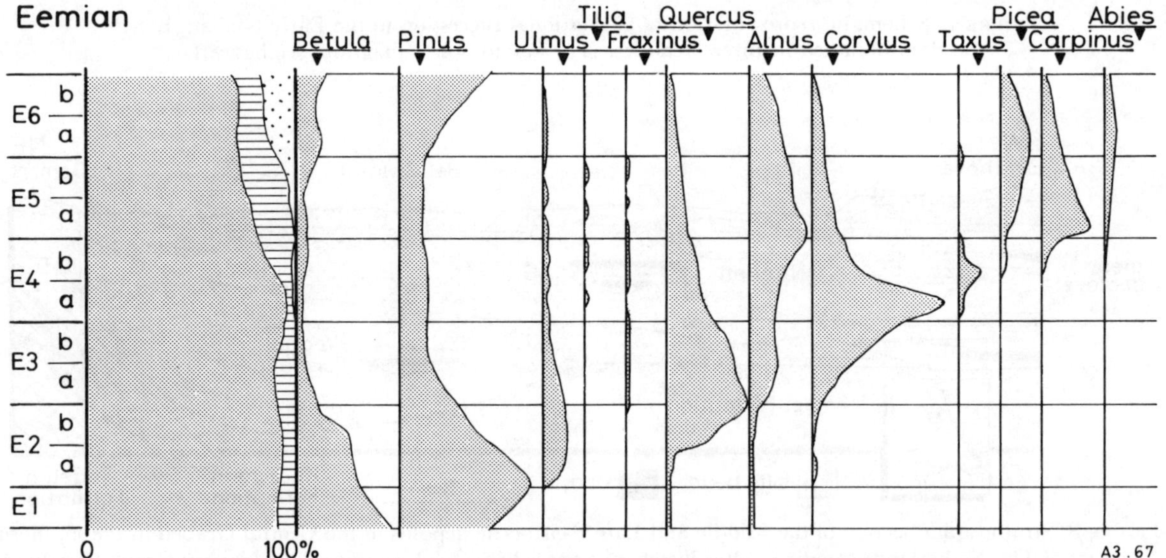

FIGURE 9. An interglacial of the Late Pleistocene. Schematic example of vegetational development of an interglacial with distinctive evidence for immigration and spreading of trees.

warmth-loving trees seem to spread and decline together. There is no clear succession in the immigration of forest elements, and the majority of the contributors seem to have been present from the very beginning (figure 10).

Besides geological and other conditions, differences in climatic conditions seem to be particularly responsible for this behaviour. Presumably, in this case, the warmth-loving trees did not have to immigrate from a very distant region. This indicates that conditions were less cold in the preceding glacial than during later glacials.

The extinction of the above-mentioned 'exotic' elements in the course of time may also be an indication of increasingly unfavourable climatic conditions in the successive glacials. Most of the genera mentioned now occur in the indigenous floras of North America and Asia under climatic conditions which may not necessarily be warmer than the present climate of The Netherlands. Furthermore, they thrive very well in parks in our regions today. It is therefore less probable that the limiting factor was cooler climatic conditions during the interglacials.

To sum up, it can be stated that during the Tiglian and Waalian the extent of the temperate forest was due mainly to climatic amelioration. Immigration seems to have been of less importance and indicates that climatic conditions were less cold in the preceding glacials than in the later part of the Pleistocene. Starting in the Bavelian, the interglacials show a characteristic succession in the appearance and expansion of warmth-loving trees. This indicates a longer immigration distance and therefore presumably colder climatic conditions in the preceding glacial. Most of the exotic elements characteristic of the Early Pleistocene interglacials could, however, maintain themselves up to the Leerdam interglacial, and *Eucommia* was still present in interglacial I of the Cromerian. This may indicate that, during the preceding glacials, these 'exotics' could maintain themselves in Europe and could immigrate once again in the next interglacial. Obviously this was no longer the case during the later glacials of the Middle Pleistocene. These data lead to a general view of increasing colder climatic conditions starting in the glacials from the Early Pleistocene up to those of the Middle Pleistocene.

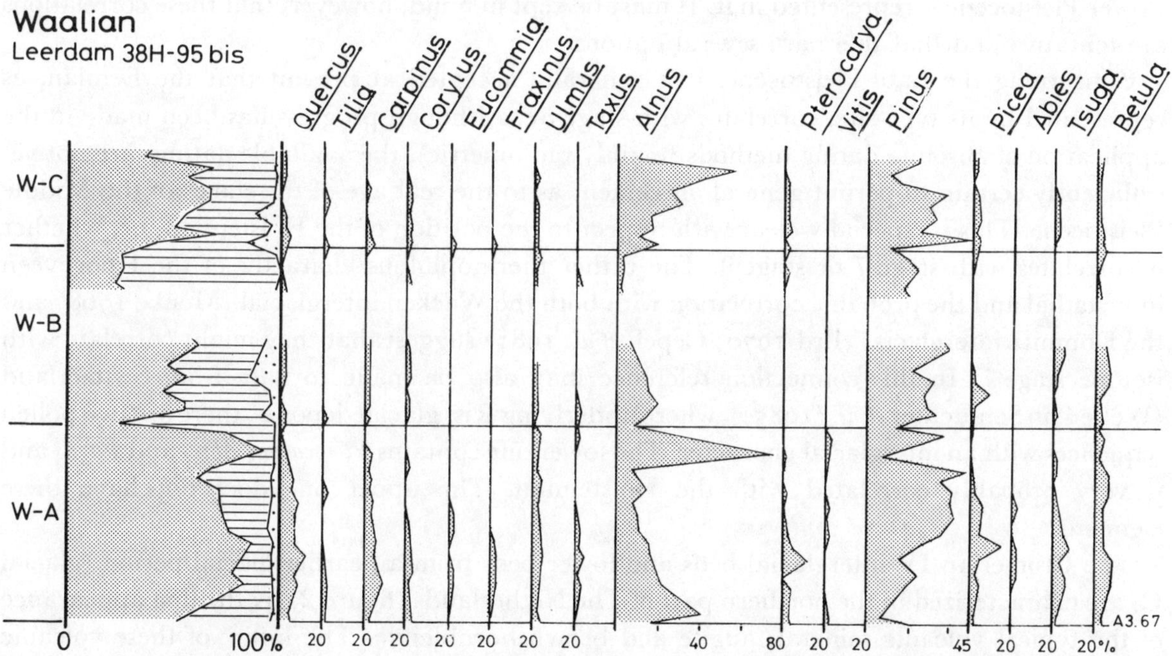

FIGURE 10. An interglacial of the Early Pleistocene. No distinct indications of systematic succession in immigration and culmination of the individual trees (after Zagwijn & de Jong 1984).

MARINE INFLUENCE

During the Neogene, the greater part of The Netherlands was part of the North Sea. At that time, and at the beginning of the Pleistocene, the coastline gradually shifted in the direction of its present position. The first regression away from the position of the recent coastline occurred in subzone TC 4c of the Tiglian. New data (T. Meijer, Geological Survey of The Netherlands, personal communication) show that, in the present coastal area of The

Netherlands too, marine influence occurs in deposits dating from the later part of the Tiglian (pollen zone TC 5) and from the Waalian (A). For the latter interval, signs of near-coastal conditions were already known (Zagwijn 1974). For the remaining part of the Early Pleistocene and the major part of the Middle Pleistocene in the present coastal area, no marine deposits are known. Starting with interglacial IV of the Cromerian 'complex', the interglacials are characterized again by a marine influence in the present coastal area. Although this behaviour can be ascribed mainly to geological circumstances, there may also have been some influence of climatic conditions.

Correlations with deep-sea isotopic records

The climatic record established for The Netherlands shows some similarities with the oxygen-isotope curve for the deep sea. A detailed correlation with a reasonable degree of certainty cannot yet be made, however. A tentative correlation with core V28-239, originating from the western equatorial Pacific (Shackleton & Opdyke 1976), has been given (figure 11). Although the resolution for the Middle Pleistocene interval is lower in this core than in core V28-238, which originates from the same area, preference is given to the former because the Lower Pleistocene is represented in it. It must be kept in mind, however, that these correlations are tentative and that there are several options.

Concerning the Late Pleistocene, it is generally accepted at present that the Eemian, as represented by its type-site, correlates with stage 5e. Although progress has been made in the application of absolute dating methods (mainly radiometric), the available datings are not yet sufficiently certain to permit general agreement as to the real age of the stages of the Middle Pleistocene. This is especially clear with respect to the position of the Holsteinian, i.e. whether it correlates with stage 7 or stage 9. The rather thermophilous character of the Hoogeveen interstadial and the probable correlation with both the Wacken interglacial (Menke 1968) and the Dömnitz interglacial (Erd 1970; Cepek *et al.* 1981) suggest that they might correlate with isotope stage 7. In this connection reference may also be made to data from Switzerland (Welten, in Schlüchter *et al.* (1985)), where underlying Riss glacial deposits, there are two pollen sequences with an interglacial character. The lower one contains *Pterocarya*, *Abies* and *Fagus* and is very probably correlated with the Holsteinian. The upper one does not have these elements.

The Cromerian IV interglacial beds and lower beds from an earlier glacial period (glacial C) are characterized in the northern part of The Netherlands (figure 2) by the first appearance of the typical volcanic minerals augite and brown hornblende. The influx of these volcanic minerals into the Rhine deposits is associated with eruptions in the Eifel area (Selbergit tuff). Although the lithostratigraphic correlation of the various deposits of the Rhine is complex, radiometric dating of the oldest known Selbergit tuff (Evernden *et al.* 1957; Frechen & Lippolt 1965) points to an age of about 400 ka BP for the first influx of these minerals (Zagwijn 1985). This means that the Cromerian must date from that time or slightly later and can therefore be correlated with isotopic stage 11, in agreement with the estimated date of that stage established by interpolation of the rate of sedimentation (Shackleton & Opdyke 1973).

A series of palaeomagnetic determinations with normal polarity of sediments initially assigned to the Waalian refers instead to deposits of the Bavel interglacial. This leads to the conclusion that palaeomagnetically the Bavel interglacial (base of the Bavelian stage) can be

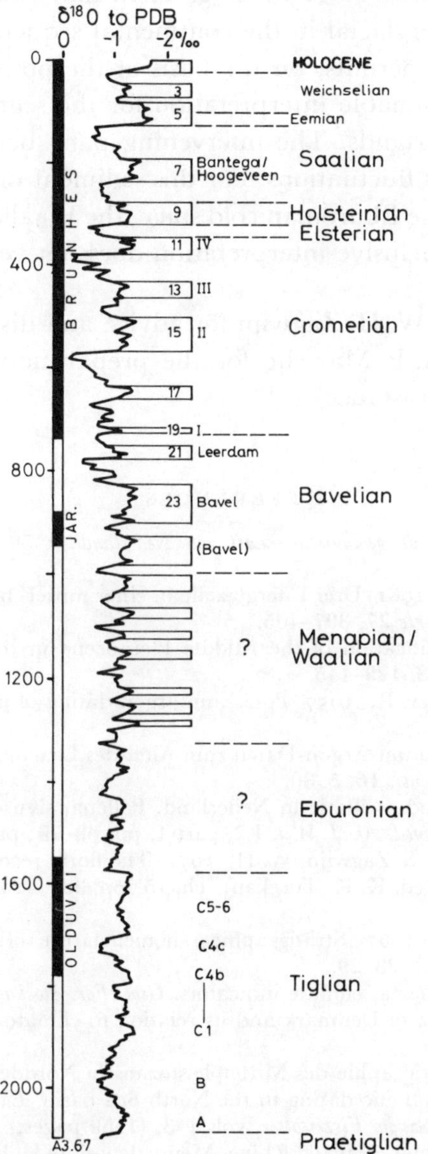

FIGURE 11. Tentative correlation of the climate curve of the Quaternary in The Netherlands with the isotope stages in the deep-sea core V28-239 (from Shackleton & Opdyke 1976).

dated in the Jaramillo (Zagwijn & de Jong 1984). This in turn may point to isotopic stage 23 (25?) for the Bavel interglacial, stage 21 for the Leerdam interglacial, and stage 19 for Cromerian I, because the deposits of the latter two interglacials are palaeomagnetically reversed. Furthermore it is assumed that the amplitude of stage 19 is sufficient for the Cromerian I interglacial. The position of stage 19 in the Matuyama reversed epoch is the case for core V28-238 with a higher resolution, but not for core V28-239.

On the basis of terrace correlations in the Lower Rhine basin in Germany, Cromerian III may correlate with Main terrace 3 (Zagwijn 1985) and could therefore be slightly older than the beginning of the Selbergit volcanism (augite and brown hornblende), that is, slightly older than *ca.* 400 ka. This leads to the suggestion that Cromerian III may correlate with stage 13 and Cromerian II presumably with stage 15, stage 17 in that case representing an unknown, or not accurately classified, interglacial in the continental sequence.

The lower part of the curve permits, on the basis of the position of the Olduvai normal palaeomagnetic interval, a reasonable interpretation for the segment below about 1590 cm, with the Tiglian of The Netherlands. The intervening part, between *ca.* 950 and 1590 cm, shows a considerable number of fluctuations. For this sediment interval, which covers at least the period of the upper part of the Eburonian cold stage, the Waalian interglacial complex, and the Menapian cold stage, a conclusive interpretation does not yet seem possible.

I am greatly indebted to Dr W. H. Zagwijn for advice and discussion of various aspects of this paper. I also thank Mr A. P. Marselje for the preparation of figures, and Miss E. H. Vreugdenburg for secretarial assistance.

References

Bisschops, J. H. 1973 *Toelichting bij de geologische kaart van Nederland* 1: 50000. Haarlem: Rijks Geologische Dienst.

Cepek, A. G., Erd, K. & Zwirner, R. 1981 Drei Interglaziale in einer mittel- bis jung pleistozaner Schichtenfolge ostlich von Berlin. *Z. angew. Geologie* **27**, 397–405.

Erd, K. 1970 Pollen-analytical classification of the Middle Pleistocene in the German Democratic Republic. *Palaeogeogr. Palaeoclim. Palaeoecol.* **8**, 129–145.

Evernden, J. F., Curtis, G. H. & Kistler, R. 1957 Potassium argon dating of pleistocene volcanics. *Quaternaria* **4**, 1–5.

Frechen, J. & Lippolt, H. J. 1965 Kalium-Argon-Daten zum Alter des Laacher Vulkanismus, der Rheinterrassen und der Eiszeiten. *Eiszeitalter Gegenw.* **16**, 5–30.

van der Hammen, T. 1949 De Allerød-oscillatie in Nederland. Pollenanalytisch onderzoek van een laatglaciale meerafzetting in Drente. *Proc. K. ned. Akad. Wet.* L2, part 1, pp. 69–76; part 2, pp. 169–176.

van der Hammen, T., Wijmstra, T. A. & Zagwijn, W. H. 1971 The floral record of the late Cenozoic of Europe. In *The late Cenozoic glacial ages*, (ed. K. K. Turekan), ch. 15, pp. 391–424. New Haven and London: Yale University Press.

van der Heide, S. & Zagwijn, W. H. 1967 Stratigraphical nomenclature of the Quaternary deposits in The Netherlands. *Meded. geol. Sticht.* **18**, 23–29.

Iversen, J. 1944 *Viscum, Hedera* and *Ilex* as climate indicators. *Geol. For. Stockh. Forh.* **66**, 463–483.

Iversen, J. 1954 The late-glacial flora of Denmark and its relation to climate and soil. *Dan. geol. Unders.* **2** (80), 87–119.

Menke, B. 1968 Beitrage zur Biostratigraphie des Mittelpleistozans in Norddeutschland. *Meyniana* **18**, 15–42.

van Montfrans, H. M. 1971 Paleomagnetic dating in the North Sea basin. *Earth planet. Sci. Lett.* **11**, 226–236.

Penck, A., Bruckner, E. 1909 *Die Alpen im Eiszeitalter*, vols 1–3. (1159 pages.) Leipzig:

de Ridder, N. A. & Zagwijn, W. H. 1962 A mixed Rhine–Meuse deposit of Holsteinian age from the south-eastern part of The Netherlands. *Geol. Mijnb.* **41**, 125–130.

Schluchter, Ch., Wegmuller, S. & Welten, M. 1985 *On Quaternary reference sections in the eastern and central Alpine Foreland of Switzerland. Guidebook to the excursions of Oct. 16 and 17, 1985; INQUA, Subcommission on European Quaternary Stratigraphy, Symposion, Zurich.* (88 pages.)

Shackleton, N. J. & Opdyke, N. D. 1973 Oxygen isotope and palaeomagnetic stratigraphy of equatorial Pacific core V28-238: oxygen isotope temperatures and ice volumes on a 10^5 year and 10^6 year scale. *J. Quat. Res.* **3** (1).

van der Vlerk, I. M. & Florschutz, F. 1950 *Nederland in het IJstijdvak.* (287 pages.) Utrecht: de Haan.

van der Vlerk, I. M. & Florschutz, F. 1953 The palaeontological base of the subdivision of the Pleistocene in The Netherlands. *Verh. K. ned. Akad. Wet.* **20** (2), 1–58.

Zagwijn, W. H. 1957 Vegetation, climate and time-correlations in the Early Pleistocene of Europe. *Geol. Mijnb.* **19**, 233–244.

Zagwijn, W. H. 1961 Vegetation, climate and radiocarbon datings in the Late Pleistocene of the Netherlands. Part I: Eemian and Early Weichselian. *Med. geol. Sticht.*, **14**, 15–45.

Zagwijn, W. H. 1973 Pollenanalytic studies of Holsteinian and Saalian Beds in the northern Netherlands. *Meded. Rijks geol. Dienst* **24**, 139–156.

Zagwijn, W. H. 1974 Vegetation, climate and radiocarbon datings in the Late Pleistocene of the Netherlands. Part II: Middle Weichselian. *Meded. Rijks geol. Dienst* **25** (3), 101–111.

Zagwijn, W. H. 1974 Palaeographic evolution of The Netherlands during the Quaternary. *Geol. Mijnb.* **53** (6), 295–468.

Zagwijn, W. H. 1975 Variation in climate as shown by pollen analyses, especially in the Lower Pleistocene of Europe. In *Ice Ages: ancient and modern*, Geological Journal special issue no. 6 (ed. A. E. Wright & F. Moseley), pp. 137–152. Liverpool: Seel House Press.

Zagwijn, W. H. & van Staalduinen, C. J. (eds) 1975 Toelichting bij geologische overzichtskaarten van Nederland, pp. 109–114. Haarlem: Rijks Geologische Dienst.

Zagwijn, W. H. & Doppert, J. W. C. 1978 Upper Cenozoic of the Southern North Sea Basin: Palaeoclimatic and Palaeogeographic evolution. *Geol. Mijnb.* **57**, 577–598.

Zagwijn, W. H. & de Jong, J. 1984 Die Interglaziale von Bavel und Leerdam und ihre stratigraphische Stellung im niederlandischen Fruh-Pleistozan. *Meded. Rijks geol. Dienst* **37** (3), 155–169.

Zagwijn, W. H. 1985 An outline of the Quaternary stratigraphy of The Netherlands. *Geol. Mijnb.* **64**, 17–24.

Phil. Trans. R. Soc. Lond. B **318**, 619–635 (1988)

Printed in Great Britain

Correlation of marine events and glaciations on the northeast Atlantic margin

By D. Q. Bowen and G. A. Sykes

Department of Geography, Royal Holloway and Bedford New College, Egham, Surrey TW20 0EX, U.K.

Stratigraphic units representing high-sea-level events in Britain, northern France, Belgium, The Netherlands, northwest Germany, Denmark, Sweden and Norway, are correlated by aminostratigraphy (*D* (alloisoleucine)/*L* (isoleucine) (ratios from *Littorina littorea*, *Macoma balthica*, *Macoma calcarea* and *Arctica islandica*). The eight sea-level events recognized are modelled with the constraints provided by the oxygen-isotope signal of sea-level variability, and available geochronometric age determinations for calibrating the *D*/*L* data. These data are used to constrain the timing and extent of glaciations in the British Isles and Scandinavia during the Middle and Late Pleistocene.

Introduction

Despite the convincing telecorrelation of continental loess ('glacials') and palaeosol ('interglacials' or 'interstadials') sequences with the oxygen-isotope signal of Brunhes time, notably in Central Europe (Kukla 1977), many uncertainties remain. In particular, the correlation of actual glacial deposits with specific oxygen-isotope stages does not yet rest on a firm stratigraphic basis. As a result much uncertainty occurs and major disagreement obtains on the timing and extent of glaciations, and especially their correlation with the oxygen-isotope signal. One view accepts the number of glaciations recognized by long-standing convention and makes correlation by a 'count from the top' (Kukla 1977) procedure down the even-numbered stages of the oxygen-isotope chronostratigraphic scale. Another has adapted an isoleucine epimerization time scale of marine events and correlated these with the conventionally recognized 'Cromerian', 'Holsteinian' and 'Eemian' Stages of Europe, by implication also dating the glaciations between those events (Miller & Mangerud 1985). Yet another view, widely represented in the IGCP Report on Project-24 'Quaternary glaciations in the Northern Hemisphere' (Sibrava *et al.* 1986) ascribes the European glaciations to the even-numbered stages of the oxygen-isotope signal for all of Middle Pleistocene time, in particular by recognizing two Elsterian Glaciations (I and II) and two (or three) Saalian Glaciations (Sibrava *et al.* 1986). But except for the Late Devensian and Late Weichselian Glaciations, which are constrained by radiocarbon dating, evidence for the age of the earlier ones is still under discussion.

Meanwhile, it is clear that progress is being influenced by pre-existing conventional classifications. The precept of the United States Geological Survey in declaring some of the American historical chronostratigraphic labels (e.g. 'Nebraskan' and 'Kansan') redundant (Richmond & Fullerton 1986) has yet to be followed in Europe despite unmistakable imperatives. For example, in a conventional sequence of three glaciations (from 'Cromerian' to present), the reclassification of a till at Esbjerg, Denmark, as 'pre-Cromerian' (formerly 'Elsterian') and overlying interglacial beds as 'Cromerian' (formerly 'Holsteinian') (Miller & Mangerud 1985) begs searching questions on the validity of correlating deposits by using these and other names elsewhere. Given such circumstances (see numerous examples in Sibrava *et al.* 1986), together with poor subsurface control and a generally inadequate knowledge of the geometry of stratigraphic units over wide areas, it is not unreasonable to suppose that, although correlation should be based on 'all possible means', the correlation and age of glacial stratigraphic units can only be convincingly based on a comprehensive geochronology from multiple dating methods. The promise of relatively new dating methods, such as amino acid epimerization, thermoluminescence dating (TL), electron spin resonance (ESR) dating, and the enhanced means of uranium-series dating (Edwards *et al.* 1986), has not yet been realized, but in combination they offer the prospect of solving many problems of correlation and dating.

The margin of the northeast Atlantic Ocean is an ideal testing ground for some of these methods because of its unique stratigraphic sequences of marine and glacial deposits; and although the 'state of the art' review shows that wider and more intensive dating programmes are desirable, the emerging correlations and age determinations already offer a stratigraphic framework superior to that of long-standing conventional classifications.

To a large and critical extent the global integrity of the oxygen-isotope signal and its unique attributes both as a chronostratigraphic and geochronological framework, and as a surrogate of ice-volume variability and sea-level change (Shackleton & Opdyke 1973), may be used to constrain the timing and patterning of sea-level change, and glaciation based on such application of multiple age determinations. But although the oxygen-isotope signal is a surrogate for global ice-volume variability it does not indicate the changing location nor the relative strength of ice centres, which is a necessary prelude to successful modelling of climatic change in the British Isles and Europe. The location of both these regions on the margin of the northeast Atlantic Ocean makes it likely that they were particularly sensitive to the climatic forcing of ice-sheet growth, especially in the case of western Britain.

The problem of relating the even-numbered stages of the oxygen-isotope signal to regional ice volume and extent is matched by the problem of the regional status of the odd-numbered stages of relatively low global ice volume: while these may rank as 'interglacial' on a global scale, provincial proxy data may be classified as 'interglacial' or 'interstadial' (see in Sibrava *et al.* 1986). Thus there is much potential for regional mis-correlation on the continents and between continents and oceans.

Successful ascription of the timing and relative extent of different ice-centres to the oxygen-isotope timescale will allow modelling of ice accumulation, notably related to patterns of precipitation availability, the circulation of the north Atlantic Ocean and, eventually, to modelling the atmosphere on timescales of 10^5 years.

Correlation

On the littoral margin of the northeast Atlantic Ocean in southwest Britain the geochronology of sea-level events and glaciations allows correlation with the oxygen-isotope signal (Bowen *et al.* 1985). It is based on the lithostratigraphy of marine and glacial stratigraphic units, which outcrop along the coastline (Bowen 1973), and its amplification by aminostratigraphy, using D-alloisoleucine: L-isoleucine ratios (D/L) from *Littorina littorea* (together with some other species which epimerize at similar rates), *Macoma balthica*, *Macoma calcarea* and *Arctica islandica*. Such shells occur in marine stratigraphic units as well as in glacial deposits. Shelly glacial deposits occur where Pleistocene glaciers have crossed parts of the continental shelf and former littoral areas, from where they incorporated the marine fauna. Many marine units, still in a geomorphological context of shore-platform and former sea-cliff, have withstood the passage of glacier ice across them (Bowen 1973). But others, notably in northwest Europe, have been brought near to the present surface along thrust planes caused by glacial tectonics: for example, in northwest Germany, 'Holsteinian' marine deposits are 'only exposed where they have been up-thrust by glacial tectonics' (Ehlers 1983). In such circumstances a regional lithostratigraphy is unlikely to be definitive however close the subsurface control may be, a view widely upheld in Sibrava *et al.* (1986), and independent age evaluation by aminostratigraphy and other dating methods is more likely to be definitive than a classification based on the use of facies-floras in 'interglacial' deposits.

Apart from some possible variability, between southern Britain and southwest Norway, it is assumed that fluctuations in Pleistocene temperatures throughout northwest Europe were uniform, thus eliminating the temperature term in epimerization kinetics. A limited number of geochronometric age determinations supports this. Similarly the species term is eliminated by using molluscs known to epimerize at the same rate. The use of overlapping species which epimerize at different rates from the same lithostratigraphic units extends the range of the method. Two data sets are used: one based on the relatively slow rates of epimerization shown by *Littorina littorea* (and other species (Bowen *et al.* 1985)), and another based on the relatively moderate rates of epimerization shown by *Macoma balthica*, *Macoma calcarea* and *Arctica islandica* (Miller & Mangerud 1985) (figure 1). The overlap between the two data sets is important, not only for use where one may be absent because of different palaeoecological conditions, but also because only *Macoma* and *Arctica* are found in shelly glacial deposits. The approach to correlation and dating, therefore, is unimpeded by conventional models of classification (Kukla 1977), and is based on independent parameters for correlating and dating stratigraphic units.

Sea-level events

Reproducible D/L ratios from molluscs in stratigraphic units throughout northwest Europe permit recognition of discrete events of high sea level. The data from *Macoma balthica*, *Macoma calcarea* and *Arctica islandica* show a larger number of sea-level events than the *Littorina* data, but the two data sets may be correlated, and eight high-sea-level events are recognized.

High-sea-level event 1 (*Macoma–Arctica* $D/L = 0.46$). This includes the fossiliferous marine sediments at Esbjerg, Denmark (Hansen 1965; Miller & Mangerud 1985), and the marine beds at Noord Bergum, The Netherlands (Miller & Mangerud 1985).

High-sea-level event 2 (*Macoma–Arctica* $D/L = 0.37$; *Littorina* $D/L = 0.26$). This includes the Vognsbøl Sand and the marine sands at Eartham (Boxgrove) in Sussex, which allows correlation with the *Littorina* D/L data (figure 1). Also included in this event are the marine sands at Hummelsbüttel ('Holsteinian'), Hamburg (Miller & Mangerud 1985).

High-sea-level event 3 (*Macoma–Arctica* $D/L = 0.29$; *Littorina* $D/L = 0.2$). This includes the upthrusted marine clays at Wacken ('Holsteinian'), which contain both *Macoma–Arctica* and *Littorina*, thus allowing correlation with the *Littorina* data (figure 1), including a reworked

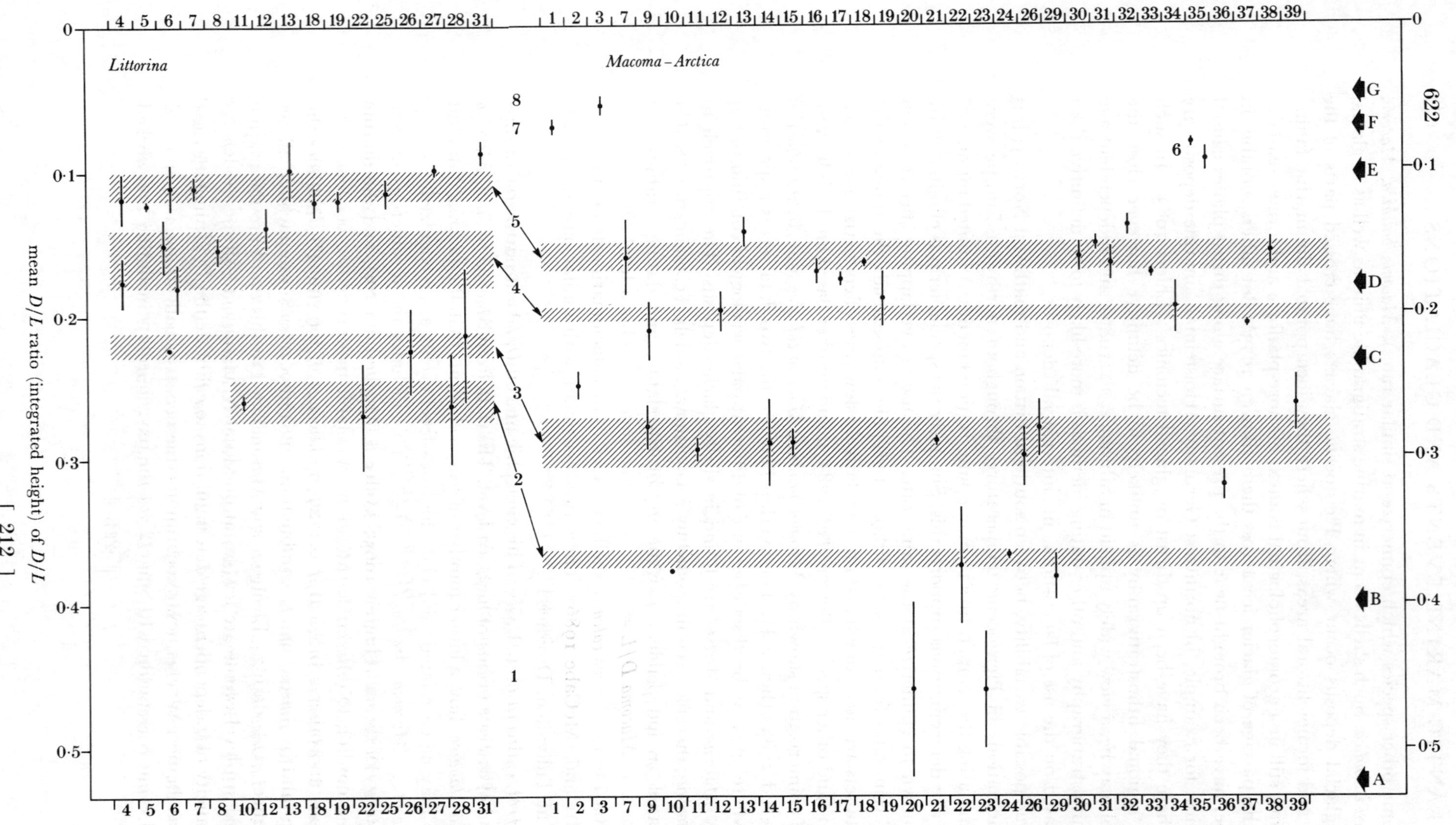

FIGURE 1. Aminostratigraphic correlation of D/L sea-level events in north-west Europe and the British Isles (based on Miller & Mangerud 1985; Bowen *et al.* 1985, unpublished data). One standard deviation for each sea-level event recognized is shown by the shading. The sea-level events are numbered 1 to 8, and glaciations lettered A to G. Two separate data points (*Arctica* and *Macoma*) are given for Wacken. The Sussex coastal plain data are from separate localities and are not stratigraphically superposed. Numbers at top and bottom indicate localities:

1, Belderg, Co. Mayo, Ireland; 2, Cleongart, Kintyre, Scotland; 3, Ardyne, Firth of Clyde; 4, Minchin Hole Cave, Gower; 5, Bacon Hole Cave, Gower; 6, southwest Britain (Pembrokeshire to Dorset); 7, Burtle Beds, Somerset; 8, Easington, Co. Durham; 9, Sussex; 10, Norton Farm, Sussex; 11, Eartham (Boxgrove), Sussex; 12, Fjøsanger, Norway; 13, Bø, Norway; 14, Herzeele, France; 15, Zeebrugge, Belgium; 16, Koolkerke, Belgium; 17, Bergen, The Netherlands; 18, Castricum, The Netherlands; 19, Zunderdorp, The Netherlands; 20, Noord Bergum, The Netherlands; 21, Scharnhorn, northwest (NW) Germany; 22, Vognsbøl, Denmark; 23, Esbjerg, Denmark; 24, Kaas Hoved, Denmark; 25, Tønder, Denmark; 26, Wacken, NW Germany; 27, Rodemis, NW Germany; 28, Halstenbek, NW Germany; 29, Hummelsbüttel, NW Germany; 30, Holnis, NW Germany; 31, Stensigmose, Denmark; 32, Mommark, Denmark; 33, Trappeskov, Denmark; 34, Ristinge, Denmark; 35, Skaerumhede, Denmark; 36, Slettenshage, Denmark; 37, Strandegaard, Denmark; 38, Stubberup, Denmark; 39, Margareteburg, Sweden.

fauna in Gower (Bowen *et al.* 1985). Included in this event are marine beds at Herzeele, some of the Norton Farm boring beach sands of the Sussex coastal plain (Lovell & Nancarrow 1983), and the marine beds at Margareteburg in Sweden. *Macoma balthica* shells from Margareteburg give slightly lower D/L ratios (Miller & Mangerud 1965), which could be explained by a small difference in the integrated temperature history of that locality.

High-sea-level event 4 (*Macoma–Arctica* $D/L = 0.2$; *Littorina* $D/L = 0.16$). This includes the Fjøsanger marine beds, as well as those at Strandegaard and Ristinge Klint in Denmark (Miller & Mangerud 1985). These European sites are correlated with stratigraphic units identified throughout southwest Britain and, in particular, with the stratotype at Minchin Hole Cave, Gower (Bowen *et al.* 1985). The slightly lower *Littorina* D/L ratios of 0.14 at Fjøsanger are explained by a small difference in the integrated temperature history of that site, which lies some 10° north of Gower.

High-sea-level event 5 (*Macoma–Arctica* $D/L = 0.16$; *Littorina* $D/L = 0.11$). The Eemian parastratotype of Castricum is included in this event, together with numerous other localities in northwest Europe. This high-sea-level event is recognized throughout southwest Britain and is represented by the Outer Gravel Beach at Minchin Hole Cave, Gower, and the marine beds at Bacon Hole Cave, Gower (Bowen *et al.* 1985). Overlapping *Macoma* and *Littorina* species in the Burtle Beds of Somerset and at Castricum and Zunderdorp in The Netherlands (Eemian parastratotypes), demonstrate the strength of this correlation. The marine beds of Bø, Norway, yield *Littorina* with somewhat lower D/L ratios (0.097) (Miller & Mangerud 1985), but are correlated with Rodemis in northwest Germany (0.098) and Stensigmose in Denmark (0.086); but it is possible that the marine beds at the two latter localities may be slightly younger.

High-sea-level event 6 (*Macoma* $D/L = 0.093$; *Arctica* $D/L = 0.082$). This is based on measurements of D/L ratios from shells in the marine sediments encountered in the Skaerumhede Borings (Miller & Mangerud 1985). These D/L ratios correlate with those measured in *Macoma* and *Arctica* shells from shelly glacial deposits in the British Isles, which show a marine transgression of that age across the British seas.

High-sea-level event 7 (*Macoma* $D/L = 0.07$). This glaciomarine high-sea-level event is shown by D/L ratios of 0.07 from *Macoma calcarea* shells collected from a glaciomarine delta at Belderg, County Mayo, Ireland (McCabe 1986). These ratios correlate with similar ones on the east coast of Ireland at Tullyallen, Drogheda.

High-sea-level event 8 (*Arctica* D/L 0.055). This event is based on D/L ratios from *Arctica* shells from the marine Clyde Beds of Scotland at Ardyne (Miller *et al.* 1987).

Geochronology

The geochronology of the eight high-sea-level events is modelled in terms of oxygen-isotope stages (figure 2) according to different assumptions, available independent geochronometric age determinations and other geological information. The age of the younger D/L sea-level events is secure and is based on a combination of radiocarbon, uranium-series and thermoluminescence (TL) dating. Electron spin resonance (ESR) ages for some of the intermediate-aged sea-level events are few or, at one locality, in conflict. All the models are limited by the paucity of independent age determinations to calibrate the D/L timescale, and modifications may be necessary as new geochronometric ages are forthcoming.

High-sea-level event 8 is ascribed to Stage 2 (of the oxygen-isotope scale) by radiocarbon

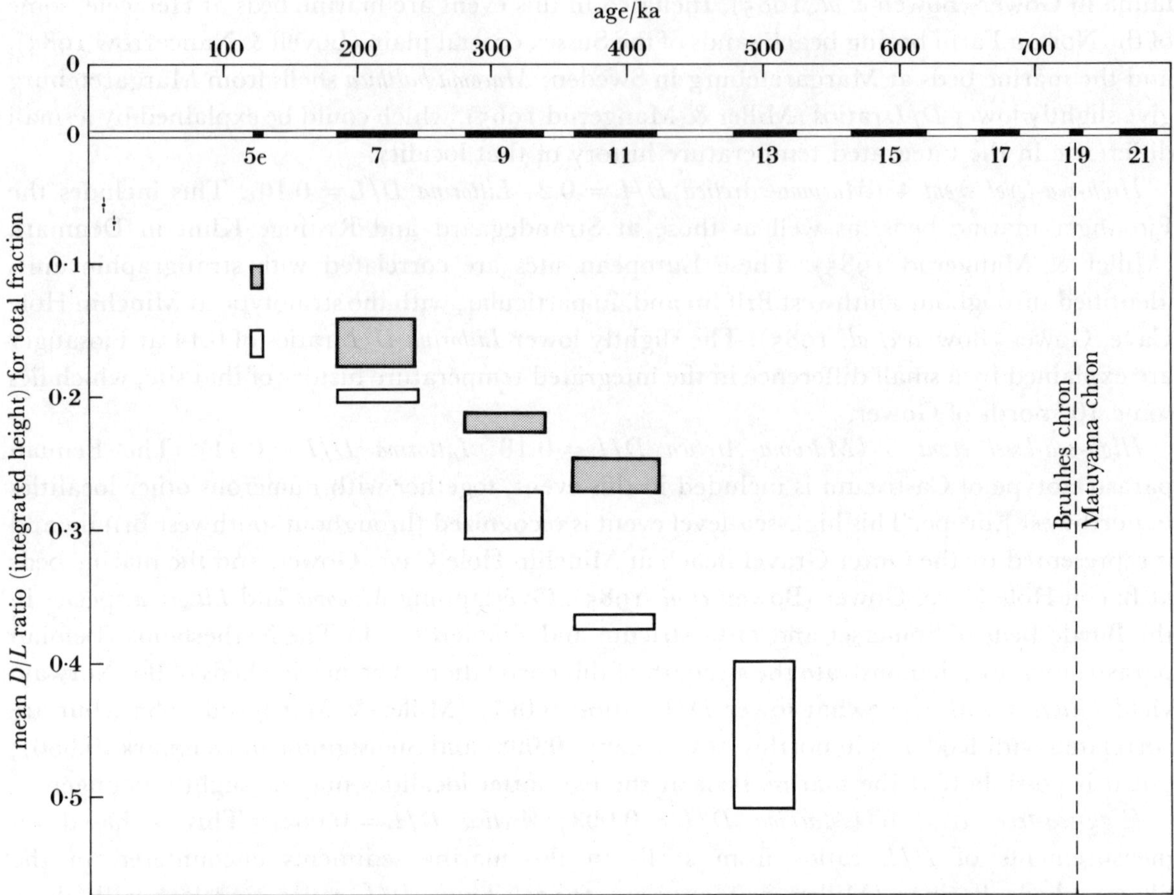

FIGURE 2. Correlation of D (alloisoleucine)/L (isoleucine) high-sea-level events with odd-numbered oxygen-isotope stages on the SPECMAP timescale of Imbrie *et al.* (1984). The rectangles show one D/L standard deviation correlated with odd-numbered oxygen-isotope stages. Shaded boxes show the *Littorina* D/L data; unshaded boxes, *Macoma–Arctica* D/L data. (Model 1 assumptions.)

dates of 11.2 and 11.6 ka BP on *Arctica* shells from Ardyne, Scotland (Miller *et al.* 1987). Sea-level event 7 is dated by radiocarbon ages of 16.9 and 17.3 ka BP on an *in situ Macoma calcarea* fauna from a glaciomarine delta, at 40 m above sea-level, at Belderg, County Mayo. Correlation by aminostratigraphy shows that the glaciomarine deposits at a similar elevation on the east coast of Ireland at Tullylallen, Drogheda, are the same age. Sea-level event 6 is not directly dated by a geochronometric age determination but is ascribed to Stage 3 because it lies between sea-level event 8 (*ca.* 17 ka BP) and sea-level event 5 (122 ka BP; see below).

Sea-level events 5 and 4 are calibrated by uranium-series and thermoluminescence dates from Gower. At Bacon Hole Cave, D/L ratios on *Littorina* of 0.122 are overlain by speleothem fragments with a mean age of 122 ka BP (Stringer *et al.* 1986). The Bacon Hole marine beds are correlated with the Outer Gravel Beach and associated beds at the adjacent cave of Minchin Hole both by aminostratigraphy and by further uranium-series dates of 116 ka BP, from a speleothem block in a similar relation to the beach deposits as at Bacon Hole Cave (Sutcliffe & Currant 1984). The outer gravel beach overlies and is separated from an inner sand beach by a thermoclastic deposit with a cold-stage fauna (Sutcliffe & Currant 1984). The inner sand beach at Minchin Hole Cave, Gower, contains *Littorina littorea* shells with a D/L

ratio of 0.176. It also has a mean TL age of 191 ka BP (Southgate 1986). Thus the beds with a D/L ratio of 0.1 are correlated with Substage 5e and those with a D/L ratio of 0.176, with Stage 7. The stratotype for both sea-level events is Minchin Hole Cave, where the lithostratigraphic sequence shows the Stage 7 beach overlain by a cold-climate head deposit, in turn overlain by the Substage 5e marine beds (Sutcliffe & Bowen 1973; Bowen 1977). The Bacon Hole succession shows no further marine event in a sequence of fossiliferous terrestrial deposits between the 5e marine beds and a stalagmite floor dated by uranium series to 81 ka BP (Stringer *et al.* 1986) at the base of Devensian head deposits. Thus along the northern side of the Bristol Channel there is no evidence for 5c or 5a sea levels. These age ascriptions have an important bearing on similarly aged events on the continent and in Scandinavia. Much rests, therefore, on the lithostratigraphy, D/L data, and geochronology of Minchin Hole Cave. The only uncertainty arises because the speleothem age determinations are on broken samples, not *in situ*.

In southwest Britain an overlap between *Littorina* (and *Patella*) and *Macoma balthica* occurs in the Burtle Beds of Somerset, where *Littorina* D/L ratios of 0.11 ± 0.007 ($n = 11$) overlap with *Macoma balthica* D/L ratios of 0.163 ± 0.025 ($n = 17$) (0.15 if possible re-worked samples are excluded from the calculation). This compares with the overlap relations shown by Miller & Mangerud (1965) for *Littorina* (0.098) and *Arctica* (cf. *Macoma*) (0.016) in Schleswig-Holstein. The D/L data from Minchin Hole Cave and Gower show a sea-level event (the Minchin Hole (D/L) Stage (Bowen *et al.* 1985)) with a mean ratio of 0.175, ascribed to Stage 7, which is apparently unrecorded in Schleswig-Holstein (table 1). After allowing for a 10° latitudinal

TABLE 1. CORRELATION OF D/L RATIOS IN *LITTORINA LITTOREA*

(B, Bø; F, Fjøsanger; asterisks indicate the stratigraphic position of the thermoclastic scree (with cold-stage fauna) at Minchin Hole Cave, Gower (Bowen *et al.* 1985).)

Minchin Hole Cave	Bacon Hole Cave	southwest Britain	Norway	Schleswig-Holstein
0.12 ± 0.018	0.12 ± 0.002	0.11 ± 0.016	0.09 ± 0.017(B)	0.098 ± 0.004
**********				
		0.153 ± 0.018[a]		
0.18 ± 0.017		0.18 ± 0.015	0.15 ± 0.03(F)[c]	
		0.21 ± 0.008[b]		0.22 ± 0.03

[a] Separate D/L sea-level event identified by Bowen *et al.* (1985) throughout SW Britain (equivalent *Patella* D/L ratio is 0.135); it is suggested that this is close in time to the 0.18 D/L sea-level event.

[b] D/L ratios from *Littorina littoralis*, *Nucella lapillus* and *Patella vulgata* standardized to *Littorina littorea* (see Bowen *et al.* 1985).

[c] It is probable that two distinct populations are represented by this mean ratio; on this assumption two populations could be represented by D/L ratios of 0.193 ± 0.01 ($n = 4$) and 0.137 ± 0.011 ($n = 8$), thus demonstrating possible reworking of older fauna and sediments.

distance between Gower and southwest Norway, with a consequent difference in integrated temperature history, it may be possible to recognize that sea-level at Fjøsanger (*Littorina* $D/L = 0.15$); the beds at nearby Bø (*Littorina* $D/L = 0.097$), although correlated by Miller & Mangerud (1985) with those at Fjøsanger, may correlate with the Pennard (D/L) Stage 0.11, which is ascribed to Substage 5e (Bowen *et al.* 1985). Thus Fjøsanger is ascribed to Stage 7 (as in one of the two models of Miller & Mangerud 1985)), and Bø to Substage 5e. Clarifying the *Arctica–Macoma* data is less straightforward. But the *Arctica* D/L ratios at Bø of 0.14 compare with those of 0.16 from the Burtle Beds in Britain and Schleswig-Holstein; and the *Arctica D/L* ratios of 0.197 from Fjøsanger could compare with ratios of *ca.* 0.21 as typical Stage 7 values elsewhere. Such an expectation is fulfilled, and other Stage 7 sea-level events may be recognized, at Ristinge (Denmark) (*Arctica* $D/L = 0.2$), Strandegaard, Denmark (*Arctica* $D/L = 0.21$), and Ham Cottages, Sussex (*Macoma* $D/L = 0.21$). All of these localities were previously considered to be Eemian or 'last interglacial'. Some stratigraphic units ascribed to Substage 5e, e.g. at Castricum and Koolkerke, include some re-worked fauna of Stage 7 age.

ESR age determinations from Hummelsbüttel ('Holsteinian'), in northwest Germany, which give Stage 7 ages (Linke *et al.* 1985), are incompatible both with the *D/L* data (*Macoma* $D/L = 0.382$) from that site (Miller & Mangerud 1985), and with a further ESR determination of 371 ka BP (Sarnthein *et al.* 1986). The *D/L* data and the ESR age determination of Sarnthein *et al.* (1986), are, however, not incompatible.

For high-sea-level events previous to those correlated with Stage 7, the *D/L* data can be modelled in at least three ways.

Model 1

The ESR determinations of Sarnthein *et al.* (1986) may be used to ascribe *D/L* high-sea-level event 3 (represented by the marine deposits at Wacken ('Holsteinian'), dated to 300 ka BP, and those at Herzeele, dated to 348 ka BP) to Stage 9. Similarly the ESR age of 371 ka BP at Hummelsbüttel ('Holsteinian') (Sarnthein *et al.* 1986) is used to ascribe *D/L* sea-level event 2 to Stage 11. Assuming that no major breaks in the stratigraphic record occur, *D/L* high-sea-level event 1 can therefore be ascribed to Stage 13. Zagwijn ascribed the Esbjerg marine beds (*D/L* sea-level event 1) to 'Cromerian IV' (in Mangerud & Miller 1985).

Model 2

It is possible, however, that the difference of 0.09 between the 0.46 and 0.37 *Macoma–Arctica D/L* groups represents more than the 55 ka between Stages 13 and 11. The oldest available *D/L* ratios (*Macoma* $D/L = 0.46$) are from Noord Bergum in The Netherlands (formerly classified as 'Holsteinian' but revised to 'Cromerian' by Ter Wee (1983)), and Esbjerg in Denmark (also formerly classified as 'Holsteinian' but revised to 'Cromerian' (Miller & Mangerud 1985). The Esbjerg marine deposits grade upwards from late-glacial beds which overlie a till, formerly classified as 'Elsterian' (Sjorring 1983). On the basis of model 1, the glaciation which deposited the till is ascribed to Stage 14. It is, however, possible that the basal till at Esbjerg is older. One possibility for correlation is with the till at the top of the North Sea Aberdeen Ground Beds Formation, which lies just above the Matuyama–Brunhes boundary (Cameron *et al.* 1987). If this correlation is correct, the basal till at Esbjerg could be time-equivalent to, for example, Stage 18, and sea-level event 1 occurred in Stage 17. Consequently slower epimerization kinetics must be assumed and sea-level event 2 could be ascribed, for example, to Stage 13 (figure 2), and sea-level event 3 to Stages 11 or 9. Equally the glacial beds could be ascribed to Stage 16; this ascription is advantageous because, in global terms, this was a time of exceptional ice-volume on the continents. On this basis the *Macoma* $D/L = 0.46$ sea-level event could be Stage 15 in age.

Model 3

A further consideration arises from contextual changes in sea-levels throughout the global ocean. It has been shown, for example, that a major 'lacuna' (Blackwelder 1981) occurs with no dated marine deposit between Stage 11 and 1.1 Ma BP. This may correspond with the lower temperatures and lower sea-levels inferred from the isotope signal for Stages 13 to 21 (see, for example, Shackleton *et al.* 1984). Sea-level data showing this are widespread and include those from the coasts of the Mediterranean basin (Hearty *et al.* 1986), California (Karrow & Bada 1980; Muhs & Szabo 1982), Atauro (Chappell & Veeh 1978), and the eastern U.S.A. (Blackwelder 1981). Yet models 1 and 2 (above) recognize pre-Stage 11 sea-levels in northwest Europe. This result may have been brought about by neotectonic movements, or it may be that the sequence is incorrectly ascribed to the isotope signal. Alternatively, if the data are modelled from Stage 11 onwards, then two of the D/L high-sea-level events need to be subsumed within the same stage. Similar problems of ascription of the sea-level events arise if the epimerization kinetic pathway is more rapid than that assumed in model 1.

The reproducibility of D/L ratios from marine molluscs is clear (figure 1) (Miller & Mangerud 1985; Bowen *et al.* 1985), although more information could modify the pattern of high-sea-level events adduced here. Similarly, the timing of these events may be modified by further geochronometric age determinations. But, with the available information, model 1 is preferred and to some extent can be used to place constraints on the extent and timing of glacial advances (table 2).

TABLE 2. CORRELATION OF D/L RATIOS IN *MACOMA* AND *ARCTICA*

(B, Bø; F, Fjøsanger; *, Arctica.)

southern Britain			
Burtle Beds	Sussex	Norway	northwest Europe
0.157±0.018		0.14±0.01*(B)	0.16±0.01*
0.21±0.004[a]	0.206±0.015[b]	0.197±0.01*(F)	
	0.30±0.02		0.28±0.01*
			0.30±0.02

[a] A reworked faunal element is assumed in the Burtle Beds and is represented by D/L ratios of 0.21.
[b] The same stratigraphic unit contains *Littorina littoralis* ($D/L = 0.144 \pm 0.006$) and *Littorina saxatilis* ($D/L = 0.132 \pm 0.006$); standardized to *Littorina littorea*, these data give a D/L ratio of 0.144 ± 0.006 ($n = 10$) (cf. table 1).

THE SEQUENCE OF GLACIATIONS

The earliest northwest European glaciations of the past 3 million years are older than the available D/L data. The oldest one inferred is based on the evidence of ice-rafting in DSDP hole 552a (Rockall), dated to 2.4 Ma BP (Shackleton *et al.* 1984), which in the broadest terms could correlate with the Scandinavian erratics found in Pliocene deposits in northwest Germany (Ehlers 1983). Subsequent glaciation may be inferred from the heavy mineral assemblage derived from a northerly source in the Baventian deposits of East Anglia (Solomon, in Funnell & West 1962) and the Yorkshire Wolds (Catt 1982), and the Menapian deposits of The Netherlands (Zagwijn 1986). But the first major glaciations, around about the

Matuyama–Brunhes boundary, are to be inferred from erratics of north Wales provenance found in the Kesgrave Formation of Essex (Rose & Allen 1977; Hey 1965). Rose (in Bowen *et al.* 1986) has inferred four glaciations of upland Britain from such evidence in the sediments of the Kesgrave Formation and its correlatives in the Middle Thames (Green *et al.* 1980). These are correlated with the Yarmouth Roads Formation of the North Sea (Balson & Cameron 1985) and the Sterksel Formation of The Netherlands (Zagwijn 1986).

Glaciation A (*Stage* 14) ('*Elster* I'?)

The oldest glacial deposit directly related to the *D/L* sea-level sequence is the till beneath the marine sediments of sea-level event 1 at Esbjerg, Denmark. On model 1 considerations it is ascribed to Stage 14 because its deposits grade upwards into the (Stage 13) Esbjerg marine beds. This till is probably the same age as the one beneath the interglacial beds at Harreskov ('Cromerian'), Denmark (Sjorring 1983), and thus could be equivalent to the pre-Voigstedt ('Cromerian') Elster I Glaciation proposed in East Germany (Cepek 1986). On model 2 assumptions, the till, correlated with the one on top of the Aberdeen Ground Beds Formation, just above the Matuyama–Brunhes boundary ('Glacial A' of Cameron *et al.* (1987)), could represent a glaciation (?Stage 18) broadly coeval with one of the early upland glaciations of western Britain. But, as previously discussed, this glaciation could be time-equivalent to Stage 16.

Glaciation B (*Stage* 12) ('*Elster* II'?, *Anglian*, '*older drift*' *Irish Sea*)

This is younger than *D/L* sea-level event 1 (Stage 13), but older than sea-level event 2 (Stage 11). On this basis the upper till at Esbjerg and the glaciolacustrine beds at Noord Bergum are ascribed to the Stage 12 (B) Glaciation. If *D/L* sea-level event 2 (Stage 11) is taken as a minimum age for the Stage 12 (B) Glaciation, the basal tills at Kaas Hoved and Hummelsbüttel are ascribed to this event. Because a Stage 10 glaciation has not been widely recognized in the northern hemisphere, including Europe, it is assumed that this relatively brief cold stage did not witness extensive glaciers; this assumption is supported by the available geological evidence (Sibrava *et al.* 1986). It follows, therefore, that the marine deposits ascribed to the *D/L* sea-level event 3 (Stage 9) may also be used as minimum ages for the Stage 12 (B) Glaciation. Thus the 'Holsteinian' deposits at Wacken may also be used in this way, although they are not the same age as the 'Holsteinian' deposits at Hummelsbüttel (as shown by both *D/L* ratios (Miller & Mangerud 1986) and ESR age determinations (Sarnthein *et al.* 1986)). In Gower the mixed faunas (sea-level event 3) of Hunts Bay can be used to fix a minimum age for a Stage 12 (B) 'Irish Sea' Glaciation of the Bristol Channel, which includes the Fremington Till of north Devon (Bowen *et al.* 1985). *D/L* measurements on non-marine shells from 'interglacial' deposits overlying the Anglian tills (Hughes 1987) are ascribed to Stages 11 and 9, thus dating the underlying tills as Stage 12 (Glaciation B). This dating is consistent with the views of Perrin *et al.* (1979) that the chalky tills of the Anglian Glaciation in midland England and in East Anglia represent the maximum extension of ice in lowland Britain, which is correlated with the post-Esbjerg interglacial (sea-level event 1) glaciation of The Netherlands, that is, with the glaciolacustrine deposits of the Elster Glaciation (Elster II ?). This view accords with those of Sarnthein *et al.* (1986), who also ascribed the Elster Glaciation to Stage 12, and Bowen (1985), who ascribed this event to Elster II. (On model 2 considerations this glaciation would be ascribed to Stage 14, the Elster I advance of Cepek (1986) and others; see Sibrava (1986).)

TABLE 3. CORRELATION OF OXYGEN-ISOTOPE STAGES, *D/L* SEA-LEVEL EVENTS AND GLACIATIONS, BASED ON MODEL 1 ASSUMPTIONS

age[a]	stage	*D/L* sea-level event *Macoma–Arctica* (*Littorina*)	geochronology ka BP	glaciation
		8 0.05	11	G Loch Lomond
	2			
		7 0.07	17	F late Devensian late Weichselian
24				
	3	6 0.085		
59				
	4			E in Scotland, Ulster and Ireland
71				
	5			
122				
	5e	5 0.16 (0.11)	122	
128				
	6			D 'Warthe'
186				
	7	4 0.2 (0.16)	191	
245				
	8			C 'Drenthe'? Paviland
303				
	9	3 0.29 (0.22)	300 348	
339				
	10			
362				
	11	2 0.37 (0.26)	371	
423				
	12			B 'Elster' (Anglian, 'Irish Sea')
478				
	13	1 0.46		
524				
	14			A 'Elster' ('Elster I'?)
565				

[a] SPECMAP age (Imbrie *et al.* 1984).

Glaciation C (*Stage* 8) (*Paviland, 'Drenthe'*)

This glaciation is fixed between *D/L* sea-level events 3 (Stage 9) and 4 (Stage 7). Ice-thrusted marine sediments of *D/L* sea-level event 3 age occur at Halstenbek, near Hamburg. At this locality the *D/L* ratios (Miller & Mangerud 1985) are in conflict with a U-series age which places the sediments in the Eemian (5e) interglacial (Sarnthein *et al.* 1986). At Slettenshage, Denmark, marine clay with *D/L* ratios of sea-level event 3 (Stage 9) occurs in till, which is thus ascribed to Stage 8 (Glaciation C). That same glaciation (C) is comparatively restricted in the British Isles. In Gower, the Paviland Moraine and its deposits are older than marine deposits of the *D/L* sea-level event 4 (Stage 7) and have been ascribed to Stage 8 (Glaciation C) (Bowen *et al.* 1985). Possible correlatives of this have been identified in the west Midlands of England at Quinton, Birmingham, which is the basis for the extent shown in figure 3. At Welton, in Lincolnshire, a till sequence has also been ascribed to the Stage 8 (Glaciation C) advance (Bowen *et al.* 1986). Because the March Gravels of the Fenland (ascribed to *D/L* sea-level event 9) are not overridden by glacial deposits they may also be used to constrain the extent of Stage 8 glaciation. In all three cases in the British Isles the limit of the Stage 8 (Glaciation C) advance lies immediately outside the limit of late Devensian glaciation. On this evidence it is most unlikely that the British and European ice-sheets were in contact at this time.

Glaciation D (Stage 6) ('Warthe')

This occurred between *D/L* sea-level events 4 (Stage 7) and 5 (Substage 5e). In Europe, possible representatives of this glaciation (D) are tills above marine deposits of *D/L* sea-level 4 at Ristinge Klint and Strandegaard in Denmark. At both localities three till units overlie marine deposits of Stage 7 age. Till units beneath Substage 5e marine beds occur at Tønder and Stubberup in Denmark (Miller & Mangerud 1985). There is no unequivocal evidence in the British Isles for glaciation during Stage 6. Certainly any ice did not extend as far south as County Durham, on the east coast of England, where the Easington raised beach (sea-level event 4, Stage 7) is overlain by Late Devensian till. Equivocal *D/L* data from shelly tills in Ireland and Scotland (Caithness and Orkney) may show Stage 6 glaciation of these areas. But an alternative interpretation is that these are Stage 4 (Early Devensian) in age (below). More data, and geochronometric calibration, may resolve this problem. In any event, it is difficult to envisage no ice cover in the British Isles at this time, which corresponds generally to the 'Warthe' Glaciation of Europe (figure 3); upland Britain, at least, would have been ice-covered at this time.

Three separate glaciations are recognized in the Devensian Stage of Britain. These are: an early Devensian Glaciation (E), the Late Devensian Glaciation (F) (Dimlington Stadial of Rose (1985)), and the Loch Lomond Glaciation (G). *D/L* data for recognizing these comes principally from the western seaways and continental shelf areas of Britain.

Glaciation E

An Early Devensian Glaciation (E) is recognized in Ireland and Scotland. In both areas, tills occur containing only molluscs ascribed to the Substage 5e sea-level event (5) or older (D. Q. Bowen, G. A. Sykes, D. G. Sutherland, J. Gordon & A. M. McCabe, unpublished data). None of these contains molluscs with a Middle Devensian (Weichselian) *D/L* signature (Bowen, Sykes, Sutherland, Gordon & McCabe, unpublished data). It follows, therefore, that glaciation occurred after Substage 5e (sea-level event 5) but before sea-level event 6 (Middle Devensian, Stage 3). In contrast, glacial deposits of the Late Devensian Glaciation contain molluscs from both the Substage 5e (sea-level event 5) and Stage 3 (sea-level event 6) marine transgressions, and sometimes, from a glacio-marine sea-level event (7) contemporaneous with Late Devensian ice. Because the sea would have occupied some areas of the continental shelf more or less continuously, especially as the result of isostatic depression during and after the glaciations, it may be difficult to discriminate precise high-sea-level events on the basis of *D/L* signature, as is evident from the spread of *D/L* ratios from molluscs in glacial deposits. But, as one interpretation of the available data shows, a post-Substage 5e and pre-Stage 3 glaciation affected parts of Caithness and Orkney in Scotland, and Ulster and the eastern coast of Ireland (figure 3). In the absence of independent geochronological control, however, it is impossible

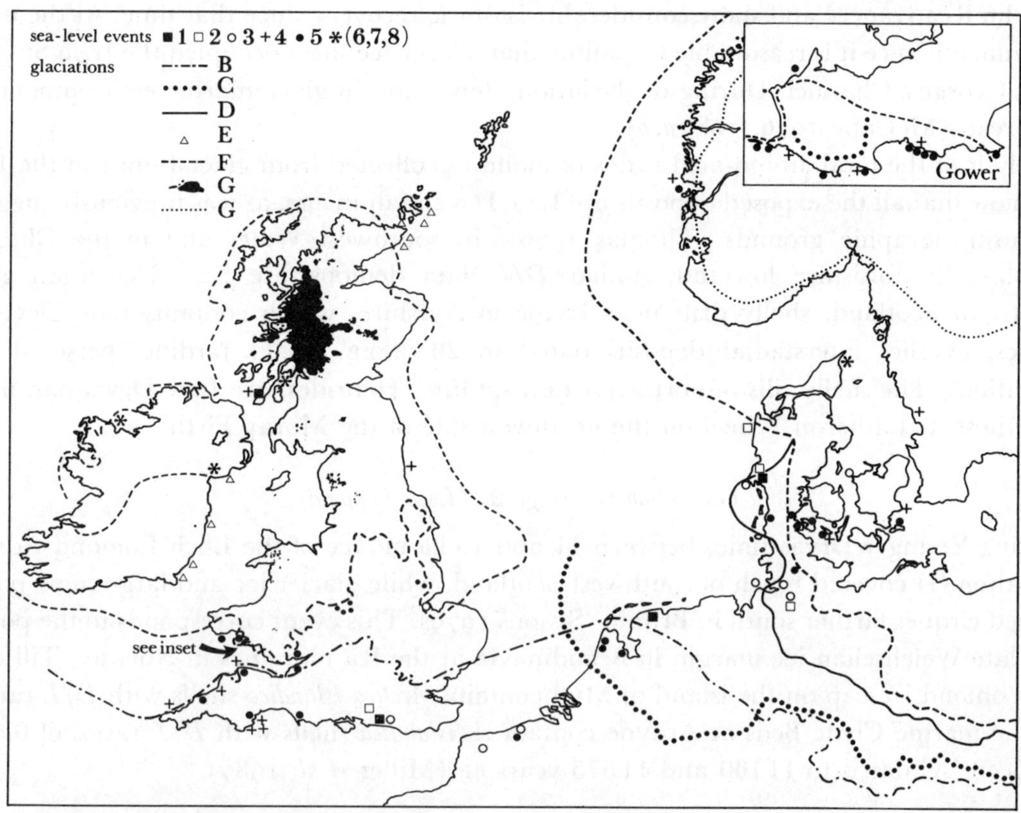

FIGURE 3. Extent of glaciations in the British Isles and northwest Europe. Sea-level events (*Macoma/Arctica* data): 1, $D/L = 0.46$ (Stage 13); 2, $D/L = 0.37$ (Stage 11); 3, $D/L = 0.29$ (Stage 9); 4, $D/L = 0.2$ (Stage 7); 5, $D/L = 0.16$ (Substage 5e); 6, $D/L = 0.83, 0.92$ (Stage 3); 7, $D/L = 0.07$ (Stage 2); 8, $D/L = 0.055$ (Stage 2). Glaciations: B, Stage 12 ('older drift' Irish Sea, Anglian, Elster' (II ?); C, Stage 8 (Paviland, 'Warthe'); D, Stage 6 ('Warthe'); E, ?Stage 6, or Substage 5d or Stage 4 (shelly glacial deposit localities); F, Stage 2 (Late Devensian, Late Weichselian); G, Stage 2 (Loch Lomond; asterisk indicates Ra and central Sweden moraines).

to ascribe this event to either Stage 4 or 5d. Independent calibration, however, may demonstrate an important temperature term in the data, and it is not impossible that some of the D/L ratios currently interpreted as 5e in age could be older, perhaps of Stage 7 age. In that eventuality, the glacial deposits of Caithness and Orkney could be older and possibly ascribed to Stage 6. As such they would find a ready parallel with the 'Warthe' advance of Denmark and northwest Germany, which also extended beyond the margin of Late-Devensian–Weichselian glaciation.

Glaciation F (*Stage* 2) (*Late Devensian–Late Weichselian*)

The establishment of the margins of the Late Devensian Glaciation (F) has been controversial (Bowen 1973) (figure 3). But Late Devensian glacial units containing marine molluscs of Stages 3 and 2 have been identified throughout the British Isles. Molluscs from glaciomarine beds at Belderg, County Mayo, Ireland, have been dated to *ca.* 17 ka BP. In elevation (*ca.* 40 m above sea-level) they compare with similar deposits on the east coast of Ireland at Tullyallen in County Meath. The beds at both localities are contemporaneous with the

'Drumlin Readvance' and show considerable isostatic recovery since that time. At the time of maximum advance it is reasonable to assume that a large ice sheet occupied the Irish Sea basin and St George's Channel. During deglaciation, deposition in glaciomarine environments was widespread (McCabe 1976, 1987*a*, *b*).

Analysis of the total amino-acid ratios of molluscs collected from glacial units in the Isle of Man show that all the exposed deposits are Late Devensian in age, as was previously suggested on lithostratigraphic grounds (Thomas 1976). In southwest Wales and in the Cheshire–Shropshire–Staffordshire lowland, similar *D/L* data demonstrate Late Devensian glacial deposits. In Scotland, shelly drift near Irvine in Ayrshire, which contains Late Devensian molluscs, overlies interstadial deposits dated to 29 ka BP (W. G. Jardine, personal communication). The shelly tills of northwest Lewis, Outer Hebrides, are Late Devensian in age, as are those at Latheron Wheel on the northwest side of the Moray Firth.

Glaciation G (Stage 2) (Loch Lomond)

During Younger Dryas time, between 11 and 10 ka BP, ice of the Loch Lomond Advance (Glaciation G) covered much of southwest Scotland, while glacier ice and large snow patches occupied cirques farther south in Britain (Sissons 1979). This event corresponds to the position of the late Weichselian ice margin in Scandinavia at the Ra moraines in Norway. Till of the Loch Lomond ice-cap on the island of Mull contains *Arctica islandica* shells with *D/L* ratios of 0.4. The marine Clyde Beds at Ardyne contain *A. islandica* shells with *D/L* ratios of 0.5 and have been ^{14}C dated to 11160 and 11575 years BP (Miller *et al.* 1987).

We thank J. Rose and N. J. Shackleton, F.R.S., for commenting on the original manuscript; Gerald Sykes; and also the following for their discussion of the problems raised in this paper: G. H. Miller, D. G. Sutherland, J. Gordon, A. M. McCabe, Sandra Hughes, Alayne Reeves, J. T. Andrews, R. J. Price, W. G. Jardine and D. D. Harkness.

References

Andrews, J. T., Gilbertson, D. D. & Hawkins, A. B. 1984 The Pleistocene succession of the Severn Estuary: a revised model based upon amino acid racemization studies. *J. geol. Soc. Lond.* **141**, 967–974.

Balson, P. S. & Cameron, T. T. J. 1985 Quaternary mapping offshore East Anglia. *Mod. Geol.* **9**, 221–239.

Blackwelder, B. W. 1981 Late Cenozoic marine deposition in the United States Atlantic Coastal Plain related to tectonism and global climate. *Paleogeogr. Paleoclim. Palaeoecol.* **34**, 87–114.

Boulton, G. S., Jones, A. S., Clayton, K. M. & Kenning, M. J. 1977 A British ice sheet model and patterns of glacial erosion and deposition in Britain. In *British Quaternary studies* (ed. F. W. Shotton), pp. 231–246. Oxford: Clarendon Press.

Bowen, D. Q. 1973 The Pleistocene succession of the Irish Sea. *Proc. Geologists' Association* **84**, 249–272.

Bowen, D. Q. 1977 The coast of Wales. *Geol. J.* special issue no. 7, pp. 223–256.

Bowen, D. Q., Sykes, G. A., Reeves, A., Miller, G. H., Andrews, G. T., Brew, J. S. & Hare, P. E. 1985 Amino acid geochronology of raised beaches in south west Britain. *Quat. Sci. Rev.* **4**, 279–318.

Bowen, D. Q., Rose, J., McCabe, A. M. & Sutherland, D. G. 1986 Correlation of Quaternary glaciations in England, Ireland, Scotland and Wales. *Quat. Sci. Rev.* **5**, 299–340.

Cameron, T. D. J., Stoker, M. S. & Long, D. 1987 The history of Quaternary sedimentation in the U.K. sector of the North Sea basin. *J. geol. Soc. Lond.* **144**, 43–58.

Catt, J. A. 1982 The Quaternary deposits of the Yorkshire wolds. *Proc. North of England Soils Discussion Group* **18**, 61–67.

Catt, J. A. & Penny, L. F. 1966 The Pleistocene deposits of Holderness, East Yorkshire. *Proc. Yorks. geol. Soc.* **35**, 375–420.

Cepek, A. G. 1986 Quaternary stratigraphy of the German Democratic Republic. *Quat. Sci. Rev.* **5**, 359–364.

Chappell, J. & Veeh, H. H. 1978 Late Quaternary tectonic movements and sea level changes at Timor and Atauro Island. *Bull. geol. Soc. Am.* **89**, 356–368.

Derbyshire, E., Foster, C., Love, M. A. & Edge, M. J. 1984 Pleistocene lithostratigraphy of north-east England: a sedimentological approach to the Holderness sequence. In *Correlation of Quaternary chronologies* (ed. W. C. Mahaney), pp. 371–384. Norwich: Geo Books.

Ehlers, J. 1983 The glacial history of north-west Germany. In *Glacial deposits in north-west Europe* (ed. J. Ehlers), pp. 229–238. Rotterdam: Balkema.

Ehlers, J., Mayer, K. D. & Steffan, H. J. 1984 Pre-Weichselian glaciations of north-west Europe. *Quat. Sci. Rev.* **3**, 1–40.

Funnell, B. M. & West, R. G. 1962 The early Pleistocene of Easton Bavents, Suffolk. *Q. Jl geol. Soc. Lond.* **118**, 125–141.

Hansen, S. 1965 The Quaternary of Denmark. In *The geologic systems: The Quaternary* (ed. K. Rankama), vol. 1, pp. 1–90. New York: Academic Press.

Hearty, P. J., Miller, G. H., Stearns, C. E. & Szabo, B. J. 1986 Aminostratigraphy of Quaternary shorelines in the Mediterranean basin. *Bull. geol. Soc. Am.* **97**, 850–858.

Hey, R. W. 1965 Highly quartzose pebble gravels in the London basin. *Proc. Geol. Ass.* **76**, 403–420.

Hughes, S. A. 1987 The aminostratigraphy of British Quaternary non-marine deposits. Ph.D. thesis, University of Wales.

Imbrie, J., Hayes, J. D., Martinson, D. G., MacIntyre, A., Micks, A. C., Morley, J. J., Pisias, N. G., Prell, W. L. & Shackleton, N. J. 1984 The orbital theory of Pleistocene climate: support from a revised chronology of the marine 018 record. In *Milankovitch and climate*, part 1 (ed. A. Berger, J. Imbrie, J. D. Hayes, G. Kukla & B. Saltzman), pp. 269–306. Dordrecht: Reidel.

Jenkins, D. G., Bowen, D. Q., Adams, C. G., Shackleton, N. J. & Brassell, S. C. 1985 The Neogene. In: *The chronology of the geological record*, ed. N. J. Snelling (Geological Society Memoir 10), pp. 199–210.

Karrow, P. F. & Bada, J. L. 1980 Amino acid racemization dating of Quaternary raised marine terraces in San Diego County, California. *Geology* **8**, 200–204.

Kukla, G. 1977 Pleistocene land/sea correlations. 1. Europe. *Earth Sci. Rev.* **13**, 307–374.

Linke, G., Katsenburger, O. & Grun, R. 1985 Description and ESR dating of the Holstenian interglaciation. *Quat. Sci. Rev.* **4**, 319–332.

Lovell, J. H. & Nancarrow, P. H. A. 1983 The sand and gravel resources of the country around Chichester and north of Bognor Regis, Sussex. Mineral Assessment Report 138, Institute of Geological Sciences, N.E.R.C.

McCabe, A. M. 1986 Late Pleistocene tide water glaciers and glacial marine sequences from north County Mayo, Republic of Ireland. *J. Quat. Sci.* **1**, 73–84.

McCabe, A. M. 1987*a* Glacial facies deposited by retreating tide water glaciers: an example from the late Pleistocene of Northern Ireland. *J. sedim. Petrol.* **56**, 880–894.

McCabe, A. M. 1987*b* Quaternary deposits and glacial stratigraphy in Ireland: a review. *Quat. Sci. Rev.* **6**, 259–300.

McCabe, A. M. 1988 Sedimentation at the margins of the late Pleistocene ice lobe terminating in shallow marine environments, Dundalk Bay, eastern Ireland. *Sedimentology* **34**. (In the press.)

Miller, G. H. & Mangerud, J. 1985 Aminostratigraphy of European marine interglacial deposits. *Quat. Sci. Rev.* **4**, 215–278.

Miller, G. H., Jull, T., Linnick, T., Sutherland, D. G., Serjup, P., Brigham, J. K., Bowen, D. Q. & Mangerud, J. 1987 Racemization derived Late Devensian temperature reduction in Scotland. *Nature, Lond.* **326**, 593–595.

Morgan, A. V. 1973 The Pleistocene geology of the area north and west of Wolverhampton, Staffordshire, England. *Phil. Trans. R. Soc. Lond.* B **265**, 293–297.

Muhs, D. R. & Szabo, B. J. 1982 Uranium series age of the Eel Point Terrace, Sante Clemente Island, California. *Geology* **10**, 23–26.

Perrin, R. M. S., Rose, J. & Davis, H. 1979 The distribution, variation and origins of pre-Devensian tills in Eastern England. *Phil. Trans. R. Soc. Lond.* B **287**, 535–570.

Petersen, K. S. 1973 Tills in dislocated drift deposits on the Rosnaes Peninsula, north-western Sjaelland, Denmark. *Bull. geol. Inst. Univ. Ups.* **5**, 41–49.

Richmond, G. M. & Fullerton, D. S. 1986 Introduction to Quaternary glaciations in the United States of America. *Quat. Sci. Rev.* **5**, 3–10.

Rose, J. 1985 The Dimlington stadial/Dimlington chronozone: a proposal for naming the main glacial episode of the late Devensian in Britain. *Boreas* **14**, 225–230.

Rose, J. & Allen, P. 1977 Middle Pleistocene stratigraphy in south-east Suffolk. *J. geol. Soc. Lond.* **133**, 83–102.

Sarnthein, M., Stremme, H. E. & Mangini, A. 1986 The Holstein interglaciation: time stratigraphic position and correlation to stable isotope stratigraphy of deep sea sediments. *Quat. Res.* **26**, 283–298.

Shackleton, N. J. & Opdyke, N. D. 1973 Oxygen isotope and palaeomagnetic stratigraphy of equatorial Pacific core V28-238: oxygen isotope temperatures and ice volumes on a 10^5 and 10^6 year scale. *Quat. Res.* **3**, 39–55.

Shackleton, N. J., Backman, J., Zimmerman, H., Kent, D. V., Hall, M. A., Roberts, D. G., Schnitker, D., Baldauf, J. G., Desprairies, A., Homrighausner, R., Huddlestone, P., Keen, J. B., Kaltenbach, A. J., Krumsiek, K. A. O., Morton, A. C., Murray, J. W. & Westberg-Smith, J. 1984 Oxygen isotope calibration of the onset of ice rafting and history of glaciation of the north Atlantic region. *Nature, Lond.* **307**, 620–623.
Shotton, F. W. 1953 Pleistocene deposits of the area between Coventry, Rugby and Leamington and their bearing on the topographic development of the Midlands. *Phil. Trans. R. Soc. Lond.* B **237**, 209–260.
Sibrava, V., Bowen, D. Q. & Richmond, G. M. (eds) 1986 Quaternary glaciations in the northern hemisphere. *Quat. Sci. Rev.* **5**, pp. 511.
Sibrava, V. 1986*a* Correlations of European glaciations and their relation to the deep sea record. *Quat. Sci. Rev.* **5**, 433–442.
Sibrava, V. 1986*b* Scandinavian glaciations in the Bohemian Massif and Carpathian Foredeep and their relation to the extra glacial areas. *Quat. Sci. Rev.* **5**, 381–386.
Sissons, J. B. 1979 The Loch Lomond stadial in the British Isles. *Nature, Lond.* **291**, 473–475.
Sjorring, S. 1983 The glacial history of Denmark. In *Glacial deposits in north west Europe* (ed. J. Ehlers), pp. 163–180. Rotterdam: Balkema.
Southgate, G. A. 1985 Thermoluminescence dating of beach and dune sands: potential of single grain measurements. *Nuclear Tracks*, pp. 743–747.
Sparks, B. W., Williams, R. B. G. & Ransome, M. 1969 Hoxnian interglacial deposits near Hatfield, Herts. *Proc. Geol. Ass.* **80**, 243–267.
Stringer, C. B., Currant, P., Schwarcz, H. P. & Collcutt, S. M. 1986 Age of Pleistocene faunas from Bacon Hole, Wales. *Nature, Lond.* **320**, 59–62.
Sumbler, M. G. 1983 A new look at the type Wolstonian glacial deposits of central England. *Proc. Geol. Ass.* **94**, 23–31.
Sutherland, D. G. 1984 The Quaternary deposits and landforms of Scotland and the neighbouring shelves: a review. *Quat. Sci. Rev.* **3**, 157–254.
Sutcliffe, A. J. & Currant, A. P. 1984 Minchin Hole cave. In *Wales: Gower, Preseli and Fforest Fawr* (ed. D. Q. Bowen & A. Henry), pp. 33–37. Cambridge: Quaternary Research Association.
Sutcliffe, A. J. & Bowen, D. Q. 1973 Preliminary report on excavations in Minchin Hole, April–May 1973. *Newsl. William Pengelly Cave Studies Trust* **21**, 12–25.
Ter Wee, M. W. 1983 The Elsterian glaciation of the Netherlands. *Glacial deposits in northwest Europe* (ed. J. Ehlers), pp. 413–416. Rotterdam: Balkema.
Thomas, G. S. P. 1976 The Quaternary stratigraphy of the Isle of Man. *Proc. Geol. Ass.* **87**, 307–324.
West, R. G. 1956 The Quaternary deposits at Hoxne, Suffolk. *Phil. Trans. R. Soc. Lond.* B **239**, 265–356.
Zagwijn, W. 1986 The Pleistocene of the Netherlands with special reference to glaciation and terrace formation. *Quat. Sci. Rev.* **5**, 341–346.

Discussion

C. Turner (*Department of Earth Sciences, The Open University, Milton Keynes, U.K.*). The reported amino-acid measurements suggest that the Holsteinian marine deposits at Hummelsbüttel in the type area at Hamburg are (*a*) of equivalent age to the deposits at Boxgrove in southern England and (*b*) distinctly older than the Holsteinian marine deposits at Wacken, Schleswig-Holstein. This proposal causes considerable stratigraphic problems. The Quaternary deposits of the Hamburg area have been investigated both at depth and more intensively than perhaps anywhere else in northern Europe. It is clear that the Holsteinian deposits in Hamburg and at Wacken post-date the youngest of the Elsterian tills and can be correlated with each other from many section and borehole records in considerable detail based on marine macro- and micro-fauna and palynology. This was well demonstrated at a recent meeting of the INQUA Sub-commission on European Quaternary Stratigraphy in Hamburg. Both Hummelsbüttel and Wacken unequivocally belong to the same interglacial stage. On palaeontological grounds, the Boxgrove site is now agreed to relate to a temperate stage older than the Anglian glacial stage (and, of course, older than the Hoxnian interglacial). The correlations proposed in this paper would, therefore, imply that the Anglian glacial period of Britain is younger than, and not to be correlated with, *any* part of the Elsterian complex in Europe, and that the Holsteinian typesite is not only older than other Holsteinian sites nearby but likewise older than the Hoxnian, with which the Holsteinian is now firmly correlated.

These deductions from amino-acid measurements fly in the face of some of the most carefully and firmly established correlations in North European stratigraphy. Such problematical and apparently aberrant results have to be looked into more carefully, even if initially repeatable (which is, in itself, not an explanation) to maintain confidence in the use of amino-acid racemization measurements as a precise and useful chronostratigraphic tool.

D. Q. Bowen. One of the principal thrusts of this paper has been to show that existing classifications of the continental Quaternary are inadequate and mostly oversimplified. The isoleucine epimerization data (amino acid geochronology) show this for 'interglacial' marine stratigraphic units, as has already been shown for non-marine 'interglacial' units (Hughes 1987). Dr Turner's comments are based on the view that classification based on assemblage floras of 'interglacial aspect' may be used as a suitable means of identifying fixed points in time. Such an approach, however, is inappropriate because it is not based on first appearances, nor extinctions of fossil taxa, but on facies–floras controlled by climate. His comments also seem to be a defence of the view that Middle Pleistocene deposits in Europe and Britain represent a 'short chronology', as opposed to a longer one inferred from signals of climatic change based on oxygen isotope variability.

Dr Turner is incorrect to suggest that an implication of the paper is that the 'Anglian' of England is younger than the 'Elsterian' of Europe. The data presented cannot be interpreted in that way. What the isoleucine epimerization on shells of *Macoma* from Hummelsbüttel (0.382) and Wacken (0.283) show, is that the latter is clearly younger than the former, a view held by many German workers on other grounds (see, for example, Sarnthein *et al.* 1986; Cepek 1986). Regardless of how these data are modelled, they show that not all 'interglacial' deposits labelled 'Holsteinian' are the same age (see also Kukla 1977).

Dr Turner comments that the Quaternary deposits of the Hamburg area have been investigated 'more intensively that perhaps anywhere else in northern Europe'; yet regional lithostratigraphy is hardly secure when 'Holsteinian' deposits are only exposed where they have been brought to the surface along thrust planes (Ehlers 1983). I agree with him that the Eartham (Boxgrove) site is older than the Hoxnian of Suffolk: what the data show is that the Boxgrove site is the same age as Hummelsbüttel.

Finally, it is misleading of Dr Turner to say that his correlations are based on 'stratigraphy': they are based on a facies–flora biostratigraphy, entirely lacking in geochronometric control, which shows a comparatively simple model of climatic history. The isoleucine epimerization data, on the other hand, show that the Middle Pleistocene history of Europe was as complex as suggested by the models of Kukla (1977), Sibrava (1986) and Cepek (1986).

Phil. Trans. R. Soc. Lond. B **318**, 637–644 (1988)

Printed in Great Britain

The glacial history of Iceland during the past three million years

By T. Einarsson and K. J. Albertsson
Department of Geosciences, Science Institute, University of Iceland, 101 *Reykjavík, Iceland*

Iceland is built up of volcanic rocks with sedimentary interbeds, which have been piled up continuously since Miocene times. In the Pleistocene rock series, sediments of fluvial, lacustrine, marine and glacial origin and soils are very common and frequently thick. A sudden climatic deterioration took place at about 3 Ma BP. The Pliocene lusitanic marine fauna was replaced by a boreal fauna. Conifers and deciduous forest vanished and the flora became similar to the present one. From 3 to 2 Ma BP, inland ice caps were common during cold spells. From then on ice sheets reaching down to sea level have covered most of the country at least 12 times during glacials.

Introduction

Iceland is situated on the active Mid-Atlantic Ridge in the North Atlantic. It forms the middle part of the extensive Brito-Arctic volcanic area. The opening of the North Atlantic Ocean in late Cretaceous and early Tertiary times was accompanied by volcanic activity in the adjacent regions; volcanic activity persists at present only in Iceland and Jan Mayen.

Iceland is almost wholly built up of volcanic rocks which have been piled up more or less continuously since Miocene times. The oldest rocks so far dated by the K–Ar method show ages of 14–15 Ma (Gale *et al.* 1966; Moorbath *et al.* 1968; Albertsson & Einarsson 1982; McDougall *et al.* 1984). The bedrock in Iceland can be divided into three major formations: the Tertiary Plateau Basalt Formation, the Grey Basalt Formation (3–0.7 Ma), and the upper Pleistocene Palagonite Formation (younger than 0.7 Ma).

Sediments of varying origin make up 2–10% of the thickness of the Tertiary strata, whereas their Quaternary counterpart constitutes, in places, up to 50% of the thickness of the Quaternary pile. Some of the sediments are good indicators of climate, i.e. the palaeosols and tillites (diamictites); others contain fossils, which are also useful for the interpretation of past climates. Volcanic rocks are in themselves evidence of past environments; for example, hyaloclastite ridges and table mountains can in most cases be considered to have been heaped up subglacially during glacials and lava flows formed during interglacials.

Present climate

The climate of Iceland is affected very much by both the movement of the polar front, i.e. the boundary between cold air masses from the polar regions and the warm air masses from the south, and also by the movements of the boundary between Arctic and Atlantic sea water. The mean temperature in southern and western Iceland is 10–11 °C in July and 0 to −1 °C in January. The corresponding figures for northern and eastern Iceland are 8–11 °C and −2 to −4 °C respectively. At present, glaciers cover 11260 km^2 or 11% of the country; the biggest is Vatnajökull, which has an area of 8300 km^2. The annual precipitation varies greatly from about 400 mm in northern Iceland, to 800 mm in the southwest, and to some 3600 mm on the southeastern coast. The precipitation on southern Vatnajökull above the snowline at 1000–1100 m is 4000 mm; on Mýrdalsjökull the precipitation is some 6000 mm. North of Vatnajökull the snowline lies above 1700 m (Eythorsson & Sigtryggsson 1971).

The Irminger Current flows clockwise around Iceland. The temperature on the south coast is 11 °C in August and 6–7 °C in March; on the north coast it is 8 °C and 2–3 °C respectively. In cold years, drift ice can occasionally block the north and east coast; this last happened during the winters of 1965 and 1968. In cold years the cold East Greenland Current can block off the Irminger Current at the northwestern peninsula, causing the north and east coasts to be dominated by the Arctic seawater (Stefánsson 1981).

The Last Glacial and Holocene as a model for past glaciations in Iceland

At the maximum of last glaciation, Iceland was almost completely covered by glaciers. End-moraines, probably of Weichselian age, have been found on the shelf, approximately 130 km off Breiðafjörður, west Iceland, at 150–250 m depth (Ólafsdóttir 1975). The ice divide of the main glacier was some 50 km south of the present water divide. According to the height of subglacially formed table mountains of Weichselian age, the surface of the ice sheet was probably no higher than 2000–2500 m at the ice divide. On the northwestern peninsula and Snæfellsnes and the highlands in northern and eastern Iceland, there were independent ice caps and valley glaciers. During the retreat some re-advances occurred, the latest being the Álftanes stage (about 12 ka BP correlated with the Older Dryas) and the Búði stage (11–10 ka BP, Younger Dryas). Marine beds of Kópasker age (i.e. Bølling) and Saurbær age (i.e. Allerød) have been found at different places in western and northern Iceland. By 8 ka BP the glaciers were probably smaller than the present ones (Einarsson 1985).

Owing to isostatic and eustatic sea-level adjustments, the greater part of the lowlands was flooded by the sea as the glaciers retreated. The highest shorelines are approximately 100 m above present sea level in south Iceland and 30–60 m elsewhere. They are probably of Saurbaer age (12–11 ka BP). The isostatic readjustment was very rapid and by 9 ka BP the shore was everywhere lower than today and probably at −20 m by 8 ka BP.

The late glacial and early Holocene marine fauna was similar to the present one. However, arctic species like *Portlandia arctica* are seen to have lived only at glacier snouts in fiords. In late-glacial times the temperature of the sea on the south and west coast of Iceland can be considered to have been only 2 °C lower than at present (Einarsson 1985). It is also worth mentioning that only two ice-wedge casts have been found in late-glacial sediments in northern Iceland; this observation indicates that the climate was not very cold during late-glacial times (Norðdahl 1983).

No plant-bearing deposits (peat or lacustrine sediments) have been found in Iceland from late-glacial times, although up to one half of the country was already ice-free during the Saurbær and Búði stages. The oldest organic sediments so far found are of Preboreal age and contain a flora similar to the present one.

Several hypotheses have been presented on the origin of the Icelandic flora, which contains about 450 species of vascular plant of which over 98% are of European affinity. One of the hypotheses assumes that one half of the flora survived the last glacial in ice-free areas, mainly in northwest, north and east Iceland. About 25% were introduced after the human settlement

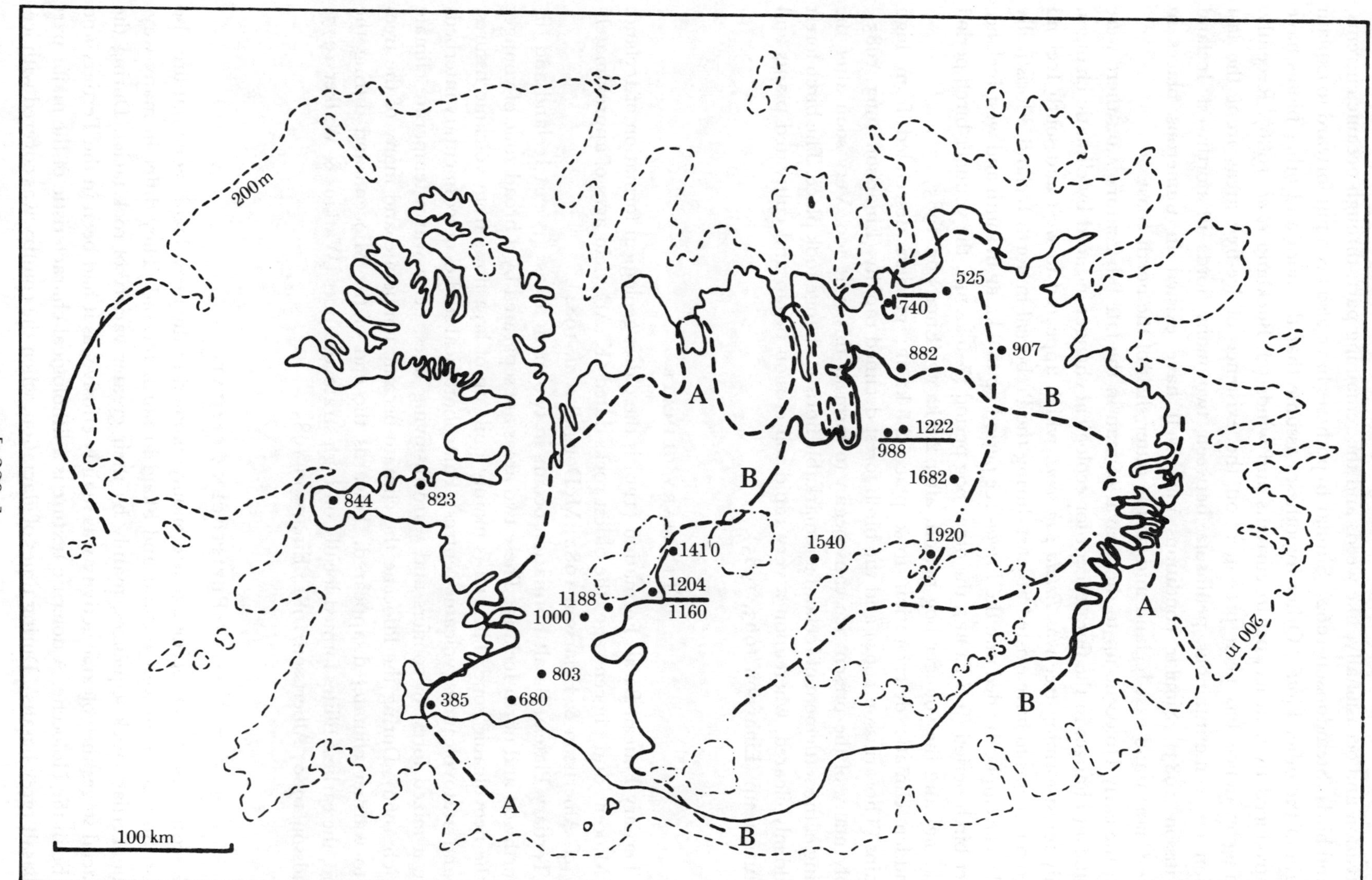

FIGURE 1. The Weichselian ice sheet in Iceland. The maximum extent (end-moraines off west Iceland) and the readvance stages Álftanes (A) and Búði (B). Numbered dots show the altitudes of Weichselian table mountains and the highest glacial striae on Saalian table mountains (outlined). Also shown is the 200 m depth contour of the shelf and ice divide in late Weichselian time.

by cultivation and occasionally, like weeds and the remaining part, through oceanic currents, winds and birds (Steindórsson 1962). Similar hypotheses have also been put forward to explain the origin of the other biotas. Other hypotheses assume that the flora and other biotas have been introduced by drift ice, ocean currents and winds (cf. Buckland *et al.* 1986). Recently, confirming evidence has been presented on the existence of ice-free areas from the last glaciation on a mountainous peninsula between two main fiords in northwest Iceland (Sigurvinsson 1983). Similar conditions may well have existed in numerous places in northwest, north and east Iceland and on the outer shelf of the north coast.

From the early Holocene up to the arrival of man in the late 9th century A.D., there were no significant changes in the flora except for ecological changes induced by climatic changes. The only forest-forming tree was *Betula pubescens*, which appeared in west and south Iceland 9 ka ago but seems to have existed at least during the Preboreal in north Iceland. Floristically the Holocene can be divided into the *Betula*-free period up to 9 ka BP in south and west Iceland, the older birch period (9–7 ka BP), the older bog period (7–5 ka BP), the younger birch period (5–2.5 ka BP) and the younger bog period, after 2.5 ka BP (Einarsson 1985).

A sudden climatic deterioration took place 2.5 ka BP: glaciers descended from high mountains (Thorarinsson 1964) and the birch forest declined rapidly (Einarsson 1963, 1985). Probably many of the present ice caps began to form at this time too. Very soon after the beginning of the settlement of Iceland a remarkable floristic change took place. The birch forest was suddenly cleared, whereupon a very rapid soil erosion began and cultivated plants and weeds turned up (Einarsson 1963, 1985).

Tertiary climate

The Tertiary Plateau Basalt Formation (TPB) is the oldest geological formation of Iceland. It is of Miocene and Pliocene age: the oldest rocks dated (K–Ar) yield ages of approximately 14 Ma (cf. Albertsson & Einarsson 1982; McDougall *et al.* 1984).

The Tertiary Plateau Basalt Formation occurs in two main areas: (i) east Iceland and (ii) west, northwest and north Iceland. These two areas are separated by a broad zone of younger rocks, the neovolcanic zone. The TPB is mainly built up of lava flows from volcanic fissures, shield- and stratovolcanoes (volcanic centres). Palaeobotanical studies of sedimentary interbeds indicate a mixed forest of conifers and warmth-loving trees, i.e. a warm-temperate climate (Friedrich 1966). During the Pliocene the climate became temperate and many of the trees suited to warmer climates disappeared. Towards the end of the Miocene and during the Pliocene, the earliest tillites formed locally on high stratovolcanoes (Watkins & Walker 1977; Sæmundsson 1980; Albertsson 1981; Einarsson 1985).

Pleistocene climate

Rock sequences of Pleistocene age are mainly exposed in the neovolcanic zone but are also found on Snæfellsnes in west Iceland and Skagi in north Iceland. They differ in many ways from the Tertiary rock sequences, mainly by their greater variety of rock facies. During the interglacial stages the volcanic activity was mainly effusive, as it had been in the Tertiary and was to be in the Holocene. A doleritic texture is a lithological characteristic of the major part of the basalt (grey basalts). During times of glaciation, when the country was covered with ice,

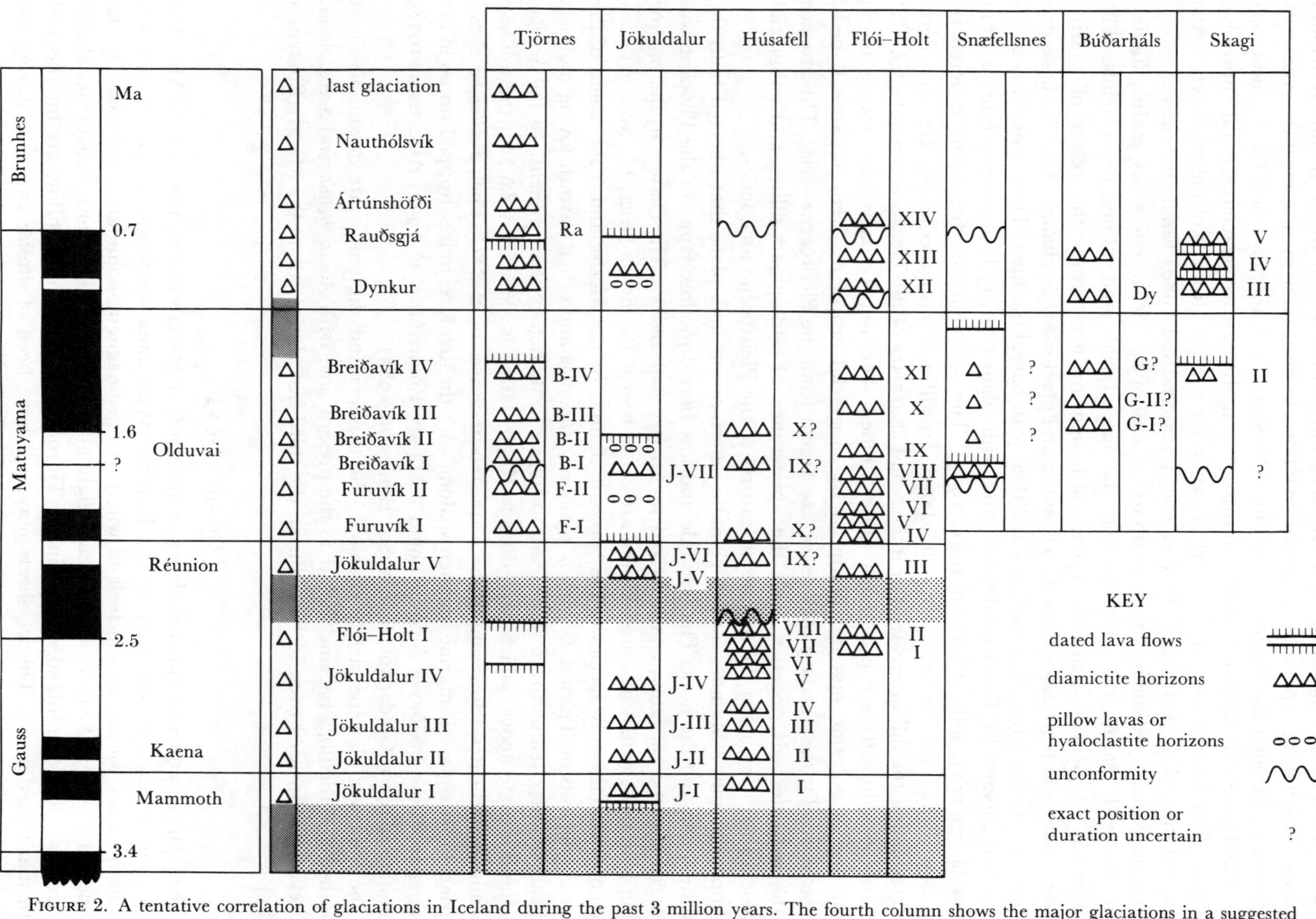

FIGURE 2. A tentative correlation of glaciations in Iceland during the past 3 million years. The fourth column shows the major glaciations in a suggested chronological order (cf. Albertsson 1976).

volcanic products were piled up over the eruption centres as tuff ridges (pillow lavas and breccias), which were capped by subaerial lava flows to form table mountains if the volcanoes were built up through the glacier sheets. The older glacial volcanoes are now for the most part eroded and buried in the pile, but those formed during the past two glacials are still impressive morphological features and can be used to define the surface altitudes of the ice sheets at the formation time of the volcanoes (Walker 1965; Sæmundsson 1980; Einarsson 1985).

In some regions, sediments comprise up to one half of the Pleistocene series, mainly fluvial, lacustrine and marine ones representing the interglacials and tillites the glacials. The sedimentary sequences are often cyclic, although mainly within the reach of marine transgressions and regressions caused by isostatic and glacioeustatic changes, i.e. till, succeeded by marine silt, first with arctic molluscs and then with boreal molluscs. These sediments are in turn often covered by fluviatile deposits and with plant-bearing lacustrine sediments, thin seams of lignite or soils (Eiríksson 1981, 1985). The sediments are frequently covered by lava flows, which can be radiometrically dated. Fossiliferous cyclic sediments are very useful in determining the full interglacial or interstadial character of the Pleistocene sequences. The Pleistocene sediments are also much more lithified then the Tertiary ones, owing to an admixture of volcanic material produced by subglacial eruptions. Red clayish soil beds, characteristic for the Tertiary rock series, are absent from the Pleistocene series. The absence of the red interbeds, together with the appearance of widespread tillites and subglacial hyaloclastites, are among the dominant features of the Pleistocene rock series.

Approximately 3 Ma BP a sudden climatic deterioration took place that affected both the marine fauna and the flora. The lusitanic and low boreal mollusc fauna of the Pliocene was replaced by a boreal fauna characterized by *Serripes groenlandicus*. The change in the marine fauna is to some extent obscured by the arrival of boreal Pacific molluscs, but younger rock sequences contain only species of the 'present' fauna. In the *Mactra* and *Tapes* zones of the Pliocene deposits on Tjörnes, *Glycimeris glycimeris*, *Abra alba* and other warmth-loving molluscs indicate a water temperature of at least 10 °C, i.e. 5 °C higher than present. In the *Serripes groenlandicus* zone above, which according to K–Ar dating is older than 2.5 Ma, the mollusc assemblage indicates a lowering of the water temperatures by 2–3 °C. Then *Portlandia arctica* appears in 'late glacial' marine sediments, along with the first foreign ice-dropped stones in the lower part of the Breiðavík deposits about 2 Ma BP (Einarsson *et al.* 1967; Albertsson 1978; Símonarson 1980; Gladenkov *et al.* 1980; Einarsson 1985).

The flora was also completely changed: the conifers and the temperate deciduous forest vanished and the flora became similar to the present one, with *Alnus*, *Betula* and *Salix*. *Alnus*, however, disappeared from Iceland during the third last glacial, about 0.5 Ma BP (Einarsson 1985).

Conclusions

The Tjörnes sequence shows evidence of at least four glacials during the past 0.7 Ma. The lowest one marks the boundary of the Brunhes–Matuyama geomagnetic epochs and is a common feature for all areas in Iceland where this geomagnetic boundary is exposed. During the period 0.7–2 Ma BP there is evidence of eight glacials in coastal sections; two further glacial horizons are found in inland sections in the Tjörnes area. Six of these tillites are interbedded with marine sediments. The ice sheets may well have been of similar extent as during the Weichselian (Einarsson *et al.* 1967; Albertsson 1978; Eiríksson 1981, 1985).

Ice caps or ice sheets formed during glacials or cold spells three to two million years ago in or adjacent to the active volcanic zones. They were probably comparable in size with glaciers in late Weichselian times, e.g. in the Húsafell area in southwest Iceland, in the Flói-Holt area in central South Iceland and the Jökuldalur area in eastern Iceland (Hopkins *et al.* 1965; McDougall & Wensink 1966; Einarsson *et al.* 1967; Eiríksson 1973; Sæmundsson & Noll 1975; Albertsson 1976). In each of these sequences there are 3–9 separate diamictite horizons, some of which have not been thoroughly studied but some of which are certainly tillites. The oldest dated tillite in these sequences is in Jökuldalur and has an age of little less than 3.1 Ma (McDougall & Wensink 1966).

The problem in recognizing the full interglacial or interstadial as well as the full glacial or glacial-stage character of the rock sequences is still not solved, except where full-scale cyclic sedimentation of interglacial–glacial character has been verified.

There is evidence of four glaciations during the past 0.7 Ma. During the period 0.7–2 Ma, at least eight glaciations, and possibly ten, occurred in the Tjörnes area. During the time interval 2–3 Ma, three to nine diamictite horizons, some of which are certainly tillites, are found in other regions. During this time the glaciers may have reached the sea in southern Iceland. According to this there have been 15–23 glaciations in Iceland during the past three million years.

References

Albertsson, K. J. 1976 K/Ar ages of Pliocene/Pleistocene glaciations in Iceland with special reference to the Tjörnes sequence, northern Iceland. Ph.D. dissertation, University of Cambridge. (268 pages.)

Albertsson, K. J. 1978 Um aldur jarðlaga á Tjörnesi. (Some notes on the age of the Tjörnes strata sequence, northern Iceland.) *Náttúrufræðingurinn* **48**, 1–8. Reykjavík.

Albertsson, K. J. 1981 On Tertiary tillites in Iceland. In *The Earth's pre-Pleistocene glacial record* (ed. W. B. Harland & M. H. Hambrey), pp. 562–565. Cambridge University Press.

Albertsson, K. J. & Einarsson, Th. 1982 Um aldur jarðlaga efst á Breiðadalsheiði. (On the age of the rocks on Breidadalsheidi, NW-Iceland.) In *Eldur er í norðri* (*Fire is up in the North*) (ed. H. Thórarinsdóttir, Ó. H. Óskarsson, S. Steinthórsson & Th. Einarsson), pp. 206–210. Reykjavík: Sögufélag.

Buckland, P. C., Perry, D. W., Gíslason, G. M. & Dugmore, A. J. 1986 The pre-Landnám fauna of Iceland: A palaeontological contribution. *Boreas* **15**, 173–184.

Einarsson, Th. 1963 *Pollen analytical studies on the vegetation and climate of Iceland in late- and postglacial times.* In *North Atlantic biota and their history* (ed. A. Löve & D. Löve), pp. 355–365. Oxford: Pergamon Press.

Einarsson, Th. 1985 *Jarðfræði* (*Geology*). Reykjavík: Mál og menning. (233 pages.)

Einarsson, Th., Hopkins, D. M. & Doell, R. R. 1967 The stratigraphy of Tjörnes, northern Iceland and the history of the Bering Land Bridge. In *The Bering Land Bridge* (ed. D. M. Hopkins), pp. 312–325. Stanford University Press.

Eiríksson, J. 1973 Jarðlagaskipan á ytra Mið-Suðurlandi. (Quaternary stratigraphy of western South-Iceland.) B.S. thesis, University of Iceland, Reykjavík. (99 pages.)

Eiríksson, J. 1981 Lithostratigraphy of the upper Tjörnes sequence, North Iceland: The Breidavík Group. *Acta nat. islandica* **29**, 1–37.

Eiríksson, J. 1985 Facies analysis of the Breidayík Group sediments, North Iceland. *Acta nat. islandica* **31**, 1–56.

Eythorsson, J. & Sigtryggsson, H. 1971 *The climate and weather of Iceland.* In *The zoology of Iceland* (ed. S. L. Tuxen), vol. **1** (3), pp. 1–62. Copenhagen: Ejnar Munksgaard.

Friedrich, W. 1966 Zur Geologie von Brjánslækur unter besonderer Berücksichtigung der fossilen Flora. *Sonderveröff. geol. Inst. Köln*, no. 10. (108 pages.)

Gale, N. H., Moorbath, S., Simons, J. & Walker, G. P. L. 1966 K/Ar ages of acid intrusive rocks from Iceland. *Earth planet. Sci. Lett.* **1**, 284–288.

Gladenkov, Yu. B., Norton, P. & Spaink, G. S. 1980 Upper Cenozoic of Iceland. Stratigraphy of Pliocene–Pleistocene and palaeontological assemblages. *Trans. Acad. Sci. U.S.S.R.* **345**, 1–116.

Hopkins, D. M., Einarsson, Th. & Doell, R. R. 1965 The stratigraphy of Tjörnes, northeastern Iceland: Its significance for the history of the Bering Land Bridge. (Abstract.) *VII INQUA Congress, Boulder, Colorado, 1965; General session*, p. 223.

McDougall, I. & Wensink, H. 1966 Paleomagnetism and geochronology of Pliocene–Pleistocene lavas in Iceland. *Earth planet. Sci. Lett.* **1**, 232–236.
McDougall, I., Kristjansson, L. & Sæmundsson, K. 1984 Magnetostratigraphy and geochronology of northwest Iceland. *J. Geophys. Res.* **89**, 7029–7060.
Moorbath, S., Sigurdsson, H. & Goodwin, R. 1968 K/Ar ages of the oldest exposed rocks in Iceland. *Earth planet. Sci. Lett.* **4**, 197–205.
Norðdahl, H. 1983 Late Quaternary stratigraphy of Fnjóskadalur, central north Iceland. *Lundqua thesis* no. 12. (78 pages.)
Ólafsdóttir, Th. 1975 Jökulgarður á sjávarbotni út af Breiðafirði. (A moraine ridge on the Iceland shelf west of Breidafjördur.) *Náttúrufræðingurinn* **45**, 31–36.
Sigurvinsson, J. R. 1983 Weichselian glacial lake deposits in the highlands of Northwestern Iceland. *Jökull* **33**, 99–109.
Símonarson, L. A. 1980 On climatic changes in Iceland. *Jökull* **29**, 44–46.
Stefánsson, U. 1981 *Sjórinn við Ísland.* (The sea around Iceland.) In *Náttúra Íslands*, pp. 397–438. Reykjavík: Almenna Bókafélagið.
Steindórsson, S. 1962 *On the age and immigration of the Icelandic flora.* Societas Scientiarum Islandiae, Rit **35**. (57 pages).
Sæmundsson, K. 1980 Outline of the Geology of Iceland. *Jökull* **29**, 7–28.
Sæmundsson, K. & Noll, H. 1975 K/Ar ages of rocks from Húsafell, W-Iceland and the development of the Húsafell central volcano. *Jökull* **24**, 40–57.
Thorarinsson, S. 1964 On the age of the terminal moraines of Brúarjökull and Hálsajökull. *Jökull* **13**, 65–75.
Walker, G. P. L. 1965 Some aspects of Quaternary volcanism. *Trans. Leicester lit. phil. Soc.* **59**, 25–40.
Watkins, N. D. & Walker, G. P. L. 1977 Magnetostratigraphy of eastern Iceland. *Am. J. Sci.* **277**, 513–584.

Phil. Trans. R. Soc. Lond. B **318**, 645–660 (1988)
Printed in Great Britain

Climatic evolution of the eastern Canadian Arctic and Baffin Bay during the past three million years

By J. T. Andrews

Department of Geological Sciences and Institute of Arctic and Alpine Research, University of Colorado, Boulder, Colorado 80309, U.S.A.

[Plates 1 and 2]

The outer east coast of Baffin Island is characterized by a series of sedimentary forelands. These contain a variety of litho- and biofacies associated with glacial marine and marine deposition into sea levels higher than those of the present. These high relative sea levels were associated with glacial isostatic loading and unloading of the crust by the NE sector of the Laurentide Ice Sheet. On the basis of amino acid epimerization ratios, eleven chronologically distinct units are delimited. The youngest unit is less than or equal to 10 ka, but all others are at or beyond the limits of radiocarbon dating. Based on biostratigraphy and amino acid data, the oldest units exposed in the forelands may be Pliocene in age. Molluscs, Foraminifera, and the palynology of buried soils and organics, indicate that the vast bulk of the exposed sequences contain floras and faunas that represent environments *warmer* than those at present. An analysis of modern and fossil pollen spectra suggests a steady decrease in low arctic conditions throughout the Quaternary.

Introduction

The Neogene palaeoclimatology, palaeoceanography, and glacial geology of Baffin Bay, the northern Labrador Sea, and the adjacent land masses (figure 1) is undoubtedly an important, if only partly understood, element in the Pliocene and Quaternary history of the Northern Hemisphere. Keen (1980) demonstrated that in the Northern Hemisphere the mean summer (June–August) temperature at 70° N is closely correlated with the summer temperature of the area between West Greenland and eastern Baffin Island (period A.D. 1951–76). This implies that the zonal average temperature is being primarily forced by the changes that occur over this small ocean basin, which links the Arctic Ocean to the main western North Atlantic (figure 1). At the hemisphere scale (Barry *et al.* 1975), the transect between West Greenland and the eastern Canadian Arctic coastline represents one of the most extreme July temperature gradients in the Arctic, with a difference in temperature of 4–6 °C across the bay (Williams & Bradley 1985).

In several publications attention has been drawn to the contrasts in climate, ice extent and nearshore oceanography between West Greenland and eastern Baffin Island (see, for example, Jacobs *et al.* 1985; Andrews *et al.* 1981) and this has been used to challenge the conventional mid-latitude orthodoxy that the cause of glaciation is a matter of a depression in temperature, with no explicit account being made of the role of snow accumulation.

The role of oceanography in the present and past climatic changes in the area is probably profound. Under present conditions (figure 2) the oceanography is dominated by a

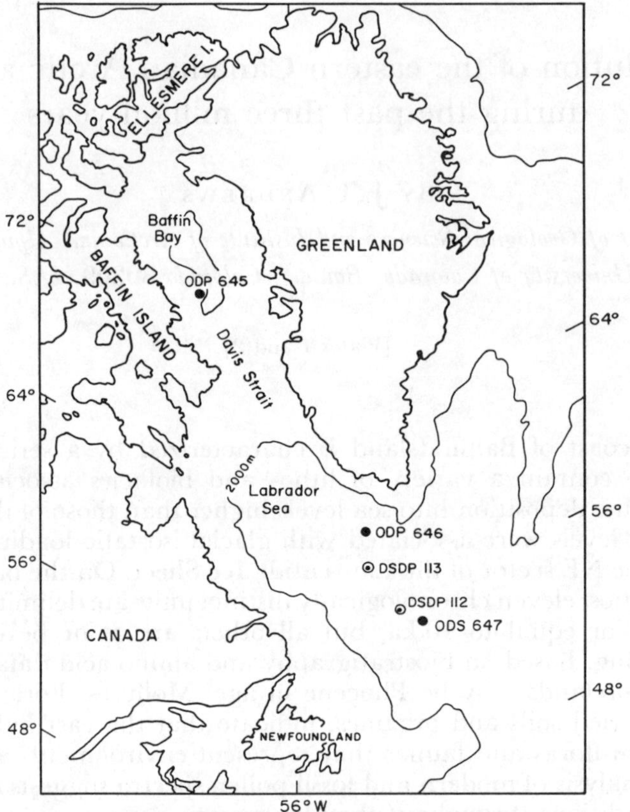

FIGURE 1. The 2000 m contour in the NE North Atlantic and the location of DSDP and ODP cores. The latter were taken on leg 105 in the autumn of 1985.

counterclockwise gyre: 'warm' water moves northward along the coast of West Greenland as the West Greenland Current (WGC). The main current turns westward in the north of Baffin Bay and descends to a depth of about 200 m, where it is held offshore against the narrow shelf of eastern Baffin Island. A shallow but cold current, the Canadian Current (CC), flows south along the west coast of Baffin Bay. The CC originates from the Arctic Ocean and moves southward through the arctic channels of the High Canadian Arctic. In the vicinity of Hudson Strait the CC joins with water from Hudson Bay and Foxe Basin and becomes the Labrador Current (LC). Both the CC and the LC are important mechanisms for the southward drift of icebergs from West and Northwest Greenland along the Canadian coastline.

Figure 2 shows the boundary between the marine arctic–subarctic and terrestrial high-arctic–low-arctic environments. This is controlled by the interactions of the WGC and CC. Diagnostic species of the low-arctic and subarctic zones are dwarf birch (*Betula* spp.) and the molluscs *Mytilus edulis* and *Macoma balthica*, among others (see Andrews 1972; Andrews *et al.* 1981; Lubinsky 1972, 1980). The average July temperature at the arctic–subarctic boundary is *ca.* 6 °C.

The west–east oceanographic contrast of Baffin Bay is also illustrated by a major contrast in the extent of glaciation between the Greenland and Canadian sides of the bay. The Greenland Ice Sheet has probably been a stable element in the geography of the region for most of the Quaternary. On the other hand, the northeastern sector of the Laurentide Ice Sheet reached tidewater along the Labrador and Baffin Island coasts several times during the same period,

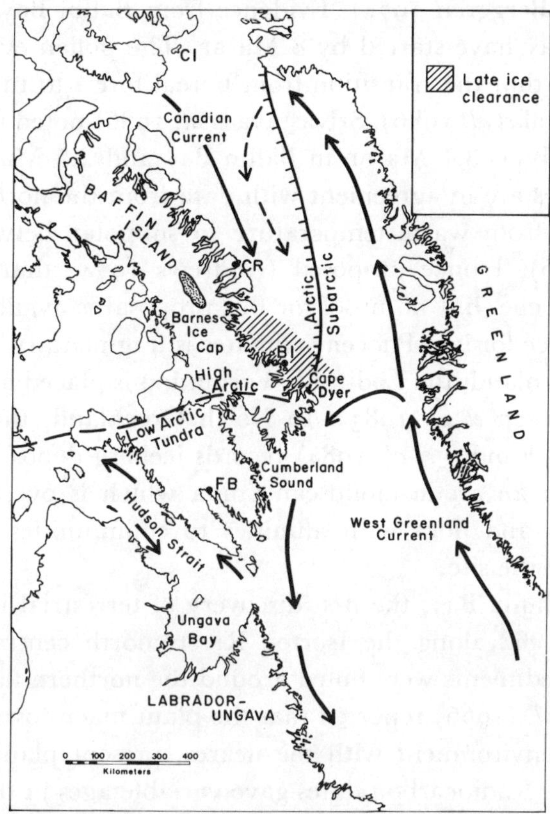

FIGURE 2. Generalized oceanographic circulation in Baffin Bay and the Labrador Sea. The arctic–subarctic zoogeographic boundary is shown (Dunbar 1968); this also corresponds with the northern limits of dwarf birch species. CR, Clyde River; BI, Broughton Island; FB, Frobisher Bay.

but today there are only a few small (less than 6000 km^2) relics of this once massive ice sheet. Despite the relatively small area of ice on Baffin Island today (37000 km^2), the remark by Tarr (1897) in the late 19th century, that Baffin Island is indeed 'wonderfully close to glaciation', is still appropriate.

Objectives

Of the many topics that might be evaluated, I wish to focus on an assessment of the chronology of glaciation and the palaeoenvironmental conditions associated with glaciation. Information from Ocean Drilling Project (ODP) leg 105, sites 645, 646, and 647 (figure 1), which were drilled in 1985, will add considerably to our knowledge of Quaternary events. The results of these surveys are just now being presented at meetings, but I would caution that evidence from these sites must be critically evaluated against the existing terrestrial evidence, much of which will be reviewed in this paper.

GLACIATION AND PALAEOENVIRONMENTS

The evidence from DSDP sites 112 and 113, between Greenland and Newfoundland (figure 1), were initially used to suggest that the onset of glaciation occurred *ca.* Ma BP (Berggren 1972). Relatively warm faunas occurred in the Labrador Sea during parts of the Pliocene (Thunnel & Belyen 1981) but in the earliest late Pliocene the Labrador Sea was invaded by polar waters (Poore & Berggren 1974). Evidence from Baffin Bay (Shipboard Party 1986) suggests that cooling may have started by 8 Ma BP. The pollen evidence from ODP site 645, in Baffin Bay, indicates that the transition from boreal forest to tundra occurred during the upper Pliocene (de Vernal *et al.* 1986). Srivastava *et al.* 1986) noted that the onset of glaciation may have begun as early as 3.4 Ma BP in Baffin Bay and 2.5 Ma BP in the western North Atlantic. These estimates are in agreement with data from the northeast Atlantic, where the surface water changed from warm-temperature to subpolar between 3.4 and 2.0 Ma BP (Loubere & Moss 1986). I once proposed (Andrews 1974) that glaciation of Greenland commenced in the Miocene, but no proof for this proposal is available.

The terrestrial evidence for late Pliocene climates is fragmentary on both Baffin Island and Greenland. In East Greenland, the Lodin Elv Formation is placed at the Pliocene–Pleistocene transition (Feyling-Hanssen *et al.* 1983). In North Greenland, the Early Quaternary Kap Kobenhavn Formation (Funder *et al.* 1984) records iceberg deposition *ca.* 1.8 Ma BP. These latter sediments contain an arctic molluscan fauna which is overlain by sediments with a boreal–low-arctic fauna and flora with affinities to communities which currently lie over 2000 km to the south of the site.

On the west side of Baffin Bay, the first discovery of terrestrial interglacial sediments was made by Andrews in 1962 along the Isortoq River, north central Baffin Island; in 1963 additional interglacial sediments were noted around the northern flank of the Barnes Ice Cap (figure 3). Terasmae *et al.* (1966) reported that the plant macrofossils and pollen assemblages indicated a low-arctic environment with the nearest present plant analogues some 600 km south of the Isortoq site. Radiocarbon dates gave variable ages in the range 14 to over 40 ka, but the younger dates were ascribed to the reworking of the sediments during the late Holocene retreat of the proto-Barnes Ice Cap. These plant-bearing beds were ascribed to the Flitaway Interglaciation, which was considered to be last-interglacial in age.

The extensive forelands that extend as low plains from the mountains toward Baffin Bay (figures 4 and 5) are a critical depository of Quaternary geological information. Only in western Spitsbergen are there comparable extensive Quaternary sections (see, for example, Boulton *et al.* 1982) along formerly glaciated coastal margins. Goldthwaite (1950), Loken (1966) and Feyling-Hanssen (1976*a*) were the first researchers to study these wave-cut exposures, which extend for more than 300 km along the outer east coast of Baffin Island (figure 4). Where cliffed by wave erosion (figure 5), the exposures consist of sediments associated with successive glacial isostatic sea-level oscillations (Feyling-Hanssen 1976*a*, *b*, 1985; Miller *et al.* 1977; Nelson 1981; Brigham 1983; Mode *et al.* 1983). The exposed sediments consisted primarily of (less than 100 m depth of water) marine silts and clays, littoral sands and gravels, some till, and peat and soils. The first ^{14}C dates (Loken 1966) indicated that the great bulk of these sediments were older than 50 ka. Research on these sections has focused on (i) the nature of the sediment cycles, (ii) their ages, and (iii) their faunas and floras.

(*a*) *Sediment cycles*

Miller *et al.* (1977), Nelson (1981) and Brigham (1983) described the succession of sediments exposed in the cliff sections (e.g. those in figures 4 and 5). From these data, Mode *et al.* (1983) presented a generalized facies model and linked the sediment sequence with the combined interactions of changes in glacier extent and glacial isostatic changes in relative sea level. Figure 6 is a schematic sketch of an idealized sequence for one glacial–interglacial cycle. The

FIGURE 4. Oblique air photograph looking northward along the outer coast of eastern Baffin Island. In the foreground is the Qivitu foreland. Numbers 1, 4 and 5 point to specific wave-cut sections: 2 locates a small raised delta some distance below the local marine limit; and 3 points to the right lateral moraine of a former ice lobe moving seaward down Narpaing Fiord.

FIGURE 5. Photograph of the exposed section at number 5 (figure 4), showing the freshly eroded face of the wave-eroded section.

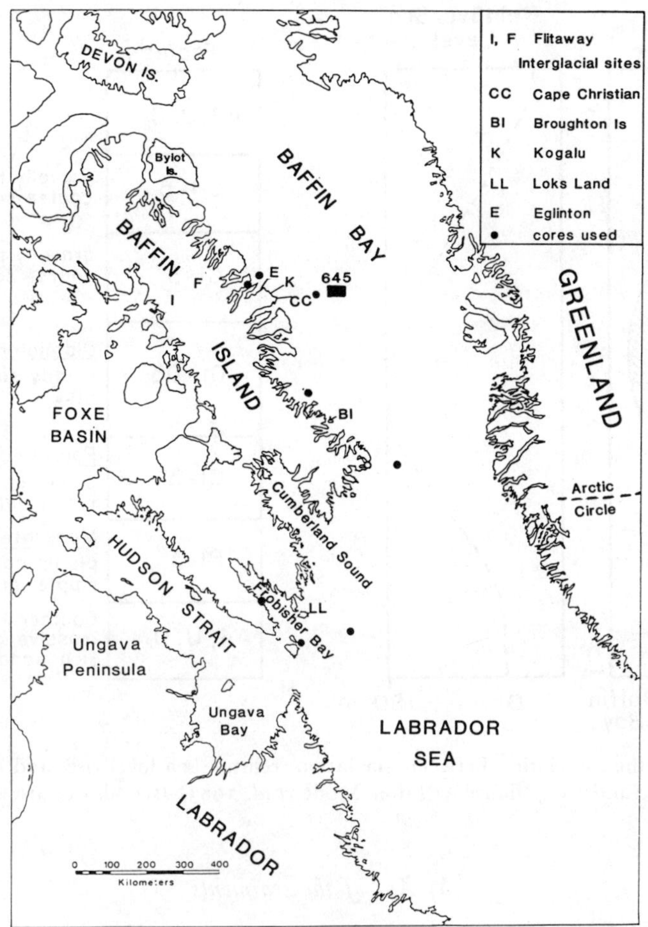

FIGURE 3. Location of typesites for some of the units described in the paper. Capital letters are used to locate the sites: I, Isortoq; F, Flitaway; CC, Cape Christian; K, Kogalu; BI, Cape Broughton; LL, Loks Land; E, Eglinton. Piston core coverage shown.

diamictons represent either basal tills or pebbly glacial marine sediments. In the eastern Baffin Island sequences, the diamictons frequently contain well-preserved foraminifera and whole bivalves, and are glacial marine sediments deposited close to ice margins that reached the outer coast and extended onto the continental shelf.

The vast majority of the forelands consist of a variety of shallow-water and littoral sediments deposited during marine transgressions and regressions. Faunas from these sediments, mainly molluscs and Foraminifera, can be used to deduce palaeoenvironmental conditions during each glacial–interglacial cycle. In some instances, organic sediments are preserved (O facies (Mode *et al.* 1983; Mode 1980, 1985)) and record periods when the sea level was at or *below* the land surface (figure 6). Such intervals may represent periods of global non-glacial climate, but periods of low sea level (and non-deposition) might also occur at times when regional glacial extent was reduced and global sea level was low. Such conditions would apply if there were an out-of-phase relationship between glaciation at high and lower latitudes (cf. Andrews & Miller 1984; Boulton *et al.* 1985; Morner 1977).

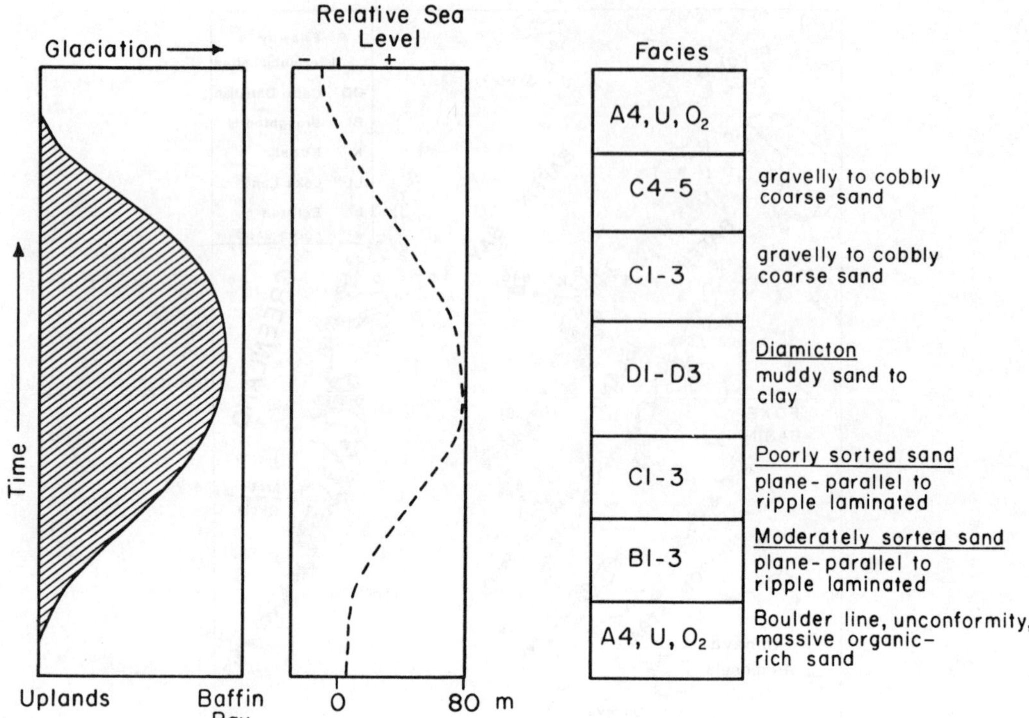

FIGURE 6. Sketch of the association between glaciation, relative sea level rise, and facies exposed in the forelands (e.g. figure 4) (after Mode *et al.* 1983) (see also figure 9).

(*b*) *Age of the sediments*

Because the bulk of the sediments lie beyond the limits of ^{14}C dating (Szabo *et al.* 1981) a variety of other dating methods have been used in an attempt to define the age, and number, of sediment packages. U-series analyses of marine molluscs proved useful in providing a series of *minimum* ages for several units. However, amino acid epimerization of common arctic bivalves, such as *Mya truncata* and *Hiatella arctica*, has been used successfully to delimit a series of aminozones along the length of the east coast of Baffin Island (see, for example, Miller *et al.* 1977; Nelson 1982; Brigham 1983; Mode 1985). This extensive body of data was reviewed by Miller (1985).

Amino acid ratios for D-alloisoleucine to L-isoleucine (aIle: Ile) in both the free (F) and total (T) fraction, increase to an equilibrium value of 1.3. The rate of epimerization is a function of both the thermal history of the site and the time since death of the organism. The aIle: Ile ratio can be used as a correlation tool for disjunct stratigraphic units. With some independent geochronological age control, the amino acid ratios can be used to estimate a range of probable ages, based on approximations of 'climate' (i.e. ground temperature). McCoy (1987) discussed errors associated with these various applications (see also Miller & Mangerud 1985).

Miller (1985, Table 14.11) evaluated the amino acid data for the late and middle Quaternary, whereas Mode (1985) presented evidence on older units (figure 7). A consideration of various thermal models suggested that the most likely effective diagenetic temperature (EDA) was -9 °C, which is associated with a mean annual temperature (MAT) of -11 to -15 °C (the present MAT is *ca.* -12 °C). The chronology shown in figure 7 is based on this model. Of note is the large number of distinct aminozones. Eleven aminozones are

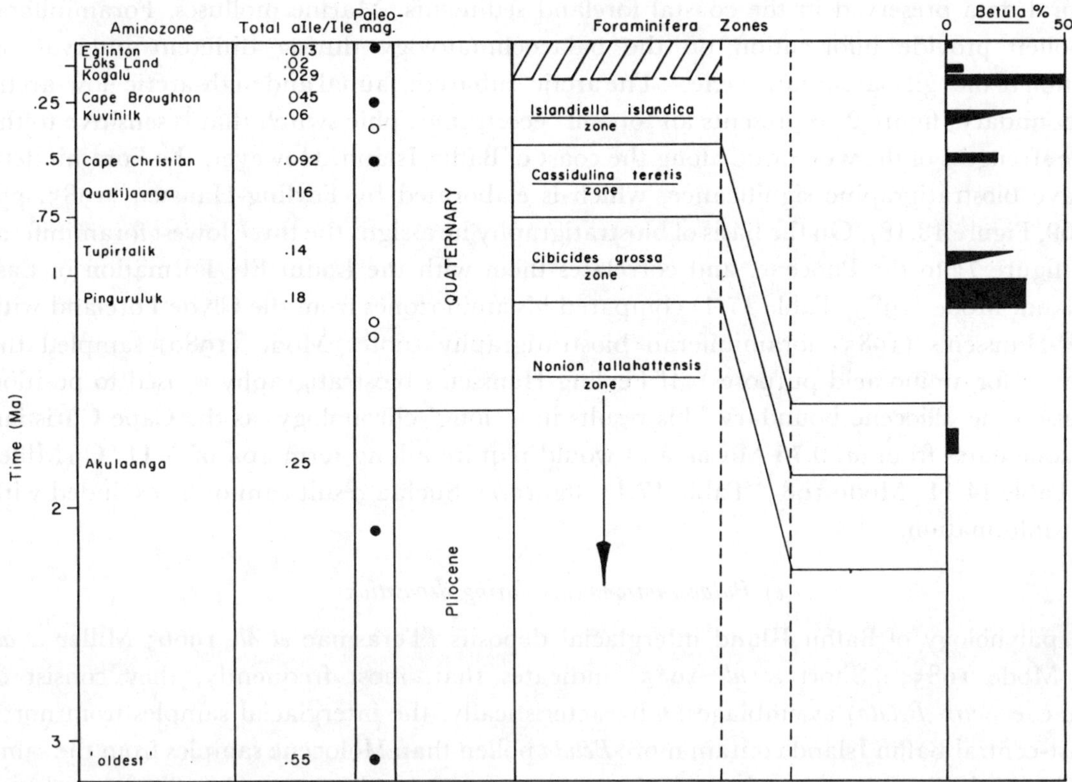

FIGURE 7. Possible geochronology for the Baffin Island aminozones (after Miller 1985) showing the limited palaeomagnetic results (solid circles represent normal inclinations). Variations in the percentages of *Betula* pollen (right-hand column, from Mode (1985)) represent a terrestrial subarctic element. Ascription of the foraminiferal zones depends on either the amino acid data (left, Miller 1985; Mode 1985) or on the biostratigraphy (Feyling-Hanssen 1985).

delimited, of which only the Eglinton and Loks Land are less than 40 ka old; this suggests that the preservation potential for these 'glacial isostatic facies' (Miller *et al.* 1977; Boulton *et al.* 1982) is remarkably high. This record in itself suggests that glacial ice was not an effective agent of sediment removal along the forelands of outer Baffin Island, although it must be stressed that the aminozones are *not* preserved anywhere in their entirety (cf. Gibbons *et al.* 1984).

In an initial chronology for eastern Baffin Island, Miller *et al.* (1977) used a U-series date to correlate the Cape Christian aminozone with the peak of the last interglaciation. Thus the Kuvinilk and Kogalu aminozones were placed in marine isotope stage 5. The Cape Christian and Flitaway deposits were correlated on the basis of their pollen assemblages. With a better understanding of both the amino acid epimerization rates, and further U-series determinations (Szabo *et al.* 1981; Brigham 1983), it was clear that the Cape Christian aminozone pre-dated stage 5e (i.e. it was much older than 125 ka).

The oldest aminozone exposed at the base of the forelands has a mean aIle: Ile ratio (T) of 0.55 (Miller 1985, p. 417). Based on the preferred EDA, this indicates an age of 3.5 Ma for this unit; but it could be as young as 1.6 Ma if an EDA of −5 °C is adopted. Limited palaeomagnetic data (Jacobs *et al.* 1985, figure 2.4) gave normal inclinations for the oldest sediment (figure 7). This result does not contradict an age of 3 Ma for the oldest aminozone, but the age estimates on figure 7 have error bars of ±50%!

Feyling-Hanssen (1985), Mode (1985), and Andrews *et al.* (1981) have reviewed the faunal and floral data preserved in the coastal foreland sediments. Marine molluscs, Foraminifera, and pollen provide information on the palaeoclimatology during different intervals of deposition of the 'glacial isostatic facies'. The arctic–subarctic faunal and high-arctic–low-arctic floral boundary (figure 2) represents an 'on–off' ecostratigraphic switch that is sensitive to the relative strengths of the WGC or CC along the coast of Baffin Island. However, the Foraminifera also have biostratigraphic significance, which is elaborated by Feyling-Hanssen (1985, pp. 366–369, Figure 13.18). On the basis of biostratigraphy he assigns the three lowest foraminiferal zones (figure 7) to the Pliocene, and correlates them with the Lodin Elv Formation of East Greenland. Mode (1985, Table 17.1) compared his aminozones from the Clyde Foreland with Feyling-Hanssen's (1985) foraminiferan biostratigraphy (note: Mode (1980) sampled the same units for amino acid purposes). If Feyling-Hanssen's biostratigraphy is used to position the Pleistocene–Pliocene boundary, this results in a 'long' chronology, as the Cape Christian aminozone dates from *ca.* 0.75 Ma BP and would require a long-term EDA of −11 °C (Miller 1985, Table 14.11; Mode 1985, Table 17.1) (figure 7). Such a result cannot be excluded with present information.

(*c*) *Palaeoenvironments during deposition*

The palynology of Baffin Island interglacial deposits (Terasmae *et al.* 1966; Miller *et al.* 1977; Mode 1985; Short *et al.* 1985) indicates that, most frequently, they consist of Gramineae–*Salix–Betula*) assemblages. Characteristically, the interglacial samples from north and east-central Baffin Island contain more *Betula* pollen than Holocene samples from the same area. On this basis, the present interglaciation is more severe than several earlier interglacial periods (see Mode 1985, Figure 17.3) (figure 7). Figure 8 is a comparison of Holocene pollen assemblages from northern Baffin Island with those from some of the older units (see figure 7). The diagram is based on a cluster analysis of the modern surface pollen rain (see, for example, Short *et al.* 1985) and the pollen spectra of peats and soils from various O facies (figure 7) within the cliffs. The north–south widths of the units (figure 8) represent the range of pollen assemblages along this 1500 km transect, whereas the width is a measure of the probability of the fossil sites being similar to the modern pollen spectra along the transect. This diagram shows that, even during the Holocene local thermal maximum, conditions were not as equable as they were during some earlier non-glacial intervals.

Of key concern is the stratigraphic position of the Flitaway and Isortoq plant-bearing beds (figures 3 and 8). These beds contain macrofossils of *Betula* and *Ledum groenlandicum*, and sufficient *Alnus* pollen to indicate that alder (probably *Alnus crispa*) grew close to the site (Terasmae *et al.* 1966). Collections from these units have been sampled for plant fossils (M. Kuc, personal communication, 1983); deposited in the National Type Collection of the Geological Survey of Canada) and beetles (Morgan *et al.* 1988). Both the moss flora and the insect fauna suggested that the sites were located close to, or even just within, the treeline. Morgan *et al.* (1982) estimated that the insect fauna indicated a July temperature 4–6 °C *above* that of the present, that is close to 10 °C. This would place the estimated July temperature 1–4 °C *above* the estimated Holocene thermal maximum for Baffin Island (Short *et al.* 1985).

When during the Neogene were temperatures this high over the Eastern Canadian Arctic? Some evidence occurs in the palynology of long piston cores from Baffin Bay and the evidence from the ODP site 645 (figures 1 and 3). de Vernal *et al.* (1986) noted that site 645 the transition

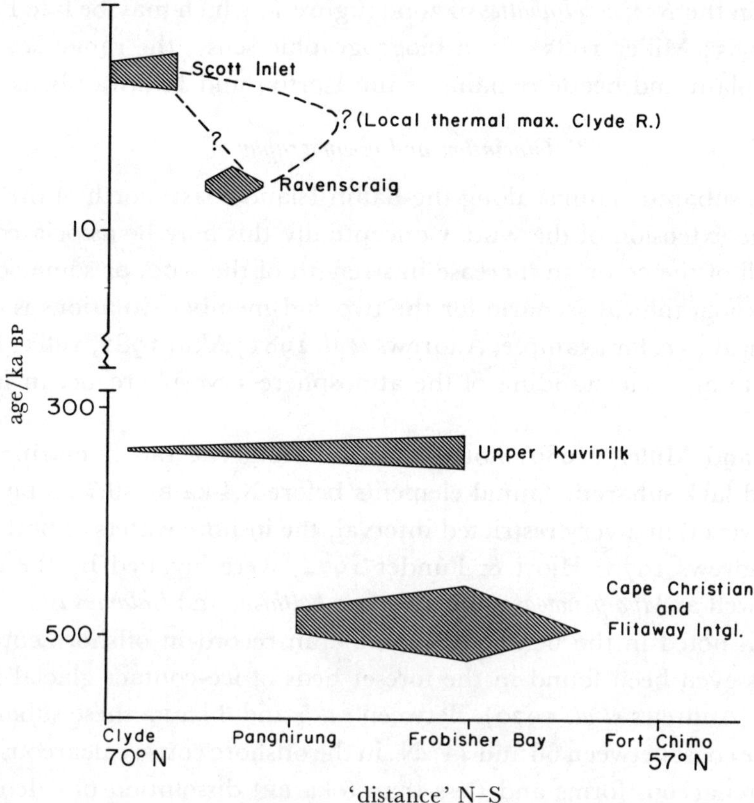

FIGURE 8. Position of fossil pollen spectra from sites on Baffin Island (figure 3) in the cliff sections near Clyde River and from the Flitaway sediments, compared with modern surface pollen rain on a transect from the boreal–tundra ecotone northwards to the Clyde River (figure 3).

from boreal–low-arctic vegetation to tundra occurred in the late Pliocene. Mudie & Short (1985, Figure 10.10) showed that, at a site south of site 645, subarctic shrub (*Betula*, *Alnus*) and conifer (*Picea* and *Pinus*) pollen occurred throughout the last 0.3 Ma and were advected to the area by either wind or ocean currents. Their graph of tree pollen percentages declines, with fluctuations, from a peak during marine-isotope stage 7 to low values within stage 1. Because pollen grains in deep-sea sediments are far-travelled, it is difficult to relate these assemblages directly to those on adjacent land masses; nevertheless, the boreal forest–tundra ecotone character of the Flitaway interglacial beds suggests considerable antiquity and certainly an age of much more than 125 ka. A late-Pliocene–early-Quaternary age cannot be ruled out, and they may be correlated with the Kap Kobenhavn floras of North Greenland (Funder *et al.* 1984).

The molluscan faunas in the raised marine sediments of eastern Baffin Island have been described qualitatively by Andrews (1972), Andrews *et al.* (1981) and Mode (1985). The northern limit of subarctic species, such as *Mytilus edulis*, occurs about 800 km farther north, in West Greenland, than it does on eastern Baffin Island (figure 2). In addition to species that extend to the limit of the subarctic zoogeographic province there are at least two other species that are currently restricted to the area of south of Hudson Strait; these are the gastropod *Colus spitsbergensis* and the pelecypod *Venericardia borealis*. The latter has only been found (one or two valves only) in some of the basal foreland sediments associated with total aIle: Ile ratios of 0.22.

This places the unit in the *Nonion tallahattensis* zone (figure 7) which may be late Pliocene in age (Feyling-Hanssen 1985; Miller 1985). In a biogeographic sense, the range is similar to that associated with the plant and beetle remains of the Isortoq and Flitaway beds (see above).

(*d*) *Glaciation and oceanography*

The occurrence of subarctic faunas along the Baffin Island coast, north of the present limit, is associated with the extension of the WGC. Conceptually this may be associated with *either* a reduction in strength of the CC or an increase in strength of the WGC, or some combination of both. The palaeoceanographical scenario for the two end-member situations is different, and this difference is critical (see, for example, Andrews *et al.* 1981; Aksu 1981, 1985; Fillon & Aksu 1985; Fillon 1985) to an understanding of the atmosphere–cryosphere–ocean interactions of this region.

Andrews (1972) and Miller (1980) noted that the Holocene raised marine sediments of eastern Baffin Island lack subarctic faunal elements before 8.4 ka BP (9.7 ka BP around outer Frobisher Bay). However, in a very restricted interval, the inshore waters of both Baffin Island and Greenland (Andrews 1972; Hjort & Funder 1974) were invaded by the common blue mussel *M. edulis*, as well as *Mya pseudoarenaria*, *Macoma balthica*, and *Chlamys islandicus*. A similar chronology has been noted in the benthic foraminiferan record in offshore cores (Osterman 1984). *M. edulis* has even been found in the foreset beds of ice-contact glacial marine deltas dated to *ca.* 8 ka BP (Andrews *et al.* 1970). Between *ca.* 5 and 3 ka BP these subarctic elements disappeared from the coast between 66 and 74° N; in the offshore cores, calcareous Foraminifera were replaced by arenaceous forms and (less than 5 ka BP) dissolution of calcium carbonate becomes clear (Osterman 1984). These developments have been associated with either the onset or the strengthening of the Canadian Current, and it is a phenomenon that can be traced for more than 3000 km from the northern Baffin Island shelf, south to the Canadian Maritimes (Scott *et al.* 1984).

The Holocene nearshore marine sequence is thus a cold–warm–cold sequence with the aquatherm being associated with the westward expansion of the WGC. This is associated with an increased advection of warm water in the North Atlantic Drift system. During the same interval, subarctic molluscs expanded along both the East Greenland and Spitsbergen coastlines (Hjort & Funder 1974; Feyling-Hanssen 1955). In the framework of our conceptual stratigraphic model (figure 6), the warm-water influx occurred early in the regressive cycle (that is, during deglaciation) whereas present conditions are marked by cold waters offshore.

Marine conditions during the onset of glaciation have figured prominently in models of glaciation. Two broad and contrasting scenarios have been proposed. In the first, glaciation is associated with relatively warm conditions so as to produce increased accumulation over the surface of the expanding snow–ice surface (Johnson & McClure 1976; Ruddiman & McIntyre 1979; Crowley 1984). In an alternative reconstruction (Denton & Hughes 1983; Denton *et al.* 1986), severe conditions during the onset on glaciation result in the freeze-up of the inner seas and channels of Arctic Canada: in other words, the glacial equilibrium line falls from present values of between 400 and 1200 m (Andrews & Miller 1972) to sea level. This constitutes an extreme form of instantaneous glaciation (Ives *et al.* 1975).

Based on the field collections from the outer coastal forelands of Baffin Island, Andrews (1984) suggested that elements of both models are required to explain the oceanography during ice sheet growth over the northeastern sector of the Laurentide Ice Sheet. In figure 6 a glacial–deglacial cycle is recorded by glacial marine sediments. As noted earlier, these contain marine fossils, preserve terrestrial organic layers (Miller *et al.* 1977; Mode *et al.* 1983) and should preserve the signal of changing nearshore conditions during the transgressive and regressive phases of glaciation. Note that the *major* assumption in this statement is the assertion that the changes in relative sea level along the outer coast are driven by glacial isostasy.

Mode (1985, Figure 17.4) examined the palaeoclimatic indicators in 8 of the 11 aminozones currently delimited on Baffin Island. His survey did not include the poorly studied Loks Land aminozone of middle Wisconsin age, nor the Holocene Eglinton aminozone (figure 7). The assessment of a 'warm' interval is based on the presence of subarctic faunal or floral elements in areas where only arctic taxa occur today. Mode (1985 , p. 515) stated: 'The marine climate was usually warmer than present during the glacial maxima because six of the eight aminozones contain subarctic species *in association with ice-proximal facies*' [italics added]. During the subsequent marine regression, conditions remained warmer than present during four of the eight aminozones. As noted by Mode (1985, p. 515), this was particularly true of the Kogalu aminozone, which contains nine subarctic molluscs in the upper (regressive) units. In addition, terrestrial evidence (*Betula* pollen percentages) also suggest environments warmer than present during this interval.

It is clear from these data that the bulk of fossiliferous raised marine sediments from eastern Baffin Island represent environments characterized by seasonal open-water conditions, terrestrial and marine temperatures warmer than present, and significant glacial meltwater production (formation of large ice-proximal deltas and extensive deposits of silty clays). Enigmatically, evidence for extreme polar climates is lacking. For example, only late Holocene niveo-aeolian sands have been described from the cliff sections. There is virtually no evidence for ice-wedge or sand-wedge casts, although these occur on the modern surface, and no-one has noted ventifacted clasts. It is possible that the vast bulk of such sediments were removed and are represented by the U (unconformities) facies (figure 6).

In the context of these findings, an important question is the age of the various aminozones. Where in the global glacial–interglacial cycles do the aminozones of eastern Baffin Island fall? (figure 9). Do they all represent the same interval within such a hypothetical cycle? Thus the chronology of the aminozones is of vital interest, but, unfortunately, one where appropriate numerical dating methods are not available to provide a solution (see, for example, Szabo *et al.* 1981).

Andrews *et al.* (1985) compared marine and terrestrial events during marine isotope stages 5 and 4 from Baffin Island, West Greenland and Baffin Bay, and concluded that the Kogalu aminozone spans the latter part of stage 5. Thus the development of the last glaciation across northeastern Arctic Canada, the Foxe Glaciation, commenced during this period. However, the Greenland Ice Sheet did not expand and advance onto the shelf until isotopic stage 3 (Kelly 1985).

When using the modern environment as a key to the past, it is important to note that the marine and terrestrial conditions of eastern Baffin Island are controlled by the presence of the Canadian Current. What would result if the channels in the High Canadian Arctic froze over and thickened to form interchannel ice shelves? The Canadian Current would be blocked from Baffin Bay and the oceanographic conditions within the bay would be controlled by the presence and vigour of the West Greenland Current.

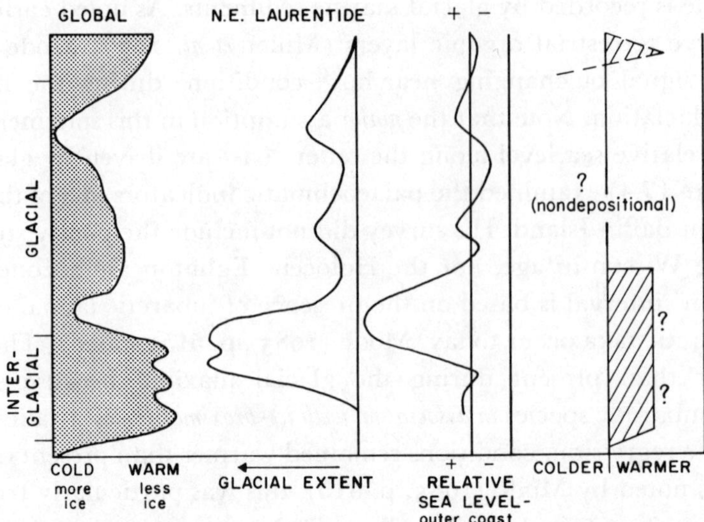

FIGURE 9. Figure 6 represents the glacial–sea-level cycle noted in the forelands (figure 4) whereas this figure portrays the history within a global interglacial–glacial cycle. In a global sense the advance of the northeastern sector on the North American ice sheet may have occurred early in the global glacial cycle. The small advances and sea-level rises associated with advances from a restricted ice sheet may explain low-level sea-level indicators, such as number 2 on figure 4. The (far) right-hand column represents climatic conditions as inferred from the palaeontology of the raised marine units.

CONCLUSIONS

There are significant differences in the interpretation of the late Quaternary palaeoceanography of the northeast and northwest arms of the Atlantic Ocean (Aksu 1985; Aksu & Mudie 1985; Fillon & Duplessy 1982; Fillon 1985; Fillon & Aksu 1985; Kellogg 1975, 1976). It is not clear whether these views represent competing models (i.e. one is right and the other is wrong), or whether both truly represent some version of reality (figure 10). The palaeoceanographic model for the northwest North Atlantic finds support from the inshore palaeontological record (Mode 1985; Andrews *et al.* 1981) and I suggest that glaciation of the northeast sector of the Laurentide Ice Sheet occurred early in a glacial cycle (figure 9). If the information from the sediment sequences, the palaeontology, and associated oceanography is pieced together, it must be concluded that the periods of major global glaciation (e.g. stage 2) are not represented in the Baffin Island raised marine record. This is because extensive regional glaciation occurs at the onset of the cycle, and glacial extent decreases progressively through the remainder of the global glacial cycle (figure 9).

If the relation of regional to global glaciation is as depicted in figure 9, it is tempting to correlate the aminozones with the marine isotopic record. Obviously at this stage, however, and with the dating problems noted earlier, such a correlation is only a hypothesis that requires rigorous testing. Possible tests include the application of new methods, such as thermoluminescence dating, and determining the number of detrital carbonate beds in site 105. On the basis of piston cores, Andrews *et al.* (1985) earlier suggested that the Cape Broughton aminozone is *ca.* 200 ka old.

Kogalu aminozone	stage 4–5
Cape Broughton aminozone	stage 6–7
Kuvinilk aminozone	stage 8–9
Cape Christian aminozone	stage 10–11.

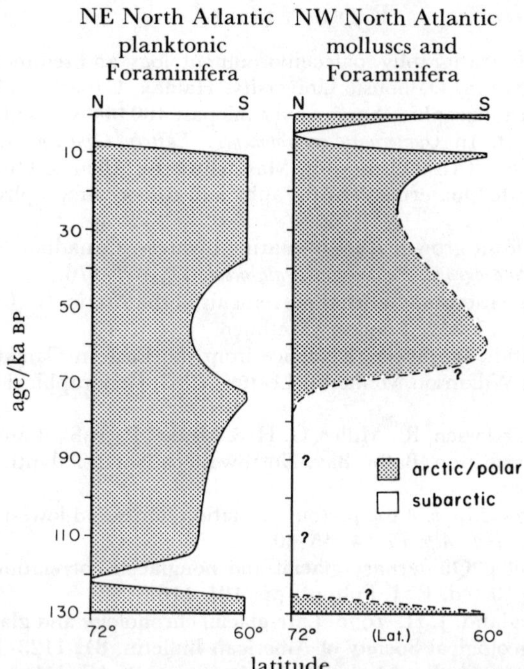

FIGURE 10. Diagrammatic comparison of changes in arctic–subarctic (polar–subpolar?) faunas and floras from the northwestern North Atlantic with those from the Greenland and Norwegian Seas (after Kellogg 1975; Fillon & Aksu 1985; Aksu 1985). Note that these diagrams are based on different biological indicators.

Such a straightforward correlation could be incorrect if there were major changes in the intensity of glaciation in different isotope stages. However, Miller (1985) noted that his age estimates (figure 7) had an 100 ka cyclicity.

The landscape of the eastern Canadian Arctic contains several elements that are rare or unique in heavily glaciated areas. These elements include: the presence of extensive sedimentary forelands that contain packages of Quaternary sediment (figures 4 and 5); the absence of coast-parallel troughs on the shelf; thick sedimentary wedges in the fiords that may pre-date the last glaciation; and the preservation of older non-glacial units on the surface and in valleys of the interior. All these elements together imply that the style of glaciation has been different from that which affected the coasts of West Greenland, Labrador or western Norway.

Marine and terrestrial fossils collected from sequences of raised marine sediments (figures 5–7) normally record intervals when the nearshore environment was warmer than present (i.e. figures 7 and 10). This situation indicates that glaciation of the northeastern sector of the Laurentide Ice Sheet was associated with a reduction in the strength of the Canadian Current and the expansion of the West Greenland Current (figures 9 and 10). Cold, full-glacial global conditions appear to be non-depositional (or erosional) events.

For the past 16 years the research reported in this paper has been supported by the National Science Foundation. I owe a considerable debt of gratitude to the many colleagues quoted in this paper for their contributions to an understanding of the Quaternary history of this pivotal region. In the context of this contribution especial thanks are owed to G. H. Miller, A. R. Nelson, W. N. Mode, A. S. Dyke and J. K. Brigham-Grette.

References

Aksu, A. E. 1981 Late Quaternary stratigraphy, paleoenvironmentology and sedimentation history of Baffin Bay and Davis Strait. Ph.D. dissertation, Dalhousie University, Halifax, Canada. (771 pages.)

Aksu, A. E. 1985 Climatic and oceanographic changes over the past 400,000 years: Evidence from deep-sea cores on Baffin Bay and Davis Strait. In *Quaternary environments: Eastern Canadian Arctic, Baffin Bay and Western Greenland* (ed. J. T. Andrews), pp. 181–209. Boston, Massachusetts: Allen & Unwin.

Aksu, A. E. & Mudie, P. J. 1985 Late Quaternary stratigraphy and paleoceanography of northwest Labrador Sea. *Mar. Micropaleont.* **9**, 537–557.

Andrews, J. T. 1972 Recent and fossil growth rates of marine bivalves, Canadian Arctic, and Late Quaternary arctic marine environments. *Palaeogeogr. Palaeoclim. Palaeoecol.* **11**, 157–176.

Andrews, J. T. 1974 Cainozoic glaciations and crustal movements of the Arctic. In *Arctic and alpine environments* (ed. J. D. Ives & R. G. Barry), pp. 277–317. London: Methuen.

Andrews, J. T. 1984 The Laurentide Ice Sheet: Evidence from the Eastern Canadian Arctic on its geometry, dynamics, and history. Norma Wilkinson Memorial Lecture 1983. Geographical Papers no. 86, University of Reading. (61 pages.)

Andrews, J. T., Aksu, A., Kelly, M., Klassen, R., Miller, G. H & Mudie, P. 1985 Land/Ocean correlations during the Last Interglacial/Glacial transition, Baffin Bay, Northwestern North Atlantic: A review. *Quat. Sci. Rev.* **4**, 333–355.

Andrews, J. T. & Miller, G. H. 1972 Maps of the present glaciation limits and lowest equilibrium line altitude for north and south Baffin Island. *Arct. Alp. Res.* **4**, 45–60.

Andrews, J. T. & Miller, G. H. 1984 Quaternary glacial and nonglacial correlations for the eastern Canadian arctic. *Geol. Surv. Can. Pap.* 84–10 (ed. R. J. Fulton), pp. 101–116.

Andrews, J. T., Buckley, J. T. & England, J. H. 1970 Late-glacial chronology and glacio-isostatic recovery, Home Bay, east Baffin Island. *Bull.* geological Society of American Bulletin, **81**, 1123–1148.

Andrews, J. T., Miller, G. H., Nelson, A. R., Mode, W. N. & Locke, W. W. III 1981 Quaternary near-shore environments on eastern Baffin Island, N.W.T. In *Quaternary Paleoclimates* (ed. W. C. Mahaney), pp. 13–44. Norwich: Geo Books, University of East Anglia.

Barry, R. G., Arundale, W. H., Andrews, J. T., Bradley, R. S. & Nichols, N 1975 Environmental change and cultural change in the eastern Canadian Arctic during the last 5000 years. *Arct. Alp. Res.* **9**, 193–210.

Berggren, W. A. 1972 Cenozoic biostratigraphy and paleobiogeography of the North Atlantic. In *Micropaleontology of oceans* (ed. F. B. Funnell & W. R. Riedel), pp. 105–149. London: Cambridge University Press.

Boulton, G. S., *et al.* 1982 A glacio-isostatic facies model and amino acid stratigraphy for late Quaternary events in Spitsbergen and the Arctic. *Nature, Lond.* **298**, 437–441.

Boulton, G. S., Smith, G. D., Jones, A. S. & Newsome, J. 1985 Glacial geology and glaciology of the last mid-latitude ice sheets. *J. geol. Soc. Lond.* **142**, 447–474.

Brigham, J. K. 1983 Stratigraphy, amino acid geochronology, and correlation of Quaternary sea-level and glacial events, Broughton Island, Arctic Canada. *Can. J. Earth Sci.* **20**, 577–598.

Crowley, T. J. 1984 Atmospheric circulation patterns during glacial inception: A possible candidate. *Quat. Res.* **21**, 105–110.

Denton G. H. & Hughes, T. J. 1983 Milankovitch theory of Ice Ages: Hypothesis of ice-sheet linkage between regional insolation and global climate. *Quat. Res.* **20**, 125–144.

Denton, G. H., Hughes, T. J. & Karlen, W. 1986 Global Ice-Sheet System interlocked by sea level. *Quat. Res.* **26**, 3–26.

Feyling-Hanssen, R. W. 1955 Stratigraphy of the marine late-Pleistocene of Billefjorden, Vestspitsbergen. *Norsk. Polarinst.* Skrifter no. 107. (186 pages.)

Feyling-Hanssen, R. W. 1976*a* The stratigraphy of the Quaternary Clyde Foreland Formation, Baffin Island, illustrated by the distribution of benthic foraminifera. *Boreas* **5**, 77–94.

Feyling-Hanssen, R. W. 1976*b* A Mid-Wisconsin interstadial on Broughton Island, Arctic Canada, and its foraminifera. *Arct. Alp. Res.* **8**, 161–182

Feyling-Hanssen, R. W. 1985 Late Cenozoic marine deposits of east Baffin Island and E. Greenland: Microbiostratigraphy, correlation, age. In *Quaternary environments: Eastern Canadian Arctic, Baffin Bay, and West Greenland* (ed. J. T. Andrews), pp. 354–393. Boston: Allen & Unwin.

Feyling-Hanssen, R. W., Funder, S. & Petersen, K. S. 1983 The Lodin Elbv Formation: A Plio/Pleistocene occurrence in Greenland. *Bull. geol. Soc. Denm.* **31**, 81–106.

Fillon, R. H., 1985 Northwest Labrador Sea stratigraphy, sand input and paleoceanography during the last 160,000 years. In *Late Quaternary environments: Eastern Canadian Arctic, Baffin Bay, and West Greenland* (ed. J. T. Andrews), pp. 210–247. Boston: Allen & Unwin.

Fillon, R. H. & Aksu, A. E. 1985 Evidence for subpolar influence in the Labrador Sea and Baffin Bay during marine isotopic stage 2. In *Late Quaternary environments: Eastern Canadian Arctic, Baffin Bay, and West Greenland* (ed. J. T. Andrews), pp. 248–262. Boston: Allen & Unwin.

Fillon, R. H. & Duplessy, J.-C. 1980 Labrador Sea bio-, tephro- and oxygen isotope stratigraphy and late Quaternary paleoceanographic trends. *Can. J. Earth Sci.* **17**, 831–854.

Funder, S., Abrahamsen, N., Bennike, O. & Feyling-Hanssen, R. W. 1984 Forested Arctic: Evidence from North Greenland. *Geology* **13**, 542–546.

Gibbons, A. B., Megeath, J. D. & Pierce, K. L. 1984 Probability of moraine survival in a succession of glacial advances. *Geology* **12**, 193–200.

Goldthwaite, R. P. 1950 Geomorphology. In *Baffin Island Expedition, 1950,* vol. 3 (*Arctic*) (ed. P. D. Baird *et al.*), pp. 139–141.

Hjort, C. & Funder, S. 1974 The subfossil occurrence of *Mytilus edulis* L in central East Greenland. *Boreas* **3**, 23–33.

Ives, J. D., Andrews, J. T. & Barry, R. G. 1975 Growth and decay of the Laurentide Ice Sheet and comparisons with Fennoscandia. *Naturwissenschaften* **62**, 118–125.

Jacobs, J. D., Andrews, J. T. & Funder, S. 1985. Environmental background. In *Late Quaternary environments: Eastern Canadian Arctic, Baffin Bay, and West Greenland* (ed. J. T. Andrews), pp. 26–68. Boston: Allen & Unwin.

Johnson, R. G. & McClure, B. T. 1976 A model for Northern Hemisphere continental ice sheet variation. *Quat. Res.* **6**, 325–353.

Keen, R. A. 1980 Temperature and circulation anomalies in the eastern Canadian Arctic summer 1946–76. *INSTAAR occasional paper* no. 34 (159 pages). Boulder: University of Colarado.

Kellogg, T. B. 1976 Paleoclimatology and paleo-oceanography of the Norwegian and Greenland seas: Glacial–interglacial contrasts. *Boreas* **9**, 115–137.

Kelly, M. 1985 A review of the Quaternary geology of western Greenland. In *Late Quaternary environments: Eastern Canadian Arctic, Baffin Bay, and West Greenland* (ed. J. T. Andrews), pp. 461–501. Boston: Allen & Unwin.

Loken, O. H. 1966 Baffin Island refugia older than 54,000 years. *Science, Wash.* **153**, 1378–1380.

Lubinsky, I. 1972 The marine bivalve molluscs of the Canadian Arctic. Ph.D. dissertation, McGill University, Montreal. (318 pages.)

Lubinsky, I. 1980 Marine bivalve molluscs of the Canadian central and eastern Arctic: Faunal composition and zoogeography. *Can. Fish. aquat. Sci. Bull.* no. 207. (111 pages.)

McCoy, W. D. 1987 Amino acid geochronology and paleothermometry: An evaluation of accuracy and precision. *Quat. Sci. Rev.* **6**, 43–54.

Meier, M. F. & Post, A. 1987 Fast tidewater glaciers. *Proceedings of Symposium on Fast Glacier Flow. J. geophys. Res.* no. 1, paper 1.

Miller, G. H. 1980 Late Foxe glaciation of southern Baffin Island, N.W.T., Canada. *Bull. geol. Soc. Am.* **91**, 39–405.

Miller, G. H. 1985 Aminostratigraphy of Baffin Island shell-bearing deposits. In *Late Quaternary Environments: Eastern Canadian Arctic, Baffin Bay, and West Greenland* (ed. J. T. Andrews), pp. 394–427. Boston: Allen & Unwin.

Miller, G. H., Andrews, J. T. & Short, S. K. 1977 The last interglacial–glacial cycle, Clyde Foreland, Baffin Island, N.W.T.: Stratigraphy, biostratigraphy and chronology. *Can. J. Earth Sci.* **14**, 2824–2857.

Miller, G. H. & Mangerud, J. 1985 Aminostratigraphy of European interglacial deposits. *Quat. Sci. Rev.* **4**, 215–278.

Mode, W. N. 1980 Quaternary stratigraphy and palynology of the Clyde Foreland, Baffin Island, N.W.T., Canada. Ph.D. dissertation, University of Colorado, Boulder. (219 pages.)

Mode, W. N. 1985 Pre-Holocene pollen and molluscan records from eastern Baffin Island. In *Late Quaternary environments: Eastern Canadian Arctic, Baffin Bay, and West Greenland* (ed. J. T. Andrews), pp. 502–519. Boston: Allen & Unwin.

Mode W. N., Nelson, A. R. & Brigham, J. K. 1983 A facies model of Quaternary glacial–marine cyclic sedimentation along eastern Baffin Island, Canada. In *Glacial-marine sedimentation* (ed. B. F. Molnia), pp. 495–534. New York: Plenum Press.

Morgan, A. V., Kuc, M. & Andrews, J. T. 1988 The beetle and plant macrofossils in the Isortoq and Flitaway sites, north-central Baffin Island. (In preparation.)

Morgan, A., Miller, R. F. & Morgan, A. V. 1982 Fossil coleoptera from arctic sites in Baffin Island and northern Alaska. *11th Annual Arctic Workshop, Boulder, Colorado; abstracts*, p. 79.

Morner, N.-A. 1977 Southward displacement of the distribution of glaciation during the three maxima of the Last Ice Age. *J. Glaciol.* **18**, 305–308.

Mudie, P. J. & Short, S. K. 1985 Marine palynology of Baffin Bay. In *Late Quaternary environments: Eastern Canadian Arctic, Baffin Bay, and West Greenland* (ed. J. T. Andrews), pp. 263–308. Boston: Allen & Unwin.

Nelson, A. R. 1981 Quaternary glacial and marine stratigraphy of the Qivitu Peninsula, northern Cumberland Peninsula, Baffin Island. *Bull. geol. Soc. Am.* **92** (1), 512–518; **92** (2), 1143–1261.

Nelson, A. R. 1982 Aminostratigraphy of Quaternmary marine and glaciomarine sediments, Qivitu Peninsula, Baffin Island. *Can. J. Earth Sci.* **19**, 945–961.

Osterman, L. E. 1984 Benthic foraminiferal zonation of a glacial/interglacial transition from Frobisher Bay, Baffin Island, Northwest Territories. *Benthos'83: 2nd International Symposium on Benthic Foraminifera (Pau, April 1983)*, pp. 471–476.

Poore, R. Z. & Berggren, W. A. 1974 Pliocene biostratigraphy of the Labrador Sea calcareous plankton. *J. foram. Res.* **4**, 91–108.

Ruddiman, W. F. & McIntyre, A. 1979 Warmth of the subpolar North Atlantic Ocean during Northern Hemisphere Ice Sheet growth. *Science, Wash.* **204**, 173–175.
Scott, D. B., Mudie, P. J., Vilks, G. & Younger, C. 1984 Latest Pleistocene–Holocene paleoceanographic trends on the continental margin of eastern Canada: Foraminiferal, dinoflagellate and pollen evidence. *Mar. Micropaleont.* **9**, 181–218.
Shipboard party 1986 End of spreading and glacial onset dated. *Geotimes* (Leg 105 Scientific Party) **31**, 11–14.
Short, S. K., Mode, W. N. & Davis, P. T. 1985 The Holocene record from Baffin Island: Holocene and fossil pollen studies. In *Late Quaternary environments: Eastern Canadian Arctic, Baffin Bay, and Western Greenland* (ed. J. T. Andrews), pp. 608–642. Boston: Allen & Unwin.
Srivastava, S. P., Arthur, M. A. & shipboard party 1986 Drilling results of Leg 105 of ODP in the Labrador Sea and Baffin Bay. Geological Association of Canada, Abstracts with program, vol. 11, p. 130.
Szabo, B. J., Miller, G. H., Andrews, J. T. & Stuiver, M. 1981 Comparison of uranium-series, radiocarbon, and amino acid data from marine molluscs, Baffin Island, Arctic Canada. *Geology* **9**, 451–457.
Tarr, R. S. 1897 Difference in the climate of the Greenland and American sides of Davis' and Baffin's Bay. *Am. J. Sci.*, ser. 4, **3**, 315–320.
Terasmae, J., Webber, P. J. & Andrews, J. T. 1966 A study of late Quaternary plant bearing beds in north-central Baffin Island, Canada. *Arctic* **19**, 296–318.
Thunell, R. C. & Belyea, P. R. 1981 Neogene planktonic foraminiferal biogeography of the Atlantic Ocean: A synthesis of DSDP legs 1–53. Geological Society of America, Abstracts with program, vol. 13, p. 567.
Vernal, de A., Mudie, P. F., Hillaire-Marcel, C. & leg 105 onboard scientists 1986 Plio-Pleistocene palynostratigraphy of ODP-Site 645, Baffin Bay: Preliminary Results. Geological Association of Canada, Abstracts with program, vol. 11, p. 63.
Williams, L. D. & Bradley, R. S. 1985 Paleoclimatology of the Baffin Bay region. In *Late Quaternary environments: Eastern Canadian Arctic, Baffin Bay, and West Greenland* (ed. J. T. Andrews), pp. 741–772. Boston: Allen & Unwin.

Discussion

H. Osmaston (*Department of Geography, University of Bristol, U.K.*). We enjoyed Professor Andrews' alternative anthropomorphic explanation of the increasing depauperization of the sub-Arctic fauna and flora in successive interglacials (that the species became increasingly browned off with the effort of returning after successive glaciations). However, would he be happy if I paraphrased it thus: that deterministic models, which predict equal returns of species after equal changes of climate, should be replaced by stochastic models, which embody for example the chance elimination of species in biological refugia?

J. T. Andrews. In my view, Dr Osmaston's assessment of the situation is correct, or at the very least is a viable alternative to the strictly deterministic association of plants and climate. In the specific case of the Baffin Island data we have to remember that migration of plants to this 500000 km^2 island involves migration across substantial water (in summer) barriers.

R. Marris (12 *East Terrace, Budleigh Salterton, Devon, U.K.*). Is it not better to regard the site in Northern Greenland as in a part of a stage in the Beaufort series of deposits until proved otherwise? The nearest on Ellesmere Island are at Yelverton Inlet and south of Alert.

J. T. Andrews. Yes, it is possible that the northern Greenland flora at Kapp Kobenhavn is an eastern outlier of the Beaufort Formation of Arctic Canada. However, the suggested age is considerably younger than the accepted Miocene age for the Beaufort. There are forest elements on Banks Island which are younger than the Beaufort and the North Greenland deposits may be correlative with those units.

Phil. Trans. R. Soc. Lond. B **318**, 661–678 (1988)
Printed in Great Britain

Evolution of marine climates of the U.S. Atlantic coast during the past four million years

By T. M. Cronin

U.S. Geological Survey, 970 *National Center, Reston, Virginia* 22092, *U.S.A.*

Marine climatic and sea-level changes in the eastern United States show two distinct modes: a gradual, directional Pliocene warming that ended with an abrupt regression, and a quasi-cyclic, high-amplitude, high-frequency middle–late Pleistocene pattern of alternating glacials and interglacials. Pliocene marine sediments of the Duplin Formation, deposited during a period of high sea level between 4.0 and 2.8 Ma BP, contain increasing percentages of tropical and subtropical ostracods, signifying a gradual warming. After maximum warm-water temperatures *ca.* 3.2–2.8 Ma BP, sea level dropped; this was followed by extensive subaerial erosion between about 2.8 and 2.0 Ma BP. This series of events reflects the emergence of the Isthmus of Panama between about 3.5 and 3.0 Ma BP, concomitant intensification of warm Gulf Stream flow along the eastern U.S.A., and initial Pliocene glaciation in the Northern Hemisphere.

In the middle–late Pleistocene, glacial–interglacial cycles occurred with a periodicity of *ca.* 100 ka. Four (possibly five) emerged interglacial marine sequences correlate with deep-sea oxygen-isotope stages 13/11, 7, 5, and 1. During some interglacials, however, climatic conditions ranged from full interglacial warmth to cool, nearly interstadial conditions; this observation indicates short-term regional climatic variability.

Introduction

The Pliocene and Pleistocene marine record of the Atlantic Coast of the United States and the adjacent continental shelf and slope is a sensitive barometer of North Atlantic oceanographic and climatic change. This paper traces the evolution of climatic variability of this region during the past four million years (Ma) and correlates it with climatic events elsewhere. A review of the stratigraphic and climatic literature on the eastern U.S.A. is beyond the scope of this paper. Consequently, with the stratigraphic record as a framework, this paper focuses on the marine palaeontological record, emphasizing temporal changes in the proportions of cryophilic and thermophilic ostracods from key stratigraphic units. Because ostracod faunal provinces on the U.S.A.'s Atlantic continental shelf parallel the distribution of modern marine climatic zones (Hazel 1970), northward and southward range expansion and contraction of temperature-sensitive species reflect oceanographic changes which themselves often signify climatic events. Each formation discussed here exhibits an important climatic and/or sea-level event covering the full range of climatic extremes: glacial, interglacial and transitional. Collectively they form a climatic framework for the past 4 Ma.

Regional stratigraphy and marine history

The emerged Coastal Plain of Delaware, Maryland, Virginia, North Carolina and South Carolina is the primary study region (figure 1). In addition, outcrops from Nantucket Island, Massachusetts, and coastal Maine, and two cores from the outer continental shelf and upper

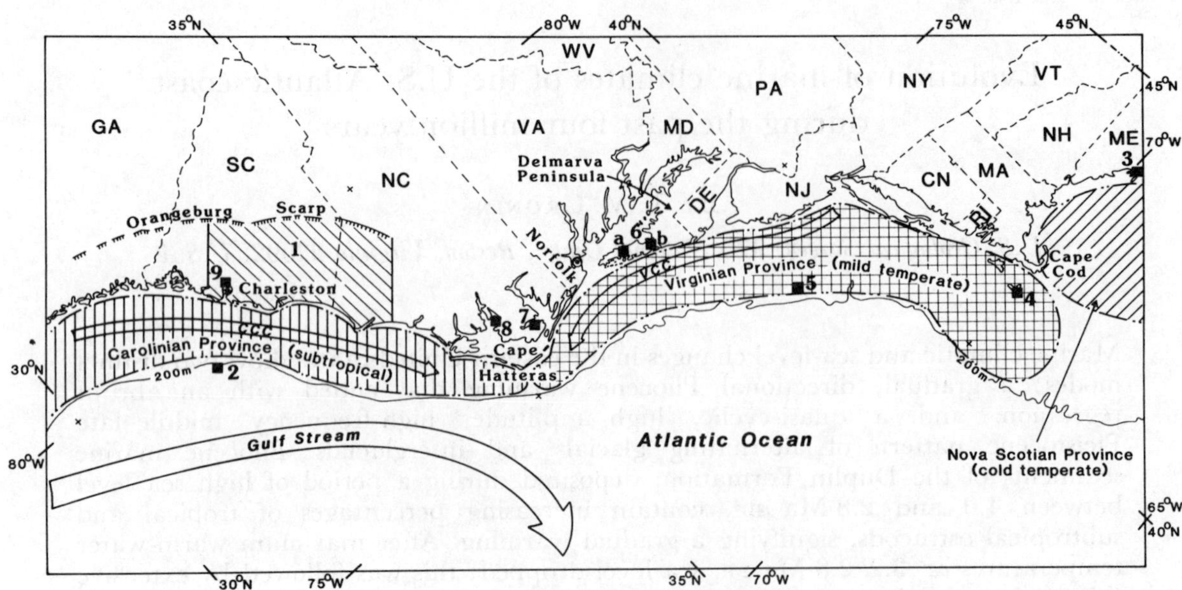

FIGURE 1. Map showing stratigraphic sequences, physiographic features, faunal provinces and associated climatic zones in parentheses. Locality: 1, general location of sections A–L of the Pliocene Duplin Formation (stippled area; discussed in detail by Cronin *et al.* (1984)); 2, AMCOR 6004; 3, Presumpscot Formation; 4, Sankaty Head, Nantucket, Massachusetts; 5, AMCOR 6010; 6, Delmarva Peninsula (a, Wachapreague Formation; b, Omar Formation); 7, Stetson core; 8, Ponzer; 9, Charleston area (a, Wando Formation; b, Ten Mile Hill beds). VCC, Virginia Coastal Current; CCC, Carolina Coastal Current.

slope off New Jersey and South Carolina (AMCOR 6010 and 6004 (Hathaway *et al.* 1979)), are included (figure 1). Table 1 lists the formations and stratigraphic sequences discussed below and gives references describing the stratigraphy and age of each.

Transgressive–regressive marine cycles, mainly reflecting changes in sea level, dominate the Plio-Pleistocene geological history of the Coastal Plain. During periods of high sea level, marine and marginal marine sediments were deposited; the onshore limit of each trangression is marked by an erosional scarp, barrier island and/or lagoonal–tidal marsh sediment (Colquhoun 1974). The local sea-level record is primarily glacio-eustatic (Cronin 1981 *a*); with important exceptions (discussed below), marine deposits contain faunas indicating interglacial climates.

Periods of low sea level account for much of the past 4 Ma and can be identified in the Coastal Plain by lithological disconformities, signifying non-deposition and fluvial erosion. This incomplete stratigraphic record poses an obvious limitation to quantifying climatic variability. None the less, many parts of the Coastal Plain have been intensely studied and mapped and provide a climatic record for particular intervals. Further, glacial oceanography can be pieced together by integrating the emerged postglacial record from northeastern North America with the record from cores and dredged samples from the outer continental shelf and upper slope, which represent times of low sea level during glacial periods.

On the basis of lithology alone, it is often impossible to correlate marine units over broad areas, owing to limited exposures and complex lateral facies changes. Sometimes it also is difficult to determine whether the apparent diachronous relations between the marine deposits of two regions reflects local tectonic subsidence or uplift, erosion and channelling, or errors in

TABLE 1. SUMMARY OF PLIOCENE AND PLEISTOCENE STRATIGRAPHY

(Figure 1 shows location of sequences 2–9; sequence 1 is a composite of 12 sections located in hatched area in figure 1; figure 3 shows climatic data from sequences 3–9.)

no.	stratigraphic sequence	formation	age	reference
1 A–L	Pliocene composite	Duplin	early–middle Pliocene	Cronin *et al.* (1984), Owens (1987)
2	South Carolina continental slope	none	late Pliocene, Pleistocene	Hathaway *et al.* (1979)
3 a–c	coastal Maine	Presumpscot	late Wisconsinan	Stuiver & Borns (1975), Bloom 1960 Cronin (1988)
4	Nantucket Island	Sankaty Sand	late Pleist.	Oldale *et al.* (1982)
5	New Jersey continental shelf	none	late Pleistocene	Hathaway *et al.* (1979)
6a, b	Delmarva Peninsula	a. Wachapreague	late Pleist.	Mixon (1985)
		b. Omar	mid. Pleist.	Mixon *et al.* (1982)
7	Stetson core, North Carolina	Core Creek sand	late Pleist.	Mixon & Pilkey (1976), York & Wehmiller (1987)
		Flanner Beach	mid. Pleist.	Miller (1985)
8	Ponzer, North Carolina	Flanner Beach	middle Pleistocene	Cronin *et al.* (1981), Miller (1985)
9	Charleston, South Carolina	a. Wando	late Pleist.	McCartan *et al.* (1984)
		b. Ten Mile Hill beds	mid. Pleist.	Weems & Lemon (1984), Szabo (1985)

dating techniques. Consequently, biostratigraphic, radiometric and palaeomagnetic data have played an important role in regional correlation of the formations discussed here (Hazel 1983; Cronin *et al.* 1984).

Ostracod zoogeography and climatic change

Marine Ostracoda – small, bivalved Crustacea common in marine and marginal marine sediments – have long been used to reconstruct palaeoceanographic conditions of eastern North America (Brady & Crosskey 1871), although much more so in the past 20 years (Hazel 1968; Cronin 1981*b*, 1988). The modern geographic distributions of ostracod species on the Atlantic continental shelf between Maine and Florida is known from about 700 modern bottom samples (Hazel 1970, 1975; Valentine 1971; Cronin 1983; Lyon 1988). These and several hundred additional Neogene and Quaternary samples described in papers mentioned below constitute the zoogeographic, taxonomic and biostratigraphic *terra firma* for this paper. Six ostracod faunal provinces, delineated qualitatively (Hazel 1970) and quantitatively (Valentine 1971; Hazel 1975), are shown in table 2 with associated climatic zones, geographic limits and bottom-water temperatures.

Boundaries between faunal provinces are regions of strong thermal gradients that limit the poleward or equatorward expansion of species. For example, at Cape Hatteras, North Carolina, the confluence of the warm, northeastward-flowing Carolina Coastal Current (figure 1) and the cool, southward-flowing Virginia Coastal Current causes cryophilic (cold-water) and thermophilic (warm-water) ostracod species to have Cape Hatteras as their southern or northern boundary. One hundred and four species, each of which is restricted to one or several

TABLE 2. OSTRACOD PROVINCES AND ASSOCIATED CLIMATIC ZONES

faunal province[a]	climatic zone	geographic range	temperature ranges (°C)
Arctic	frigid	circumpolar, Greenland	−2–4
Labradorian	subfrigid	S. Greenland–Gulf of St. Lawrence	−2–12
Nova Scotian	cold temperate	Gulf of St. Lawrence–Cape Cod	2–20
Virginian	mild temperate	Cape Cod–Cape Hatteras	3–26
Virginian–Carolinian[b]	warm temperate	Cape Cod–Cape Canaveral	12–26
Carolinian	subtropical	Cape Hatteras–Cape Canaveral	15–28
Caribbean	tropical	Caribbean, south Florida, Bahamas	20–30

[a]Hazel (1970) described ostracod provinces along the eastern United States.
[b]The warm-temperate zone does not exist today, owing to strong isothermal convergence at Cape Hatteras, but is recognized in the late Pleistocene (Valentine 1971; Cronin 1979). Annual temperature ranges are approximate; see text for references.

climatic zones, were divided into six groupings of species with similar temperature tolerances. From coldest to warmest these are: frigid zone only, subfrigid–cold-temperate, cold-temperate–mild-temperate, mild-temperate–subtropical, subtropical only, and tropical (Appendix 1). Changes in the proportions of each species-group in fossil assemblages were used as a measure of change in water temperature and regional climate.

The use of modern zoogeography to estimate species' temperature tolerances and nearshore oceanographic conditions is analogous to the use of planktonic Foraminifera from deep-sea core tops to infer surface-water conditions and climatic patterns from deep-sea sediments (Imbrie & Kipp 1971). In each case it must be assumed that, in lieu of direct experimental evidence, species' ecological limits can be reasonably estimated from their association in Holocene sediments (and plankton tows, for planktonic foraminiferans) with water temperatures and other physical and chemical parameters. Unlike planktonic foraminiferans with global distributions, most ostracods used here (except circumpolar Arctic species) are endemic to the western Atlantic, the Gulf of Mexico and the Caribbean and can only be applied to a regional oceanography.

In addition to latitudinal zonation, ostracods also show a bathymetric gradient from coastal to deep-sea environments (Cronin 1979, 1983); the percentages of species characteristic of inner-shelf, outer-shelf and upper-slope environments in Pleistocene sediments are also used to identify bathymetric changes related to eustatic sea-level fluctuations.

EARLY PLIOCENE CLIMATIC WARMING

During an interval of high sea level in the early and middle Pliocene, the Duplin Formation of South Carolina and southern North Carolina and the Yorktown Formation of Virginia and northern North Carolina were deposited. During this period, the Orangeburg Scarp, the highest, most conspicuous geomorphic shoreline feature in the Coastal Plain (figure 1), was occupied for the final time (Oaks & DuBar 1974; Colquohoun 1986). In some areas this scarp is 80 m above sea level at its crest, but it appears to be tectonically warped (Winker & Howard 1977) so it is not clear to what extent it represents a eustatic sea-level position (Cronin 1981*b*).

Thirty-six samples of the Duplin Formation (from 12 outcrops and cores within the stippled region in figure 1), when placed in the stratigraphic framework of DuBar *et al.* (1974) and Owens (1988), provide a representative Pliocene palaeoclimatic record. The Duplin Formation shows no obvious breaks in deposition and major hiatuses can be excluded.

The Duplin Formation contains the upper part of planktonic foraminiferan zone N19, zone N20, and perhaps the lowermost part of zone N21 (figure 2) (Cronin *et al.* 1984). Its lower part contains *Globorotalia puncticulata*, *Globigerina nepenthes*, *Neogloboquadrina acostaensis*, *Globoquadrina altispira*, and *Sphaeroidinellopsis* spp., an assemblage that restricts the age to between about 4.8 and 3.7 Ma. In South Carolina, the nannofossil *Pseudoemiliania lacunosa* is absent in the lower Duplin but appears in the upper Duplin (figure 2). Based on these data and that available from the correlative Yorktown Formation, the base of the sequence in figure 2 is estimated at approximately 4.0 Ma BP. In the middle part of the Duplin Formation, *Globorotalia puncticulata*, *Globoquadrina altispira*, *N. acostaensis*, *Globigerina apertura*, *Sphaeroidinellopsis* spp. and *Globigerinoides obliquus* occur; *Globigerina nepenthes* is absent. In the upper Duplin, *P. lacunosa* occurs with *Globorotalia inflata* and *Globigerina woodi*; these species indicate a maximum age of *ca.* 3.4 Ma and a minimum of 2.8 Ma, but perhaps as young as 2.5 Ma. The uppermost samples (figure 2, sections A, B, C) contain no diagnostic species and are probably only slightly younger than 2.8 Ma. The Yorktown Formation in Virginia and North Carolina contains a similar climatic

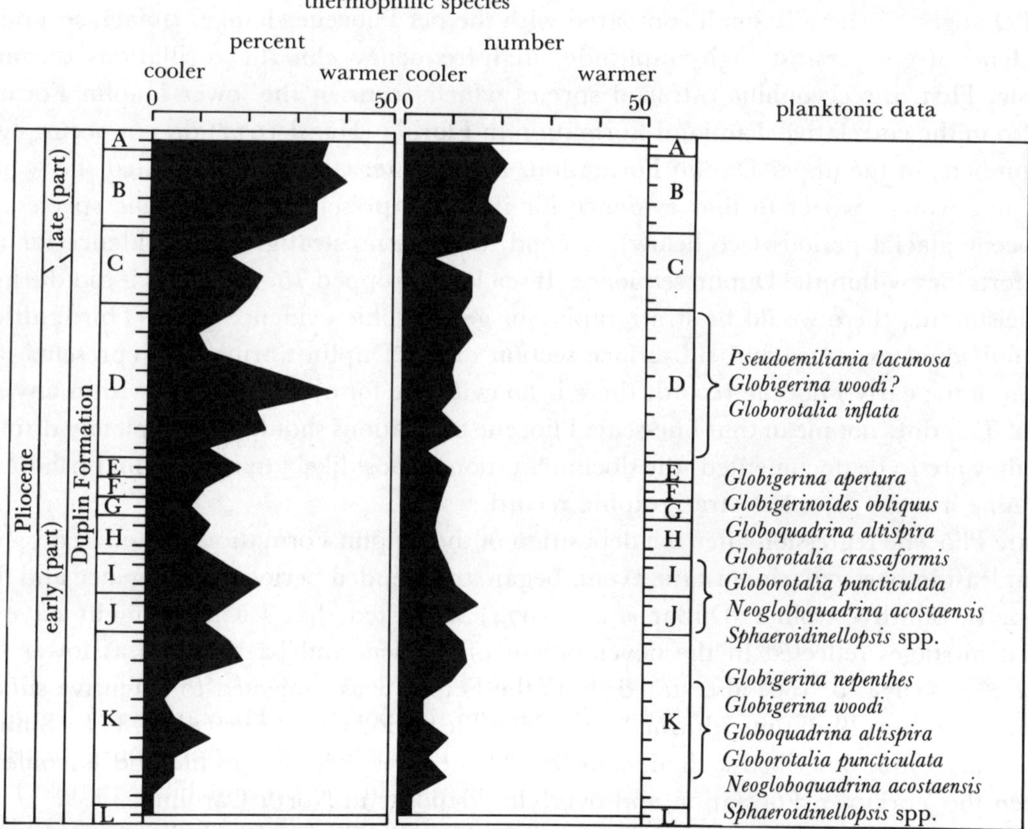

FIGURE 2. Composite sequence of the Pliocene Duplin Formation in North and South Carolina showing percentages and numbers of species of tropical ostracods. Age-diagnostic foraminifers and nannofossils shown on the right; sections A–L shown on the left; tick marks denote samples.

record to that of the Duplin and is about the same age (Akers & Koeppel 1973; Snyder *et al.* 1983; Hazel 1983; Gibson 1983).

The most significant climatic change during the interval 4.0–2.8 Ma BP is a general warming from temperate to near-tropical conditions. Figure 2 shows the numbers of extant subtropical and tropical ostracods in the Duplin sequence and their percentages out of the total assemblage in each sample. From three species in the lower Duplin, where many cryophilic temperate species occur in abundance, the number of thermophilic species increases to between 17 and 21 in the upper Duplin; this figure represents between 35 and 40% of assemblages. Among the abundant thermophilic species are *Radimella confragosa* (Edwards, 1944) and *Orionina vaughani* (Ulrich & Bassler, 1904), two species very common in the Caribbean. Species not accounted for in the 35–40% include subtropical species that live off the southeastern U.S.A. today and extinct species that belong to subtropical–tropical genera. The increase in thermophilic species in the upper Duplin Formation is accompanied by decreasing numbers of cryophilic species. Tropical molluscs also dominate the upper part of the Duplin, and water temperatures were warmer than during any middle or late Pleistocene interglacial. It is noteworthy that a parallel pattern of climatic warming was documented by Hazel (1971) for the correlative Yorktown Formation of Virginia and northern North Carolina.

Some short-term environmental changes are suggested by the staggered pattern in figure 2; however, without a single continuous section and known sedimentation rates, it is difficult to assess them. Nonetheless, with the exception of the spike in section D, the magnitude of these faunal changes is generally small compared with the net Pliocene change. In fact, several lines of evidence argue against high-amplitude, high-frequency climatic oscillations during the Pliocene. First, the cryophilic ostracod species which occur in the lower Duplin Formation and also in the correlative Tamiami Formation in Florida (Hazel 1977) do not occur, even in low numbers, in the upper Duplin Formation. If cool intervals had punctuated the warming trend, one would expect to find evidence for it in the presence of cryophilic species, as in Pleistocene glacial periods (see below). Second, there is no stratigraphic evidence for major unconformities within the Duplin sequence. If sea level dropped 75–100 m, as it did during the late Pleistocene, there would be stratigraphic or geomorphic evidence for it. Third, although each individual exposure and subsurface section of the Duplin Formation represents only a fraction of the early Pliocene record, there is no evidence for cyclic patterns within any single section. This does not mean that fine-scale Pliocene oscillations should be completely dismissed, but if they are to be documented, the documentation is most likely to come from offshore cores containing a more complete stratigraphic record.

A late Pliocene regression after the deposition of the Duplin Formation, considered by many to be at least in part a glacio-eustatic event, began an extended period of emergence and fluvial erosion. In South Carolina, DuBar *et al.* (1974) suggested that sea level might have been lowered in stages reflected in the development of barriers and barrier-flats at lower levels. DuBar *et al.* (1974, p. 156) also stated that 'the Duplin was subjected to extensive subaerial erosion in pre-Bear Bluff time' during 'a considerable period of emergence' (p. 171). Similarly, Riggs *et al.* (1982) described a significant late Pliocene weathering profile and unconformity between the Yorktown Formation and overlying deposits in North Carolina.

There is a remarkable correspondance between Pliocene Coastal Plain chronology and land and sea events in the Caribbean–Central-American region and the northern North Atlantic. The emergence of the Isthmus of Panama first blocked deep-water circulation *ca.* 3.6 Ma BP and then shallow-water circulation (Keigwin 1978), eventually isolating the Caribbean and Pacific by about 3.0 Ma BP (Jones & Hasson 1985). Increased erosion and winnowing during the early Pliocene in the Straits of Florida directly reflects enhanced flow of the Florida Current resulting from the Isthmus' closing (Brunner 1979). The age equivalence of these events and the northward migration of tropical species during the deposition of the Duplin suggests a causal relation between Coastal Plain warming and the emergence of the Isthmus. Further, evidence from deep-sea cores in the North Atlantic led Berggren (1972) and Shor & Poore (1979) to suggest a link between the closure of the Isthmus, intensification of the Gulf Stream, and Pliocene Northern Hemisphere glaciation. Shackleton *et al.* (1984) studied DSDP site 552a and determined that a North Atlantic bottom-water mass similar to that of today began by at least 3.5 Ma BP, and that major ice rafting began 2.5 Ma BP, intensifying in scale about 2.4 Ma BP. The early–middle Pliocene Coastal Plain warming and subsequent extended regression correlate well with the deep-sea chronology. Indeed, the concordance of tectonic and oceanographic events in Central America, the Atlantic Coastal Plain and the North Atlantic all seem to reflect a major oceanographic and climatic change signalling the onset of Pleistocene-type climatic cycles.

Pleistocene climatic oscillations

The middle–late Pleistocene record of the Coastal Plain consists of four, possibly five, major interglacial periods during which relatively high sea level deposited marine sediments containing climatically diagnostic species. The Canepatch Formation unconformably overlies the early Pleistocene Waccamaw in South Carolina (DuBar *et al.* 1984) and contains corals that have been dated by the uranium-series method at about 460 ± 100 ka BP (Szabo 1985). A middle-Pleistocene age for this formation is supported by its normal magnetic signal, biostratigraphic data, and its stratigraphic position (Cronin *et al.* 1984). The Canepatch probably corresponds with oxygen-isotope stage 13 or 11. The Bermont Formation of Florida and unnamed deposits in North Carolina are possible correlatives of the Canepatch but these have not been studied in detail.

Overlying the Canepatch in northeastern South Carolina, the Socastee Formation is a candidate as a correlative with isotope stage 9, but its age is unknown and it requires further study (McCartan *et al.* 1982).

A more detailed climatic record is available for the last two interglacial periods, corresponding with isotope stages 7 and 5, and the last glacial period, isotope stage 2. Figure 3 summarizes these results in a composite for the region; age data are used to place each section in the context of the deep-sea oxygen-isotope stratigraphy. Figures 4 (*a–d*), 5 (*a*, *b*), and 6 plot diagnostic ostracod percentages for six key stratigraphic sequences. Each depicts a particular type of climatic situation: interglacial, glacial and/or transitional periods.

Interglacial climates

The penultimate interglacial is represented by the Omar Formation of the Delmarva Peninsula (Mixon 1985), the Flanner Beach Formation, North Carolina (Cronin *et al.* 1981; Miller 1985) and the Ten Mile Hill beds described by Weems & Lemon (1984) in South Carolina (McCartan *et al.* 1982; Corrado *et al.* 1986). Uranium-series ages on solitary corals (Szabo 1985) and amino-acid epimerization ages on molluscs (Wehmiller & Belknap 1982;

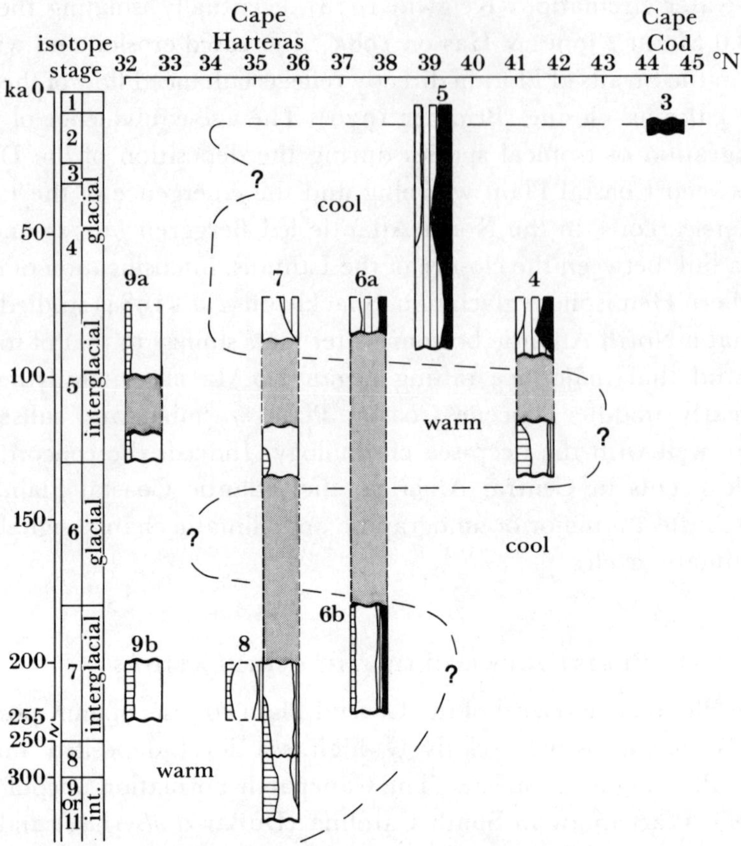

FIGURE 3. Pleistocene composite for eight sequences showing relations between latitude, inferred climate and isotope stages. Stippled portions indicate that no record was available. (Locality numbers as for figure 1; see figure 4 for explanation of temperature patterns.)

Corrado *et al.* 1986) indicate that these deposits are about 220–200 ka old and correlate with oxygen-isotope stage 7 in deep-sea cores (Shackleton & Opdyke 1973).

Figure 4*a* shows a high percentage of warm-temperate species and lower percentages of mild- and cold-temperate ostracod species in the Accomack Member of the Omar Formation. Cold-temperate species do not occur in corresponding deposits of the Ponzer and Stetson sections (figure 4*b*, *c*) and a faunal provincial boundary related to water temperature during isotope stage 7 was probably located near southeastern Virginia. Near Charleston, South Carolina, only subtropical and warm-temperate species occur in the Ten Mile Hill beds (figure 3, section 9b); no species living exclusively north of Cape Hatteras were found. The interglacial period *ca.* 200 ka BP was thus characterized by slightly cooler water than today in the Delmarva region and perhaps a steeper thermal gradient between Virginia and South Carolina.

The last interglacial period, which corresponds to isotope stage 5, shows a complex oceanography along the eastern U.S.A., including intervals of warm water conditions and cooler climatic intervals (see below). Peak warmth during the last interglacial, corresponding with isotope substage 5e, is illustrated in the Sankaty Head section, Massachusetts (figure 5*b*), where the ostracod assemblage from the lower Sankaty Sand consists primarily of subtropical

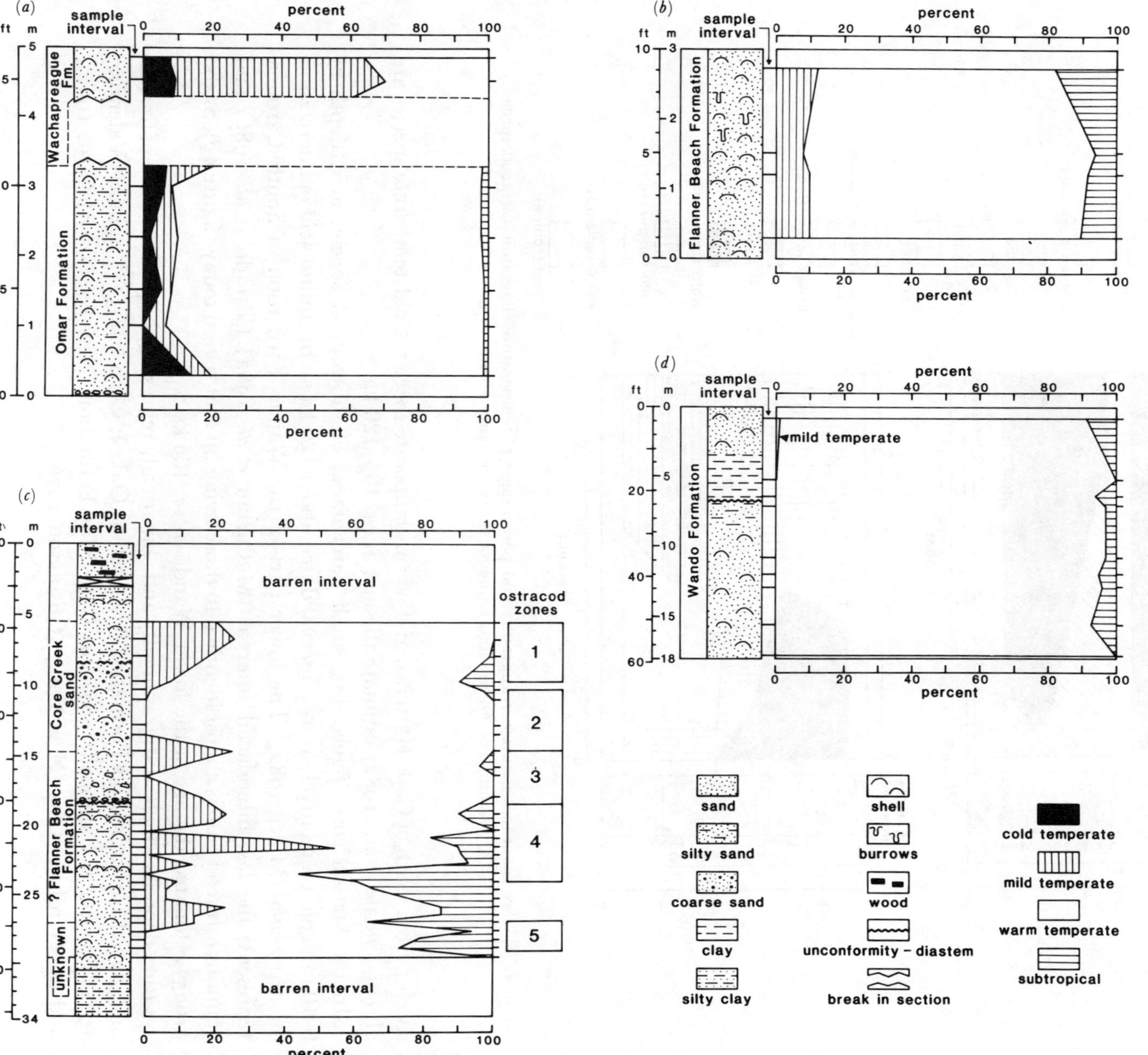

FIGURE 4. Four sequences showing stratigraphy and percentages of temperature-diagnostic ostracod species. (*a*) Delmarva Peninsula, sequence 6; (*b*) Ponzer, sequence 8; (*c*) Stetson core, sequence 7; (*d*) Charleston area, sequence 9a.

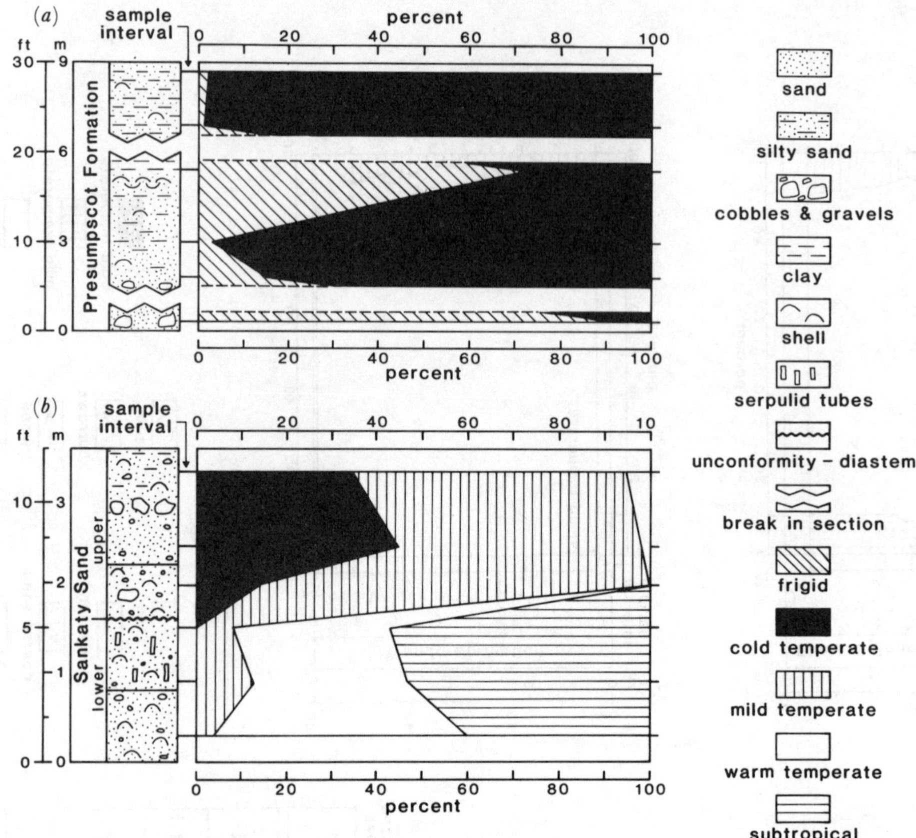

FIGURE 5. Two sequences showing stratigraphy and percentages of temperature-diagnostic ostracod species. (*a*) Presumpscot Formation, sequence 3; (*b*) Sankaty Head, sequence 4.

species, from south of Cape Hatteras, and warm-temperate species. Cold-temperate species are absent. Oldale *et al.* (1982) estimate the age at *ca.* 135–120 ka.

In the Stetson core (figure 4*c*), small percentages of subtropical species at a depth of 9–10 m suggest a relatively warm interval dated about 120 ka BP by amino acid epimerization on *Mercenaria* (Miller 1982). The lower part of the Wando Formation of South Carolina represents the last interglacial interval (McCartan *et al.* 1982; Corrado *et al.* 1986) and contains ostracod faunas almost identical to those living off that region today (figure 4*d*). Szabo (1985) estimated the age of the lower Wando at *ca.* 125 ka.

Other deposits which stratigraphically and climatically represent the peak warmth of the last interglacial are the Norfolk Formation described by Oaks & Coch (1963) (Virginia) (Valentine 1971; Spencer & Campbell 1987), the Cape May Formation (New Jersey) (Richards 1962), and the Gardiners Clay (New York) (Gustavson 1976).

GLACIAL CLIMATES

Oceanography along the eastern United States during the last glacial period was pieced together from several sources of information (figure 7). Ostracods from radiocarbon-dated postglacial deposits of the northeast document frigid–subfrigid marine conditions as far south as coastal Massachusetts and Maine between about 14.5 and 12.5 ka BP (Cronin 1981*b*,

1988). The subfrigid–cold temperature climatic boundary, now at about 48° N, near Nova Scotia, was therefore at least 5–6° farther south during the late Wisconsinan age. As the Laurentide ice sheet retreated from coastal Maine, frigid ostracod faunas of the Presumpscot Formation described by Bloom (1960) were replaced *ca.* 12.0–11.6 ka BP with the cold-temperate species that characterize the region today (figure 5*a*). A similar transition at 11–10 ka BP occurs in Champlain Sea deposits as the ice sheet retreated from New York, Vermont and southern Quebec (Cronin 1981*b*).

Evidence for Wisconsinan palaeoclimates also comes from sediments deposited on the outer shelf and upper slope during low stands of sea level, which along the eastern U.S. were about −100 m (Cronin 1987). In AMCOR 6010 (figure 6), the first appearance of *Emiliania huxleyi*

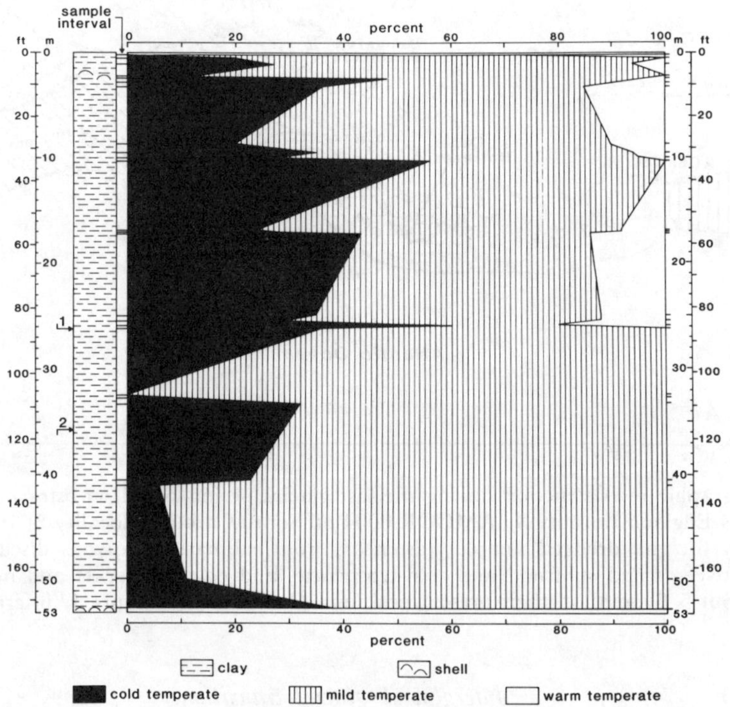

FIGURE 6. Stratigraphic section of AMCOR 6010 (sequence 5) illustrating percentages of temperature-diagnostic ostracod species. Data points on the left of the stratigraphic column are: 1, FAD of *Emiliania huxleyi*; 2, LAD of *Pseudoemiliania lacunosa.*

occurs at a core depth of *ca.* 86 m. If this is its acme zone, 75–85 ka BP (Berggren *et al.* 1980), it suggests a Wisconsinan age for the upper part of the core, where cold-temperate species account for as much as 40–60 % of some assemblages.

Hazel (1968) described glacial ostracods in rock samples dredged from ten submarine canyons between Maine and Virginia and provides convincing evidence that during glacial periods, cold temperate ostracods migrated far south of their present range. Hazel estimated the boundary between the cold-temperate and mild-temperate provinces, which now stands at about 42° N, to be about 37.5° N during the late Pleistocene. Ostracods characteristic of modern cold- and mild-temperate zones have yet to be found in Pleistocene deposits south of Cape Hatteras; there is no direct evidence off the southeastern U.S.A. for late Pleistocene marine climates that were significantly colder than today (although Wisconsinan continental climates in the southeast were colder (Watts & Stuiver 1980).

In summary, glacial climates were characterized by a southward shift of frigid, subfrigid and cold-temperate climatic zones of about 5–6° latitude, presumably owing to the proximity of the Laurentide Ice Sheet and cold-water currents off the northeastern U.S.A. Subtropical zones apparently remained warm, under continued influence of the Gulf Stream, and were not shifted southward as were temperate zones.

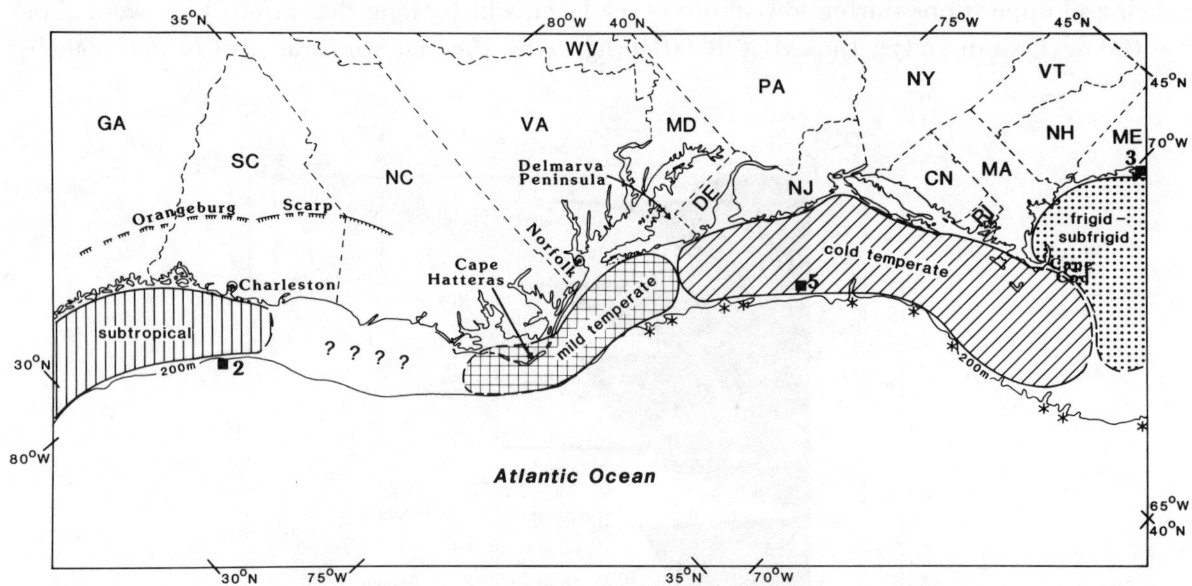

FIGURE 7. Oceanographic reconstruction during the last glacial period based on ostracod data from postglacial deposits in New England (locality 3), AMCOR 6004 off South Carolina (locality 2) and AMCOR 6010, New Jersey (locality 6), and dredged samples (asterisks) from submarine canyons discussed by Hazel (1968). Boundaries between frigid–subfrigid and cold-temperate, and cold-temperate and mild-temperate, climatic zones were about 5–6° south of their present positions. Compare with figure 1 (interglacial oceanography).

Interglacial–glacial transition

Abundant lithological, radiometric and palaeontological evidence indicates that relatively cool marine and continental climates existed at different latitudes along the eastern U.S.A. during one or two periods of high sea level corresponding to oxygen-isotope substages 5a and 5c. This cool interval is recognized across at least 7° of latitude in the youngest marine Pleistocene deposits in many areas (figure 3, sections 4, 6a, 7 and 9a). Lithological and paleontological evidence for a diastem between these sediments and underlying substage 5e sediments, containing warm-climate faunas, indicates breaks in deposition within the last interglacial. Uranium-series ages on corals from the upper sediments, bearing cool-climate faunas, suggest a correlation with oxygen-isotope substages 5c and 5a.

At Sankaty Head, the ostracode fauna in the upper sand is completely distinct from that in the lower sand. It contains as much as 30–40 % cold-temperate species and high percentages of mild-temperate species (figure 5*a*); water temperatures were much cooler than those in the region today. The contact with the underlying lower Sankaty Sand, containing warm-temperate faunas, consists of a ventifacted pebble horizon, signifying subaerial exposure (Oldale *et al.* 1982).

Cool marine climates also characterize the upper part of the Wachapreague Formation on the Delmarva Peninsula, which contains 8–10% cold-temperate and 50% mild-temperate species (figure 4*a*). According to Mixon (1985) the stratigraphic relation of the Wachapreague to the underlying Joynes Neck Sand (not represented in figure 4*a*) is unclear but some evidence suggests that the Wachapreague truncates the Joynes Neck; this evidence suggests a brief depositional hiatus. The ostracod and pollen evidence shows a distinct cooling trend for the Wachapreaue, approaching interstadial conditions.

In the Stetson core, North Carolina (figure 4*c*), the percentage of mild-temperate species in the Core Creek sand described by Mixon & Pilkey (1976) increases in the uppermost samples to about 20%; no cold-temperate species occur in the Stetson section. Ager & Shaw (1987) found cool-temperate pollen assemblages in the uppermost sediments at the Stetson section. In North Carolina and southeastern Virginia, Cronin *et al.* (1981) and Szabo (1985) reported several uranium-series ages on corals from several late Pleistocene marine units dated at *ca.* 70–75 ka BP. One coral is from an outcrop at the Stetson section, dating the uppermost marine deposits there at 72±4 ka BP (zone 1 of figure 4*c*). Lithological and faunal changes at 8 m in the Stetson section suggest a brief depositional break between sediments yielding the 120 ka BP amino acid epimerization date on *Mercenaria* and the overlying sediments yielding the 72 ka BP coral data (York & Wehmiller 1987; T. M. Cronin, unpublished observations).

Finally, near Charleston, the upper part of the Wando Formation shows a slight trace of mild-temperate species at the end of the last interglacial (figure 4*d*). Strong evidence for a break between the lower and upper Wando (figure 3, section 9) was provided by Lyon (1987) who found low-diversity, brackish-water ostracods near the contact. Szabo (1985) suggested that the upper and lower Wando Formation correlated with the isotope substages 5e and 5c, respectively, on the basis of uncorrected ^{230}Th dates on corals.

The ages of Upper Pleistocene Coastal Plain marine units have been the source of controversy (Mixon *et al.* 1982; Stearns 1984; Cronin 1983, 1987; Corrado *et al.* 1986) because most models of eustatic sea level (Bloom *et al.* 1974; Harmon *et al.* 1983; Cronin 1983; Chappell & Shackleton 1986) call for a single high stand above present sea level during the isotope substage 5e. The 70–75 ka BP Coastal Plain ages indicating correlation with substage 5a are anomalous because most estimates of eustatic sea level for this time are about 9–20 m below present sea level (see Aharon & Chappell 1986; Chappell & Shackleton 1986). This discrepancy led Mixon *et al.* (1982) to suggest the Coastal Plain units in the Chesapeake Bay area dated at 75 ka BP, and those near Charleston, dated about 100 ka BP, were actually 125 ka BP (isotope substage 5e) but that extratropical corals yielded anomolously younger ages the farther north they occurred.

The interpretation adopted here, discussed in more detail in Cronin (1987), is that the ages accurately date at least one and possibly two marine units deposited during substages 5a and 5c. Three lines of evidence support this. First, the concordance of faunal events from Massachusetts to North Carolina indicate that cool oceanic conditions were not localized, but rather a regional phenomenon, signifying a change in nearshore oceanography from those indicated from underlying deposits. Second, while this cool interval could represent the end of isotope substage 5e, this alternative seems unlikely given stratigraphic, lithic and palaeontological evidence for hiatuses between underlying deposits containing warm-water faunas, believed to represent isotope substage 5e, and overlying deposits containing cool-water faunas. Third, uranium-series ages are consistent with the stratigraphy and any post-depositional alteration of corals would have had to alter specimens equally from a broad area.

The concordance between ^{230}Th and ^{231}Pa and uranium trend age estimates on surrounding sediment also support the ages (Szabo 1985).

This hypothesis implies that cool conditions existed immediately before Wisconsinan glaciation but before a major sea-level drop occurred at the isotope stage 5–4 transition, dated by Kukla & Briskin (1982) at 75 ka BP. If we assume tectonic stability (i.e. 15–20 m of uplift has not occurred), it still remains to reconcile these data with eustatic models. The existence of cool marine climates with a high sea level might involve the presence of marine-based ice shelves, which would affect the oxygen-isotope signal in deep sea foraminiferans but would not lower sea level (Williams *et al.* 1981) but Chappell & Shackleton (1986) dismiss this, at least for the early part of the last interglacial. The issue remains unresolved; however, our poor knowledge of mechanisms and rates of vertical tectonic and isostatic movements, not only in the Coastal Plain but in all coastal regions, is such that we should reassess assumptions about 'stable' areas and estimates of uplift rates that form the foundation of eustatic models.

Conclusions

The Pliocene record of the Atlantic Coastal Plain shows a period of increasing oceanic warmth that was abruptly halted by a major regression and extended period of erosion. Since this event, the frequency and amplitude of climatic oscillations increased significantly such that in the middle and late Pleistocene an approximate cycle of 100 ka is recognized for major glacial–interglacial cycles. Although the east coast stratigraphic record is too discontinuous and dating techniques too course to discern a higher frequency signal, there is some indication that short-term climatic changes occurred during the last two interglacial periods.

I thank J. E. Hazel for his invaluable discussions on ostracods; B. J. Szabo, J. P. Owens, T. A. Ager, and J. Wehmiller for help with Coastal Plain geology; and T. Gibson, R. Z. Poore, and N. F. Sohl for comments on the manuscript. E. E. Compton-Gooding helped prepare the text and the figures.

References

Ager, T. A. & Shaw, E. G. 1987 Pollen record of Quaternary interglacials from the Stetson core, Dare County, North Carolina. *Abstracts, 12th INQUA Congress, Ottawa, Canada.*

Aharon, P. & Chappell, J. 1986 Oxygen isotopes, sea level changes and the temperature history of a coral reef environment in New Guinea over the last 10^5 years. *Palaeogeogr. Palaeoclim. Palaeoecol.* **56**, 337–379.

Akers, W. H. & Koeppel, P. E. 1973 Age of some Neogene formations, Atlantic Coastal Plain, United States and Mexico. *Soc. Econ. Paleont. Mineral., Calcareous Nannoplankton Symposium, Houston, Texas* (ed. L. A. Smith & J. A. Hardenbol), pp. 80–92.

Berggren, W. A. 1972 Late Pliocene-Pleistocene glaciation. In *Init. Repts DSDP* (ed. A. S. Laughton *et al.*), vol. **12**, pp. 953–964.

Berggren, W. A., Burckle, L. H., Cita, M. B., Cooke, H. B. S., Funnell, B. M., Gartner, S., Hays, J. D., Kennett, J. P., Opdyke, N. D., Pastouret, L., Shackleton, N. J. & Takayanagi, Y. 1980 Towards a Quaternary time scale. *Quat. Res.* **13**, 277–302.

Bloom, A. L. 1960 Late Pleistocene changes in sea level in southwestern Maine. *Maine Geological Survey, Department of Economic Development Report*, pp. 1–140.

Bloom, A. L., Broecker, W. S., Chappell, J., Matthews, R. K. & Mesolella, K. J. 1974 Quaternary sea level fluctuations on a tectonic coast: New ^{230}Th/^{234}U dates from the Huon Peninsula, New Guinea. *Quat. Res.* **4**, 185–205.

Brady, G. S. & Crosskey, H. W. 1871 Notes on fossil Ostracoda from the post-Tertiary deposits of Canada and New England. *Geol. Mag.* **8**, 60–65.

Brunner, C. A. 1979 Pliocene and Quaternary history of the Florida Current. *Geol. Soc. Am. Abstr. with Prog.* **11**, 395.

Chappell, J. & Shackleton, N. J. 1986 Oxygen isotopes and sea level. *Nature, Lond.* **324**, 137–140.
Colquohoun, D. J. 1974 Cyclic surficial stratigraphic units of the middle and lower Coastal Plains, central South Carolina. In *Post-Miocene stratigraphy, Central and Southern Atlantic Coastal Plain* (ed. R. Q. Oaks & J. R. DuBar), pp. 179–190. Logan, Utah: Utah State University Press.
Colquohoun, D. J. 1986 The geology and physiography of the Orangeburg Scarp. *Geol. Soc. Am. Centennial field guide, southeastern section*, vol. 6, pp. 321–322.
Corrado, J. C., Weems, R. E., Hare, P. E. & Bambach, R. K. 1986 Capabilities and limitations of applied aminostratigraphy, as illustrated by analysis of *Mulinia lateralis* from the late Cenozoic marine beds near Charleston, South Carolina. *S. Carol. Geol.* **30**, 19–46.
Cronin, T. M. 1979 Late Pleistocene marginal marine ostracodes from the southeastern Atlantic Coastal Plain and their paleoenvironmental implications. *Geogr. phys. quat.* **33**, 121–173.
Cronin, T. M. 1981*a* Rates and possible causes of neotectonic vertical crustal movements of the emerged southeastern U.S. Atlantic Coastal Plain. *Bull. geol. Soc. Am.* **92**, 812–833.
Cronin, T. M. 1981*b* Paleoclimatic implications of late Pleistocene marine ostracodes from the St. Lawrence Lowlands. *Micropaleontology* **27**, 384–418.
Cronin, T. M., Szabo, B. J., Ager, T. A., Hazel, J. E. & Owens, J. P. 1981 Quaternary climates and sea levels: U.S. Atlantic Coastal Plain. *Science, Wash.* **211**, 233–240.
Cronin, T. M. 1983 Rapid sea level and climatic change: Evidence from continental and island margins. *Quat. Sci. Rev.* **1**, 177–214.
Cronin, T. M., Bybell, L. M., Poore, R. Z., Blackwelder, B. W., Liddicoat, J. C. & Hazel, J. E. 1984 Age and correlation of emerged Pliocene and Pleistocene deposits, U.S. Atlantic Coastal Plain. *Palaegeogr. Palaeoclim. Palaeoecol.* **47**, 21–51.
Cronin, T. M. 1987 Quaternary sea level studies of the eastern United States of America: A methodological perspective. In *Sea level changes* (ed. M. J. Tooley & I. Shennan). *Institute of British Geographers Special Paper*, pp. 225–248. Oxford: Basil Blackwell.
Cronin, T. M. 1988 Paleozoogeography of post-glacial Ostracoda from northeastern North America. In *The Champlain Sea* (ed. N. R. Gadd). *Geol. Ass. Can. Spec. Pap.* (In the press.)
DuBar, J. R., Johnson, H. S., Thom, B. & Hatchell, W. O. 1974 Neogene stratigraphy and morphology, south flank of the Cape Fear Arch, North and South Carolina. In *Post-Miocene stratigraphy, Central and Southern Atlantic Coastal Plain* (ed. R. Q. Oaks & J. R. DuBar), pp. 139–173. Logan, Utah: Utah University Press.
Gibson, T. G. 1983 Key foraminifera from upper Oligocene to lower Pleistocene strata of the central Atlantic Coastal Plain. In *Geology and paleontology of the Lee Creek Mine, North Carolina*, vol. 1 (ed. C. E. Ray). *Smithson. Contr. Paleobiol.* **53**, 355–454.
Gustavson, T. C. 1976 Paleotemperature analysis of the marine Pleistocene of Long Island, New York, and Nantucket Island, Massachusetts. *Bull. geol. Soc. Am.* **87**, 1–8.
Harmon, R. S., Mitterer, R. M., Kriausakul, N., Land, L. S., Schwarcz, H. P., Garrett, P., Larson, G. J., Vacher, H. E. & Rowe, M. 1983 U-series and amino acid racemization geochronology of Bermuda: Implications for eustatic sea-level fluctuation over the past 250,000 years. *Palaeogeogr. Palaeoclim. Palaeoecol.* **44**, 41–70.
Hathaway, J. C., Poag, C. W., Valentine, P. C., Miller, R. E., Schultz, D. M., Manheim, F. T., Kohout, F. A., Bothner, M. H. & Sangrey, D. A. 1979 U.S. Geological Survey core drilling on the Atlantic Shelf. *Science, Wash.* **206**, 515–527.
Hazel, J. E. 1968 Pleistocene ostracode zoogeography in Atlantic coast submarine canyons. *J. Paleont.* **42**, 1264–1271.
Hazel, J. E. 1970 Ostracode zoogeography in the southern Nova Scotian and northern Virginian faunal provinces. *U.S. Geol. Surv. Prof. Pap.* 529-E.
Hazel, J. E. 1971 Paleoclimatology of the Yorktown Formation (upper Miocene and lower Pliocene) of Virginia and North Carolina. *Bull. Cent. Rech. SNPA, Pau*, **5** (suppl.), 361–375.
Hazel, J. E. 1975 Ostracode biofacies in the Cape Hatteras, North Carolina area. *Bull. Am. Paleont.* **65**, 463–487.
Hazel, J. E. 1977 Distribution of some biostratigraphically diagnostic ostracodes in the Pliocene and lower Pleistocene of Virginia and northern North Carolina. *U.S. Geol. Surv. Jl Res.* **5**, 373–388.
Hazel, J. E. 1983 Age and correlation of the Yorktown (Pliocene) and Croatan (Pliocene and Pleistocene) formations at the Lee Creek Mine. In *Geology and paleontology of the Lee Creek Mine, North Carolina*, vol. 1 (ed. C. E. Ray), *Smithson. Contr. Paleobiol.* **53**, 81–199.
Imbrie, J. & Kipp, N. 1971 A new micropaleontological method for quantitative paleoclimatology: Application to a late Pleistocene Caribbean core. In *Late Cenozoic glacial ages* (ed. K. K. Turekian), pp. 71–181. New Haven: Yale University Press.
Jones, D. S. & Hasson, P. F. 1985 History and development of the marine invertebrate faunas separated by the Central American Isthmus. In *The great American biotic interchange* (ed. F. G. Stehli & S. D. Webb), pp. 325–356. New York: Plenum Press.
Keigwin, L. D. 1978 Pliocene closing of the Isthmus of Panama based on biostratigraphic evidence from nearby Pacific Ocean and Caribbean Sea cores. *Geology* **6**, 630–634.

Kukla, G. & Briskin, M. 1983 The age of the 4/5 isotopic stage boundary on land and in the oceans. *Palaeogeog., Palaeoclim., Palaeoecol.* **42**, 35–45.

Lyon, S. K. 1988 Biostratigraphic and paleoenvironmental interpretations from late Pleistocene Ostracoda, Charleston, South Carolina. *U.S. geol. Surv. Prof. Pap.* 1367. (In the press.)

McCartan, L., Owens, J. P., Blackwelder, B. W., Szabo, B. J., Belknap, D. F., Kriausakul, N., Mitterer, R. M. & Wehmiller, J. F. 1982 Comparison of amino acid racemization geochronometry with lithostratigraphy, biostratigraphy, uranium-series coral dating, and magnetostratigraphy in the Atlantic Coastal Plain of the southeastern United States. *Quat. Res.* **18**, 337–359.

McCartan, L., Weems, R. E. & Lemon, E. M. Jr 1984 Geologic map of the area between Charleston and Orangeburg, South Carolina. *U.S. geol. Surv. Misc. Invest. Map* I-1472.

Miller, W. III 1982 The paleoecologic history of late Pleistocene estuarine and marine fossil deposits in Dare County, North Carolina. *Southeastern Geol.* **23**, 1–13.

Miller, W. III 1985 The Flanner Beach formation (middle Pleistocene) in eastern North Carolina. *Tulane Studs. Geol. Paleont.* **18**, 93–122.

Mixon, R. B. 1985 Stratigraphic and geomorphic framework of uppermost Cenozoic deposits in the southern Delmarva Peninsula, Virginia and Maryland. *U.S. geol. Surv. Prof. Pap.* 1067-G.

Mixon, R. B. & Pilkey, O. H. 1976 Reconnaissance geology of the submerged and emerged Coastal Plain Province, Cape Lookout area, North Carolina. *U.S. geol. Surv. Prof. Pap.* 859.

Mixon, R. B., Szabo, B. J. & Owens, J. P. 1982 Uranium-series dating of mollusks and corals, and age of Pleistocene deposits, Chesapeake Bay area, Virginia and Maryland. *U.S. geol. Surv. Prof. Paper* 1067-E.

Oaks, R. Q. & DuBar, J. R. (eds) 1974 *Post-Miocene stratigraphy, Central and Southern Atlantic Coastal Plain.* Logan, Utah: Utah University Press.

Oaks, R. Q. & Coch, N. K. 1963 Pleistocene sea levels, southeastern Virginia. *Science, Wash.* **140**, 979–983.

Oldale, R. N., Valentine, P. C., Cronin, T. M., Spiker, E. C., Belknap, D. F., Wehmiller, J. F. & Szabo, B. J. 1982 Stratigraphy, structure, absolute age and paleontology of the upper Pleistocene deposits at Sankaty Head, Nantucket Island, Massachusetts. *Geology* **10**, 246–252.

Owens, J. P. 1988 Geology of the Cape Fear Region, North Carolina (Florence and Georgetown 2° Sheets). *U.S. Geol. Surv. MF.* (In the press.)

Richards, H. G. 1962 Studies of the marine Pleistocene. *Trans. Am. phil. Soc.* (N.S.) **52**, 1–141.

Riggs, S. R., Lewis, D. W., Scarborough, A. K. & Snyder, S. W. 1982 Cyclic deposition of Neogene phosphorites in the Aurora area, North Carolina, and their possible relationship to global sea level. *Southeastern Geol.* **23**, 189–204.

Shackleton, N. J., Backman, J., Zimmerman, H., Kent, D. V., Hall, M. A., Roberts, D. G., Schnitker, D., Baldauf, J. G., Desprairies, A., Homrighausen, R., Huddlestun, P., Keene, J. B., Kaltenback, A. J., Krumsiek, K. A. O., Morton, A. C., Murray, J. W. & Westburg-Smith, J. 1984 Oxygen isotope calibration of the onset of ice-rafting and history of glaciation in the North Atlantic. *Nature, Lond.* **307**, 620–623.

Shackleton, N. J. & Opdyke, N. D. 1973 Oxygen isotope and paleomagnetic stratigraphy of equatorial Pacific core V28–238: Oxygen isotope temperatures and ice volumes on a 10^5 year and 10^6 year scale. *Quat. Res.* **3**, 39–55.

Shor, A. N. & Poore, R. Z. 1979 Bottom currents and ice rafting in the North Atlantic: Interpretation of Neogene depositional environments of Leg 49 cores. In *Init. Repts DSDP* (ed. B. P. Luyendyk *et al.*), vol. **49,** pp. 859–872.

Snyder, S. W., Mauger, L. L. & Akers, W. H. 1983 Planktonic foraminifera and biostratigraphy of the Yorktown formation, Lee Creek Mine. In *Geology and paleontology of the Lee Creek Mine, North Carolina*, vol. 1 (ed. C. E. Ray). *Smithson. Contr. Paleobiol.* **53**, 455–482.

Spencer, R. S. & Campbell, L. D. 1987 The fauna and paleoecology of the late Pleistocene marine sediments of southeastern Virginia. *Bull. Am. Paleont.* **92** (327), pp. 1–124.

Stearns, C. E. 1984 Uranium-series dating and the history of sea level. In *Quaternary dating methods* (ed. W. C. Mahaney). *Devs. paleont. Stratigr.* **7**, 53–66.

Stuiver, M. & Borns, H. W. 1975 Late Quaternary marine invasion in Maine: its chronology and associated crustal movement. *Bull. geol. Soc. Am.* **86**, 99–104.

Szabo, B. J. 1985 Uranium-series dating of fossil corals from marine sediments of southeastern United States Atlantic Coastal Plain. *Bull. geol. Soc. Am.* **96**, 398–406.

Valentine, P. C. 1971 Climatic implications of a late Pleistocene ostracode assemblage from southeastern Virginia. *U.S. geol. Surv. Prof. Pap.* 683–D.

Watts, W. A. & Stuiver, M. 1980 Late Wisconsin climate of northern Florida and the origin of species-rich deciduous forest. *Science, Wash* **210**, 325–327.

Weems, R. E. & Lemon, E. M. Jr 1984 Geologic map of the Mount Holly quadrangle, Berkeley and Charleston Counties, South Carolina, with text. *U.S. geol. Surv. Map* GQ-1579 (1:24000).

Wehmiller, J. F. & Belknap, D. F. 1982 Amino acid age estimates, Quaternary Atlantic Coastal Plain: Comparison with U-series dates, biostratigraphy, and paleomagnetic control. *Quat. Res.* **18**, 311–346.

Williams, D. F., Moore, W. S. & Fillon, R. H. 1981 Role of glacial Arctic Ocean ice sheets in Pleistocene oxygen isotope and sea level records. *Earth planet. Sci. Lett.* **56**, 157–166.

Winker, C. D. & Howard, J. D. 1977 Correlation of tectonically deformed shorelines on the southern Atlantic coastal plain. *Geology* **5**, 123–127.
York, L. L. & Wehmiller, J. F. 1987 Aminostratigraphy and lithostratigraphy of a subsurface Quaternary section at Stetson Pit, Dare County, North Carolina. *Abstracts, 12th INQUA Congress, Ottawa, Canada.*

Appendix 1. List of 90 temperature-sensitive ostracod species used in Coastal Plain climatology

Group 1: *frigid*

Cytheromorpha macchesneyi (Brady & Crosskey, 1871)
Finmarchinella logani (Brady & Crosskey, 1871)
Normanicythere leioderma (Norman, 1869)

These three species are the only frigid ostracods in the Presumpscot formation. Other frigid species occur in other postglacial deposits of North America (see Cronin 1988).

Group 2: *subfrigid–cold-temperate*

Acanthocythereis dunelmensis (Norman, 1865)
Baffinicythere emarginata (Sars, 1865)
Cythere lutea (O. F. Müller, 1785)
Cytheropteron angulatum Brady & Robertson, 1872
C. inflatum Brady, Crosskey & Robertson, 1874
C. nodosum Brady, 1868
C. nodosoalatum Neale & Howe, 1973
Elofsonella concinna (Jones, 1857)
Hemicythere villosa (Sars, 1865)
Hemicytherura clathrata (Sars, 1866)
Heterocyprideis fascis (Brady & Norman, 1889)
Munseyella mananensis Hazel & Valentine, 1969
Palmenella limicola (Norman, 1865)
Patagonocythere dubia (Brady, 1868)
Sarsicytheridea bradii (Norman, 1865)
S. punctillata (Brady, 1865)
Semicytherura undata (Sars, 1865)

Group 3: *cold-temperate–mild-temperate*

Actinocythereis dawsoni (Brady, 1870)
Bensonocythere americana Hazel, 1967
B. arenicola (Cushman, 1906)
B. sp. B of Valentine (1971)
Cytheridea sp. A
Cytheromorpha fuscata (Brady, 1868)
Cytherura wardensis Howe & Brown, 1935
Eucythere declivis (Norman, 1865)
Finmarchinella finmarchica (Sars, 1865)
Hulingsina sp. A
Leptocythere angusta Blake, 1935
Loxoconcha cf. *L. impressa* (Baird, 1850)
L. sperata Williams, 1966
L. aff. *L. granulata* Sars, 1865
Muellerina canadensis (Brady, 1870)
Muellerina ohmerti Hazel, 1983
Neolophocythere subquadrata Grossman, 1967
Perissocytheridea sp. A
Pseudocytheretta edwardsi Cushman, 1906
Puriana rugipunctata (Ulrich & Bassler, 1904)

Group 4: *mild–temperate to subtropical* (*warm-temperate*)

Actinocythereis captionis Hazel, 1983
Bensonocythere hazeli Cronin, 1988, in the press
B. valentinei Cronin, 1988, in the press
B. whitei (Swain, 1951)
Campylocythere laeva (Edwards, 1944)
Cyprideis mexicana Sandberg, 1964
C. salebrosa Bold, 1963
Cytheromorpha curta Edwards, 1944
C. newportensis Williams, 1966
Cytherura forulata Edwards, 1944

Group 4 (*cont.*)

C. neusensis Cronin, 1988, in the press
Cushmanidea seminuda (Cushman, 1906)
Hulingsina americana (Cushman, 1906)
H. rugipustulosa (Edwards, 1944)
H. sp. B of Valentine (1971)
H. sp. C of Valentine (1971)
Leptocythere nikraveshae Morales, 1966
Malzella floridana (Benson & Coleman, 1963)
Peratocytheridea bradyi (Stephenson, 1938)
P. setipunctata (Brady, 1869)
Proteoconcha gigantica (Edwards, 1944)
P. nelsonensis (Grossman, 1967)
Protocytheretta cf. *P. sahnia* (Puri, 1952)
Puriana mesacostalis (Edwards, 1944)
Tetracytherura choctawhatcheensis (Puri, 1954)
T. norfolkensis (Cronin, 1979)
T. sp. A

Group 5: *subtropical*

**Cytherura reticulata* Edwards, 1944
C. sandbergi Garbett & Maddocks, 1979
**Hulingsina glabra* (Hall, 1965)
Loxoconcha matagordensis Swain, 1955
Neocaudites atlantica Cronin, 1979
**Paracytheridea altila* Edwards, 1944
P. hazeli Cronin, 1988, in the press
**Paracytheroma stephensoni* Puri, 1954
**Pellucistoma magniventra* Edwards, 1944
Perissocytheridea brachyforma Swain, 1955
**Proteoconcha multipunctata* Edwards, 1944
**P. tuberculata* Puri, 1960
**Puriana carolinensis* Hazel, 1983
**P. convoluta* Teeter, 1975

* These species also inhabit tropical environments and are included in the totals for thermophilic species in the Duplin Formation (figure 2).

Group 6: *tropical*

Acuticythereis laevissima (Edwards, 1944)
Bairdia laevicula Edwards, 1944
Bairdoppilata triangulata Edwards, 1944
Cytherelloidea umbonata Edwards, 1944
C. sp.
Cytherura nucis Garbett & Maddocks, 1979
Eucythere gibba Edwards, 1944
Hemicytherura sp.
Hermanites ascitus Hazel, 1983
Kangarina sp.
Loxoconcha purisubrhomboidea Edwards, 1953
L. reticularis Edwards, 1944
Neocaudites angulatus Hazel, 1983
N. triplistriatus (Edwards, 1944)
N. variabilus Hazel, 1983
Orionina vaughani (Ulrich & Bassler, 1904)
Paracytheridea rugosa Edwards, 1944
Paradoxostoma delicata Puri, 1954
Paranesidea sp.
Perissocytheridea subpyriforma Edwards, 1944
Protocytheretta reticulata Edwards, 1944
Radimella confragosa (Edwards, 1944)
Xestoleberis sp.

Phil. Trans. R. Soc. Lond. B **318**, 679–688 (1988)
Printed in Great Britain

The evolution of oceanic oxygen-isotope variability in the North Atlantic over the past three million years

BY N. J. SHACKLETON[1], F.R.S., J. IMBRIE[2] AND N. G. PISIAS[3]

[1] *Godwin Laboratory for Quaternary Research, University of Cambridge, Free School Lane, Cambridge CB2 3RS, U.K.*

[2] *Department of Geological Sciences, Brown University, Providence, Rhode Island* 02912, *U.S.A.*

[3] *School of Oceanography, Oregon State University, Corvallis, Oregon* 97331, *U.S.A.*

Throughout the past three million years, variability in the oxygen-isotopic composition of the ocean, caused by changing ice-sheet mass on the continents, has been concentrated at the frequencies associated with changes in the earth's orbital geometry. The amplitude of variability has increased towards the present. An increase in variability associated with changes in the obliquity of the Earth's rotational axis (period 41 ka) during the early Pleistocene was followed by an increase in power related to the precession cycle (23 ka) and associated ellipticity cycle (*ca.* 100 ka) during the past million years. Although deep-sea sediments are the best place to observe this evolution in climatic variability, we will not be able to understand it without more data from other geological sources.

INTRODUCTION

The oxygen-isotope record in benthic Foraminifera from deep-sea sediments reflects both the history of ocean isotopic composition (which varies with the quantity of isotopically light ice stored on the continents) and small changes in deep-water temperature. In principal, this record provides the clearest single indicator of the evolution of Plio-Pleistocene climatic variability.

The first Pleistocene oxygen-isotope records from the North Atlantic were obtained by Emiliani (1955), who analysed planktic Foraminifera. In particular, Swedish Deep-Sea Expedition core 280, from which Emiliani analysed the species *Globorotalia inflata*, revealed in detail the now-familiar succession of the past two glacial–interglacial cycles. It is generally agreed that, in showing his findings as surface-temperature records, Emiliani underestimated the contribution that changing ice-volume makes to the overall record (Olausson 1965; Shackleton 1967). This is clear from the very similar character of oxygen-isotope records derived from analyses of benthic Foraminifera, which inhabit the cold abyssal depths where the possibility of further refrigeration during glacial times is severely restricted (Shackleton & Opdyke 1973; Ninkovich & Shackleton 1975). On the other hand, it is not recognized that, in the North Atlantic Deep Water, where the present-day temperature is around 3 °C, changes in deep-water temperature have not been negligible (Duplessy *et al.* 1980; Shackleton *et al.* 1983; Chappell & Shackleton 1986).

Probably the only area where oxygen-isotope analyses of benthic Foraminifera might yield a pure ice-volume record is the Norwegian Sea, where deep water forms at about −1 °C. Labeyrie *et al.* (1988) have obtained an important record by stacking several fragmentary sequences, but it seems unlikely that a truly detailed record will ever be available from that region. Thus a record from the open North Atlantic is at present the best means we have for examining the development of glacial–interglacial variability.

Deep-Sea Drilling Project site 552A

DSDP site 552A was cored on the flank of Rockall Bank by using the then newly developed hydraulic piston-corer (HPC) during DSDP leg 81 (Roberts *et al.* 1985). The site was located at 56° 03′ N, 23° 14′ W in 2311 m depth of water. A good magnetostratigraphic record extending through the Brunhes, Matuyama and Gauss and into the Gilbert magnetochrons was obtained by D. Kent; this, together with J. Backman's nanofossil biostratigraphy, provides a very reliable basic chronology for the lithostratigraphic and stable-isotope records (Shackleton *et al.* 1984; Shackleton & Hall 1985; Zimmerman *et al.* 1985; Backman 1985).

Lithologically the sediments from the past 2.5 Ma consist, in alternating units, of white nanofossil–foraminiferal ooze (deposited during interglacials) and greyish sandy sediment containing Foraminifera and ice-rafted detritus (deposited during glacials (Zimmerman *et al.* 1985)). Before 2.5 Ma BP, only white ooze was deposited. It is clear from the virtual absence of nanofossils in the glacial sediments that the polar front moved south of the site during glacial episodes back to 2.4 Ma BP. Apart from a very brief and minor pulse of ice-rafting at *ca.* 2.50 Ma BP, there were no ice-rafted grains reported from sediments older than this transition. It seems certain that, at least at the latitude of Britain, this transition just after 2.5 Ma BP constituted a major climatic break.

Oxygen-isotope measurements were made on benthic Foraminifera by using standard methods (Shackleton & Opdyke 1973; Shackleton *et al.* 1983). Raw data are listed in Shackleton & Hall (1984). In the present paper, as in Shackleton *et al.* (1983), they are shown after applying adjustment factors for those species that exhibit a systematic departure from isotopic equilibrium.

Oxygen and carbon isotope measurements were made at 10 cm intervals in the top 12 cores, covering the past 3.6 Ma, excluding core 7 which was badly disturbed in the coring process and was not considered worth sampling in detail. Although this provides a favourable average sampling interval of 2.5 ka, the reality is that the sampling interval represents widely varying time intervals owing to the large variations in the sediment accumulation rate. Obtaining a detailed timescale for the past 0.9 Ma was straightforward, because the record can easily be correlated with other published isotope records. In figure 1 this part is shown on the SPECMAP timescale (Imbrie *et al.* 1984).

For the section older than 0.9 Ma BP, little has been done so far to develop a detailed chronology because there have been few high-resolution data sets available to work with. Pisias & Moore (1981) examined the distribution of variance in the oxygen-isotope record of core V28-239 (Shackleton & Opdyke 1976) and concluded that, in the portion older than 0.9 Ma, variance was concentrated at a frequency of *ca.* 2.5×10^{-4} cycles ka^{-1}, corresponding to the 40 ka period associated with variations in the tilt of the Earth's rotational axis (obliquity). They proposed a detailed timescale for core V28-239 by preserving a constant phase relation between ^{18}O and tilt (as calculated by Berger (1978)). However, the degree of detail preserved in core V28-239 (accumulated at only about 1 cm ka^{-1}) is not sufficient to permit DSDP site 552A to be correlated accurately to it, especially in view of the gap in 552A resulting from the disturbance of core 7. More recently, Ruddiman *et al.* (1986) published oxygen-isotope data from DSDP site 607 and again showed that variance is concentrated at the tilt period. By using their data, it is possible to achieve a satisfactory solution for the detailed chronology of site 552A back to the Olduvai chron at 1.6 Ma BP.

In the lower Matuyama we have developed a provisional timescale based only on tuning the 552A record to orbital variations. The difficulty in doing this with satisfactory accuracy is that the accumulation rate has been so variable at this site. The records from sites such as DSDP 607, which is south of the southernmost excursions of the polar front, will ultimately provide a more accurate chronology for this interval also.

Time-series analysis

Table 1 lists the age control points that we have used to construct the time series in figure 1. Ages of samples between these were estimated by linear interpolation. To obtain estimates at uniform time intervals of 3 ka for spectral analysis, we used a Gaussian weighting method which has the effect of providing some smoothing as well as interpolation. The smoothing used is equivalent to that of a Gaussian window with a halfwidth of 3 ka. Interpolation points for which data are not available are not used in the spectral estimation.

Table 1. Age control points used for DSDP site 552A

depth/m	age/Ma	depth/m	age/Ma	depth/m	age/Ma
00.0	0.0	13.20	0.689	35.50	1.953
00.67	0.024	13.60	0.711	35.98	1.996
01.62	0.065	14.13	0.726	37.70	2.102
01.63	0.078	14.50	0.736	37.99	2.134
02.00	0.107	15.89	0.797	38.43	2.170
02.40	0.128	17.60	0.874	39.01	2.201
02.80	0.157	19.90	0.980	39.60	2.235
03.50	0.194	21.00	1.045	40.40	2.287
04.50	0.249	21.88	1.081	41.10	2.334
05.00	0.269	22.69	1.120	41.40	2.358
05.30	0.297	23.95	1.200	43.14	2.482
05.55	0.329	29.10	1.476	43.90	2.569
06.25	0.371	29.33	1.494	45.40	2.720
06.46	0.405	29.80	1.529	46.30	2.800
06.90	0.423	31.00	1.598	47.50	2.906
08.40	0.471	31.80	1.652	48.30	2.955
09.15	0.513	32.50	1.706	50.50	3.075
09.60	0.540	33.10	1.748	53.40	3.232
10.80	0.620	34.10	1.823	55.60	3.364
12.00	0.656	34.60	1.874	56.10	3.401
12.60	0.668	34.92	1.896	57.20	3.470

Spectra were calculated by using a modification of standard Fourier analysis methods (Jenkins & Watts 1968). The principal modification is that we have calculated variance with a constant bandwidth on a logarithmic frequency scale. (In most studies the bandwidth is specified as a constant frequency interval, a practice that yields very different degrees of resolution across the spectrum and makes it difficult to draw quantitative inferences).

In figure 1, spectra are shown for overlapping time intervals of 0–1, 0.5–1.2, 1.5–2.5, 2.0–3.0

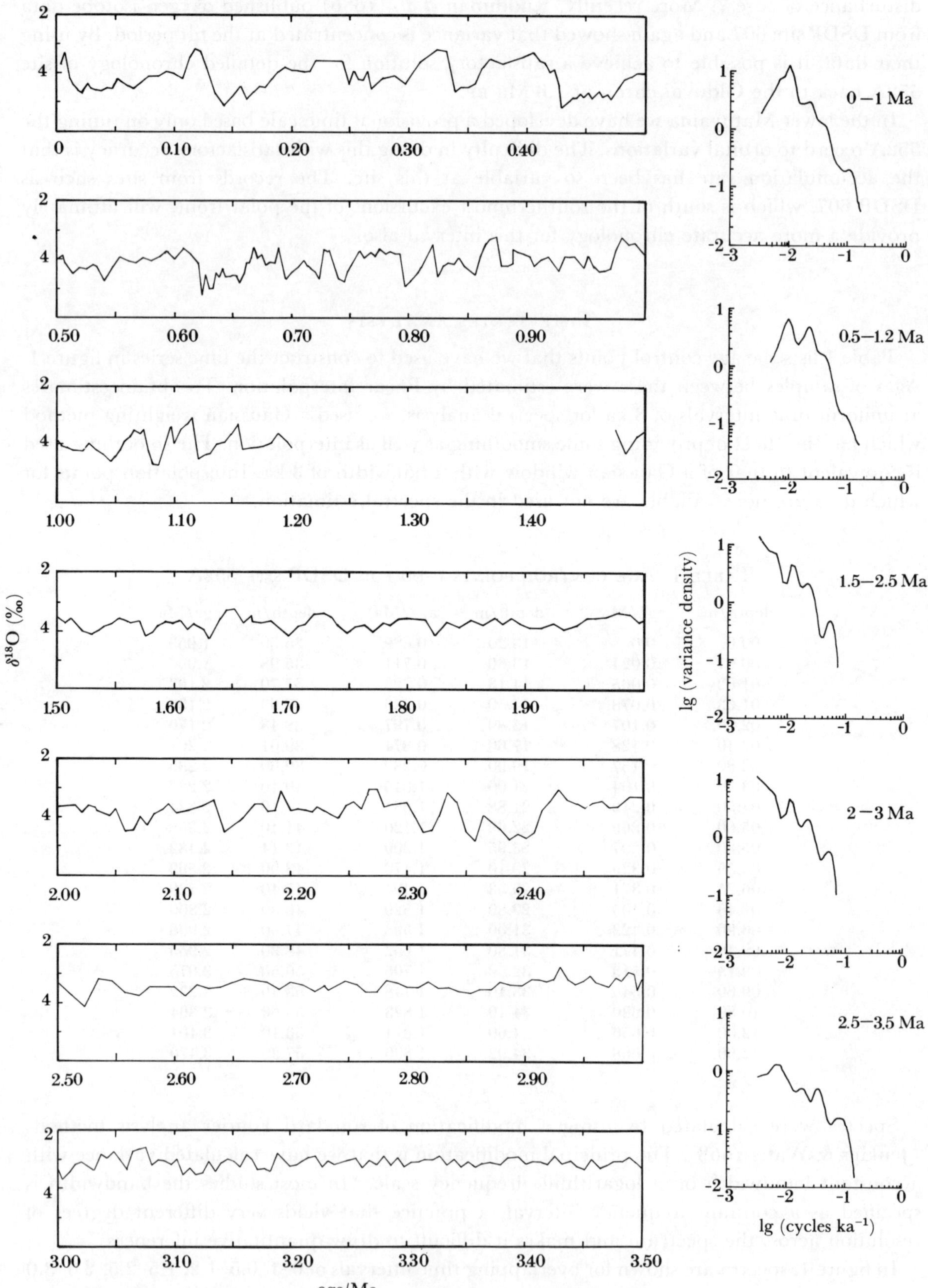

FIGURE 1. Oxygen-isotope record of DSDP site 552A; timescale from table 1. Spectra to the right are calculated with a constant bandwidth in log F for the interval; variance-density scale is (‰)^2 (cycles/ka)$^{-1}$.

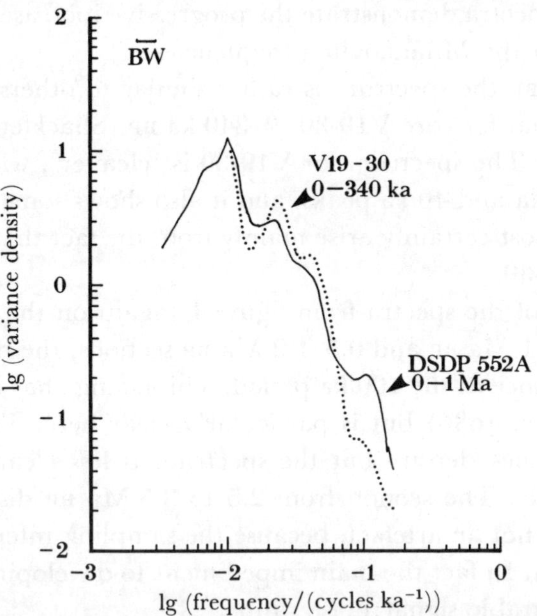

FIGURE 2. Spectrum for 0–1 Ma BP in DSDP 552A compared with that of 0–340 ka BP in piston core V19-30. Both are records of ^{18}O variations in benthic Foraminifera (proxy for global ice volume).

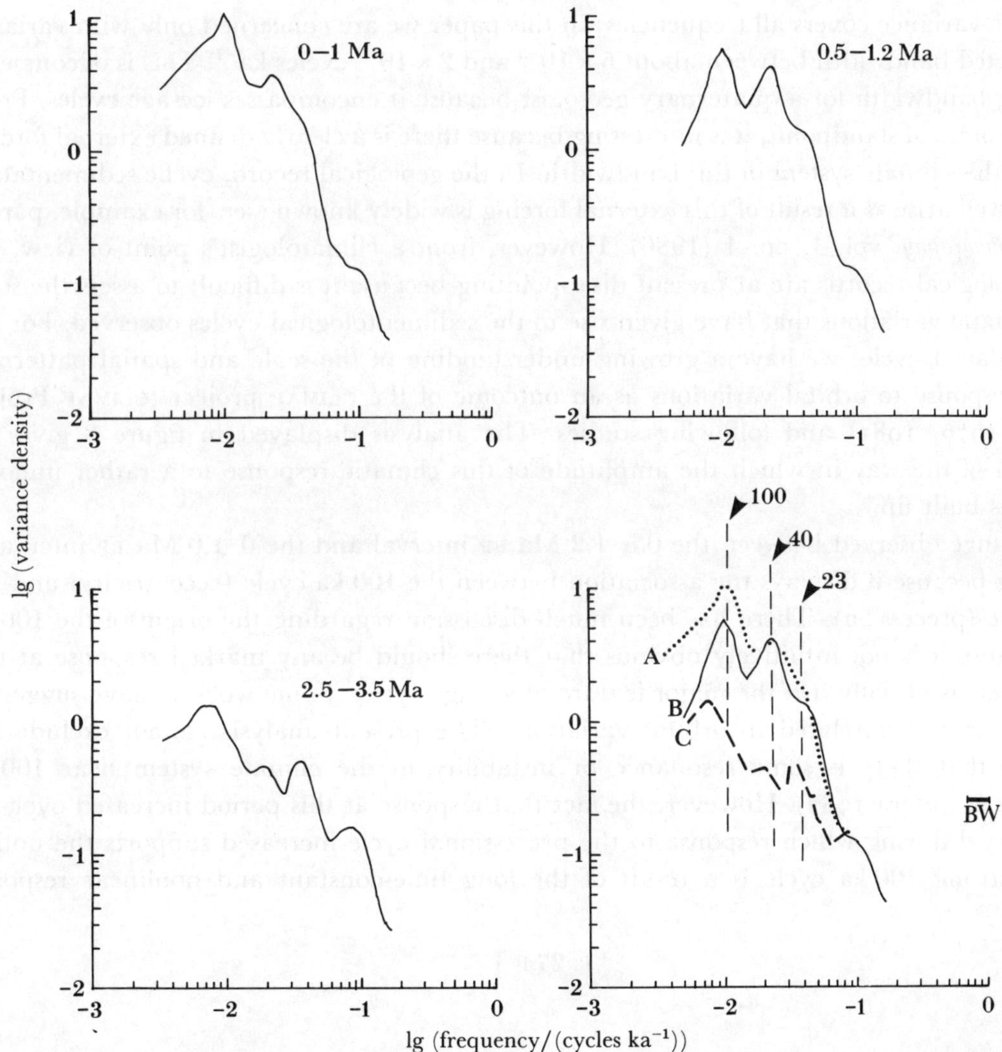

FIGURE 3. Spectra for the 0–1 Ma BP, 0.5–1.2 Ma BP and 2.5–3.5 Ma BP sections of DSDP 552A (from figure 1) compared, to demonstrate the growth in variance associated with earth-orbital variations. In the fourth sector these are superimposed with identical scales for variance density and labelled A (0–1 Ma), B (0.5–1.2 Ma) and C (2.5–3.5 Ma).

and 2.5–3.5 Ma BP. These spectra demonstrate the progressive increase towards the present in the variance associated with the Milankovitch frequencies.

For the interval 0–1 Ma BP the spectrum is rather similar to others for the same interval. Figure 2 compares it with that for core V19-30 (0–340 ka BP (Shackleton & Pisias 1985)) the same variance-density scale. The spectrum for V19-30 is 'cleaner', with a more pronounced minimum between the 100 ka and 40 ka peaks, and it also shows somewhat more variance at 23 ka. These differences almost certainly arise mainly from the fact that the sampling interval is three times closer in V19-30.

Figure 3 compares three of the spectra from figure 1 (again on the same variance-density scale). If we compare the 0–1 Ma BP and 0.5–1.2 Ma BP sections, the most striking changes is in the concentration of variance at the 100 ka period. This change has been noted before (see, for example, Ruddiman *et al.* 1986) but is particularly clear here. The section from 1.5 to 2.5 Ma BP has similar variance density but the spectrum is less clear, probably because of uncertainties in the timescale. The section from 2.5 to 3.5 Ma BP displays significantly less variance (which is certainly not an artefact, because the sampling interval is still adequate to reveal variance in this band). In fact the main impediment to developing a good timescale for this interval is the less favourable signal:noise ratio.

Discussion

Climatic variance covers all frequencies; in this paper we are concerned only with variance in a restricted bandwidth between about 5×10^{-3} and 2×10^{-1} cycles ka^{-1}. This is of course an interesting bandwidth for a quaternary geologist because it encompasses ice-age cycles. From a more theoretical standpoint, it is interesting because there is a clearly defined external forcing to the Earth's climate system in this bandwidth. In the geological record, cyclic sedimentation that may well arise as a result of this external forcing is widely known (see, for example, papers in *Paleoceanography*, vol. 1, no. 4 (1986). However, from a climatologist's point of view, the earlier geological records are at present disappointing because it is difficult to assess the scale of the climatic variations that have given rise to the sedimentological cycles observed. For the past few glacial cycles we have a growing understanding of the scale and spatial pattern of climatic response to orbital variations as an outcome of the CLIMAP project (CLIMAP Project members 1976, 1981) and following studies. The analysis displayed in figure 3 gives an impression of the way in which the amplitude of this climatic response to a rather uniform forcing has built up.

The change observed between the 0.5–1.2 Ma BP interval and the 0–1.0 Ma BP interval is interesting because it displays the association between the 100 ka cycle (eccentricity) and the 23 ka cycle (precession). There has been much discussion regarding the origin of the 100 ka cycle because it is not intuitively obvious that there should be any marked response at this period whereas visually it is the major feature of ice-age cycles. Some workers have suggested that the origin is unrelated to orbital variations. The present analysis cannot exclude the possibility that there is some resonance or instability in the climate system near 100 ka (Saltzman & Sutera 1987). However, the fact that response at this period increased over the same interval during which response to the precessional cycle increased supports the notion that the strong 100 ka cycle is a result of the long time-constant and nonlinear response associated with massive continental ice sheets. As these extended to lower latitudes, the precession effect became more and more important in controlling the distribution of solar insolation and hence in controlling their mass balance.

Passing now to the older part of the record, it is clear from figure 3 that it will be necessary to improve the signal: noise ratio of stable-isotope records before the magnitude of any response to orbital forcing in this type of record can be accurately described. Considerable subtlety will be needed if we are to build up a quantitative picture of the climatic response to orbital forcing in preglacial times.

DSDP site 552A has recovered a record of changes in globally integrated ice volume; of heavy deposition of iceberg-transported material at the latitude of Britain; and of oscillations in the position of the polar front. The location of the site is ideal for recording these aspects of climatic variability in the past, and as correlations are improved we may expect an ever-increasing similarity between this record and those reconstructed from northwest Europe. At present it is perhaps only the vegetational records of The Netherlands that offer the promise of comparability in detail, despite the fact that all the areas surrounding the North Atlantic Ocean must have experienced local variants of the same pattern of climatic change. It is appropriate to consider the question: in areas where the prospect of reconstructing long records that are suitable for comparison with the deep-sea records is remote, is there any palaeoclimatic justification for continued Quaternary research?

The answer to this question must surely be in the affirmative; there are many questions which can be answered better by terrestrial than by deep-sea records. One example concerns the interglacials. A cursory examination of the oxygen-isotope record suggests a general trend towards more positive values over the past three million years, but a more careful examination shows that this is only a result of the trend towards more positive values at the glacial extremes. If the Earth's climate system has become gradually more sensitive to glacial tendencies over this interval, one would surely expect some trend in the climate of the interglacials, unless the mechanism involved in this gradual change is very tightly tied to the actual ice sheets. Perhaps vegetational records would give us crucial information regarding the trend in interglacial climates through the past three million years if examined with this question in mind. Soil profiles, and the floras and faunas preserved in interglacial marine highstand deposits, are other sources of information which may be very valuable in this respect, despite the fact that they provide necessarily incomplete records.

At the opposite climatic extreme, the oxygen-isotope record portrays only the gross ice volume recorded at successive glacial maxima; the spatial distribution of ice masses can only be reconstructed by field observations. The question of the location of the ice sheets that came and went between 2.5 Ma BP and 1 Ma BP with a 40 ka periodicity must be answered if we are to understand why their response was different from that of the more recent ice sheets. It is to be hoped that the discussions at this meeting will help workers in different fields to exploit the full potential of their contributions from the point of view of their significance for the understanding of climatic variability. This should provide the scientific rewards that arise from interacting with a wider research community and especially with climate modellers.

We are especially grateful to Angeline Duffy for developing the program used to generate our spectral density plots. This work was supported by NERC grant GR3/3162 and NSF grants ATM-8319372 and ATM-8516140. The work would have been impossible without the support of the Ocean Drilling Programme as well as the core curatorial facility of Lamont Doherty Geological Observatory, which is supported by NSF grant OCE85-00232 and ONR grant N00014-84-C-0132.

References

Backman, J. 1985 Cenozoic calcareous nannofossil biostratigraphy for the Northeastern Atlantic Ocean- Deep Sea Drilling Project Leg 81. In *Initial Reports of the Deep Sea Drilling Project*, vol. 81 (ed. D. G. Roberts, D. Schnitker *et al.*), pp. 403–428. Washington: U.S. Government Printing Office.

Berger, A. L. 1978 Long-term variations of daily insolation and Quaternary climatic changes. *J. atm. Sci.* **35**, 2362–2367.

Chappell, J. & Shackleton, N. J. 1986 Oxygen isotopes and sea level. *Nature, Lond.* **324**, 137–140.

CLIMAP project members 1976 The surface of the ice-age earth. *Science, Wash.* **191**, 1131–1137.

CLIMAP project members 1981 *Seasonal reconstructions of the Earth's surface at the last glacial maximum.* Geological Society of America Map and Chart Series no. MC-36.

Duplessy, J.-C., Moyes, J. & Pujol, C. 1980 Deep water formation in the North Atlantic during the last ice age. *Nature, Lond.* **286**, 479–481.

Emiliani, C. 1955 Pleistocene temperatures. *J. Geol.* **63**, 538–578.

Imbrie, J., Hays, J. D., Martinson, D. G., McIntyre, A., Mix, A. C., Morley, J. J., Pisias, N. G., Prell, W. L. & Shackleton, N. J. 1984 The orbital theory of climate; support from a revised chronology of the marine $\delta^{18}O$ record. In *Milankovitch and climate* (ed. A. L. Berger *et al.*), pp. 269–305. Amsterdam: D. Riedel.

Jenkins, G. M. & Watts, D. G. 1968 *Spectral analysis and its applications* (525 pages.) San Francisco: Holden-Day.

Labeyrie, L., Duplessy, J.-C. & Blanc, P. L. 1987 Variations in mode of formation and temperature of oceanic deep waters over the past 125,000 years. *Nature, Lond.* **327**, 477–482.

Ninkovich, D. & Shackleton, N. J. 1975 Distribution, stratigraphic position and age of ash layer 'L', in the Panama Basin region. *Earth planet. Sci. Lett.* **27**, 20–34.

Olausson, E. 1965 Evidence of climatic changes in deep-sea cores, with remarks on isotopic palaeotemperatures. *Progr. Oceanogr.* **3**, 221–252.

Pisias, N. G. & Moore, T. C. 1981 The evolution of Pleistocene climate: a time series approach. *Earth planet Sci. Lett.* **52**, 450–458.

Roberts, D. G., Schnitker, D. *et al.* 1985 *Initial Reports of the Deep Sea Drilling Project*, vol. 81. Washington: U.S. Government Printing Office.

Ruddiman, W. R., Raymo, M. & McIntyre, A. 1986 Matuyama 41,000-year cycles: North Atlantic Ocean and Northern hemisphere ice sheets. *Earth planet. Sci. Lett.* **80**, 117–129.

Saltzman, B. & Sutera, A. 1987 The mid-Quaternary climatic transition as the free response of a three-variable dynamical model. *J. atm. Sci.* **44**, 236–241.

Shackleton, N. J. 1967 Oxygen isotope analyses and Pleistocene temperatures re-assessed. *Nature, Lond.* **215**, 15–17.

Shackleton, N. J., Backman, J., Zimmerman, H., Kent, D., Hall, M. A., Roberts, D. G., Schnitker, D., Baldauf, J. G., Desprairies, A., Homrighausen, R., Huddlestun, P., Keene, J. B., Kaltenbach, A. J., Krumsiek, K. A. O., Morton, A. C., Murray, J. W. & Westberg-Smith, J. 1984 Oxygen isotope calibration of the onset of ice-rafting and history of glaciation in the North Atlantic region. *Nature, Lond.* **307**, 620–623.

Shackleton, N. J. & Hall, M. A. 1985 Oxygen and carbon isotope stratigraphy of Deep-Sea Drilling Project Hole 552A: Plio-Pleistocene glacial history. *Init. Rep. Deep-sea Drilling Project* **81**, 599–609.

Shackleton, N. J., Hall, M. A. & Boersma, A. 1984 Oxygen and carbon isotope data from Leg 74 Foraminifera. *Init. Repts Deep-sea Drilling Project* **74**, 599–612.

Shackleton, N. J., Imbrie, J. & Hall, M. A. 1983 Oxygen and carbon isotope record of East Pacific core V19-30: implications for the formation of deep water in the late Pleistocene North Atlantic. *Earth planet. Sci. Lett.* **65**, 233–244.

Shackleton, N. J. & Opdyke, N. D. 1973 Oxygen isotope and palaeomagnetic stratigraphy of equatorial Pacific core V28-238: oxygen isotope temperatures and ice volumes on a 10^5-year and 10^6-year scale. *Quat. Res.* **3**, 39–55.

Shackleton, N. J. & Opdyke, N. D. 1976 Oxygen isotope and palaeomagnetic stratigraphy of Pacific core V28-239, Late Pliocene to latest Pleistocene. *Mem. geol. Soc. Am.* **145**, 449–464.

Shackleton, N. J. & Pisias, N. G. 1985 Atmospheric carbon dioxide, orbital forcing and climate. In *The carbon cycle and atmospheric* CO_2*: natural variations Archaean to Recent* (*Geophys. Monogr.* **32**), pp. 303–317.

Zimmerman, H. B., Shackleton, N. J., Backman, J., Kent, D. V., Baldauf, J. G., Kaltenbach, A. J. & Morton, A. C. 1985 History of Plio-Pleistocene climate in the Northeastern Atlantic: DSDP Site 552A. *Init. Repts Deep Sea Drilling Project* **81**, 861–875.

Discussion

J. Rose (*Department of Geography, Birkbeck College, University of London, U.K.*). It would be difficult to leave a meeting such as this without an assessment of relation of the main elements of the terrestrial record of northwest Europe with main elements the oceanic record. The forces driving the global atmospheric system are now well known and the dominant components of these forces during different parts of the Quaternary have been identified (Ruddiman & Raymo, this symposium). I wish to draw attention to the parallel between the main elements of the ocean record and the main elements of the terrestrial and offshore record in northwestern Europe when considered at the scale of geomagnetic chrons rather than conventional stages.

In the ocean record the large-amplitude $\delta^{18}O$ signature which is dominant throughout most of the Brunhes chron (0–0.6 Ma BP) (Ruddiman & Raymo, this symposium) is considered to reflect the build-up of continental ice masses. In northwestern Europe this is reflected in the glacial–interglacial cycles during which ice extended into midland and southern Britain and across part of the north European plain. The period of maximum expansion of glaciers appears to have been during the Anglian in Britain and the Elsterian of northern Europe, which are equated with oxygen-isotope Stage 12 (Bowen *et al.* 1986). This same stage has been identified from oceanic evidence as the period of maximum global cooling (Shackleton 1987, this symposium). It is probable that traces of earlier continental glaciation during the Brunhes have been obscured by the effects of the Anglian and Elsterian ice sheets. Subsequent glacial advances are almost as obscure because of the extent of the late Devensian–late Weichselian ice sheets. Nevertheless, this part of Quaternary time in northwestern Europe saw the expansion of continental ice sheets during the cold stages.

During the later part of the Matuyama chron (0.6–1.7 Ma BP) the amplitude of the $\delta^{18}O$ oceanic signal is reduced relative to the Brunhes but is still distinct (Ruddiman & Raymo, this symposium). In northwestern Europe this atmospheric force appears to be represented by glaciation in the mountains and extensive fluvial deposition in the lowlands. Typically, rivers such as the Thames and the Rhine transported large bodies of fluvial sediment, which are known respectively as the Kesgrave and Sterksel Formations of East Anglia and the Low Countries. In the southern North Sea, where these bodies of sediments coalesce, they are known as the Yarmouth Roads Formation. As far as is understood at present, this pattern of environmental change in Britain occurred between pre-Pastonian A and the Beestonian, including at least four glacial events in the Welsh uplands (Bowen *et al.* 1986); in the Netherlands it encompasses the period referred to as the Menapian, 'Bavel Complex' and early part of the 'Cromer Complex' during which time there is evidence for at least five glaciations in the headwater mountains (Zagwijn 1986).

The reason for this environmental pattern in northwestern Europe at the later part of the Matuyama chron can be explained in terms of the lower magnitude in the variation of the atmospheric forcing, with climatic severity sufficient only to maintain glaciation in western uplands and high mountains, but not sufficient to cause glaciation at a continental scale. The reasons for the exceedingly large bodies of fluvial sediments, such as have not been produced in the Quaternary either before or since, can be explained by the effectiveness of glacial processes during early stages of glaciation acting on a landscape that is not adjusted to glacial flow patterns. In these circumstances erosion would be enhanced by positive feedback until relatively adjusted large-scale glacial erosional landforms were produced and a self-regulating system was established in which erosional processes would be dampened down. (Professor Andrews made this same point with regard to North America in response to a question by Professor Sugden on the origin of fiords.)

Finally, in the period of the Matuyama chron before 1.7 Ma BP, the atmospheric variation is, in general, of a much lower magnitude. In northwestern Europe the response is to be seen in the relatively fine-grained sediments transferred to the North Sea Basin by rivers such as the Thames and Rhine, with catchments smaller even than at the present day. In the southern North Sea these deposits are represented by the Westkapelle Ground Formation, the IJmuiden Ground Formation and the Winterton Shoal Formation (Balson & Cameron 1985). In East Anglia they are represented by the Red Crag and Norwich Crag Formations (Bowen *et al.* 1986), and in The Netherlands by the Oosterhaut and Maassluis Formations. In The Netherlands, lower-energy fluvial equivalents exist in the form of the Scheemda Formation, Kieseloolite Formation, Tegelen Formation, Harderwijk Formation and Kedichem Formation (Zagwijn 1986).

References

Balson, P. S. & Cameron, T. D. J. 1985 Quaternary mapping offshore East Anglia. *Mod. Geol.* **9**, 221–239.

Bowen, D. Q., Rose, J., McCabe, A. M. & Sutherland, D. G. 1986 Correlation of Quaternary Glaciations in England, Ireland, Scotland and Wales. *Quat. Sci. Rev.* **5**, 299–340.

Shackleton, N. J. 1987 Oxygen isotopes, ice volume and sea level. *Quat. Sci. Rev.* **6**, 183–190.

Zagwijn, W. H. 1986 The Pleistocene of the Netherlands with special reference to glaciation and terrace formation. *Quat. Sci. Rev.* **5**, 341–345.